DIY Science

Illustrated Guide to Home Forensic Science Experiments

All Lab, No Lecture

First Edition

Robert Bruce Thompson and Barbara Fritchman Thompson

Illustrated Guide to Home Forensic Science Experiments

All Lab, No Lecture

by Robert Bruce Thompson and Barbara Fritchman Thompson

Published by Maker Media, Inc., 1160 Battery Street East, Suite 125, San Francisco, California 94111.

Maker Media books may be purchased for educational, business, or sales promotional use. Online editions are also available for most titles (*safaribooksonline.com*). For more information, contact our corporate/institutional sales department: 800-998-9938 or *corporate@oreilly.com*.

Print History
July 2012
First Edition

Editor: Brian Jepson
Production Editor: Melanie Yarbrough
Copy Editor: Bob Russell, Octal Publishing, Inc.
Proofreader: Linley Dolby
Indexer: Bob Pfahler
Cover Designer: Mark Paglietti
Cover Photograph: Robert Bruce Thompson
Interior Designer: Ron Bilodeau
Illustrator: Rebecca Demarest

ISBN: 978-1-449-33451-2
[LSI] [2016-10-21]

To Edmond Locard (1877 - 1966), often called the French Sherlock Holmes, who, as a professor of forensic medicine and criminology at the University of Lyons, in 1910 established the world's first police crime laboratory. Locard's lab occupied two attic rooms staffed by two assistants provided grudgingly by the Lyons police department, and was initially less well equipped than the home forensics lab we used in writing this book. Despite these limited resources, Locard's results soon convinced police departments worldwide, including Scotland Yard and the FBI, to found their own crime labs.

Locard was the first to state the fundamental principle of forensic science, now known as Locard's Exchange Principle: "Wherever he steps, whatever he touches, whatever he leaves, even unconsciously, will serve as a silent witness against him. Not only his fingerprints or his footprints, but his hair, the fibers from his clothes, the glass he breaks, the tool mark he leaves, the paint he scratches, the blood or semen he deposits or collects. All of these and more bear mute witness against him. This is evidence that does not forget. It is not confused by the excitement of the moment. It is not absent because human witnesses are. It is factual evidence. Physical evidence cannot be wrong, it cannot perjure itself, it cannot be wholly absent. Only human failure to find it, study and understand it, can diminish its value."

Contents

Preface

You're reading this preface, so it's a fair assumption that you're interested in forensic science. You're in good company. For more than 100 years, forensic science has fascinated a lot of people. Popular interest in forensic science started with the detective stories of Edgar Allen Poe and Wilkie Collins in the mid-19th century, and got a major boost in 1887 when Arthur Conan Doyle published the first of his immensely popular series of Sherlock Holmes stories. Its popularity continued to build through the early- to mid-20th century with the publication of hundreds of forensic-based mystery novels by such bestselling Golden Age authors as Agatha Christie, R. Austin Freeman, and many others. Forensic-themed novels from such authors as Patricia Cornwell, Kathy Reichs, and Tess Gerritsen continue to top the bestseller lists today.

Hollywood recognized the popular interest in forensic science and has produced hundreds of films in which forensic science—sometimes accurately portrayed, but more often not—plays a central role. Sherlock Holmes has been featured in many films, as have other fictional forensic experts such as Freeman's Dr. John Evelyn Thorndyke. Nor were television producers unaware of this popular fascination with forensic science. In 1965, the television series *The F.B.I.* premiered on ABC. Based loosely on the 1959 film, *The FBI Story*, this long-running series was the first television program that portrayed forensic science realistically and regularly. Even better, it generally got the science right, which may be no small part of why it became a top-10 series.

The F.B.I. was soon followed by a television series that did more than simply feature aspects of forensic science. In 1976, NBC introduced *Quincy, M.E.*, a television series with forensic science at its very core and a forensic pathologist as the lead character. Like *The F.B.I.* before it, *Quincy, M.E.* quickly became a top-10 hit. It lasted well into the 1980s, and set the stage for a plethora of forensic-based television programs, from cable series such as *Dexter* and *Waking the Dead* to mainstream network series like *Bones*, *Crossing Jordan*, *NCIS*, and the *CSI* franchise.

If your only knowledge of forensic science comes from watching *CSI* and similar programs, you may wonder whether modern forensic science is just a matter of white-smocked acolytes and hard-bodied assistants awaiting answers from expensive high-tech instruments, which answers they invariably get in time to solve the crime before the closing credits roll. The reality is far different. Sherlock Holmes with his magnifying glass and Dr. John Evelyn Thorndyke with his microscope and lab bench are much more realistic representations of actual day-to-day forensic science work.

Here's a startling fact: the vast majority of forensic work, even today, is done with low-tech procedures that would be familiar to a forensic scientist of 100 years ago. For every suspect illicit drug sample that is analyzed on a $100,000 spectrometer, hundreds of such samples are analyzed by using presumptive color spot tests, a technology that dates back to the 19th century. For every specimen examined with a $1,000,000 scanning electron microscope, hundreds or thousands of specimens are examined with ordinary optical microscopes.

That's not to say that all of that expensive equipment is useless. Far from it. Instrumental analysis allows today's forensic scientists to do things that were unimaginable just a few years ago, laying bare secrets that formerly would have remained forever hidden. A forensic scientist from 100 years ago would have regarded today's instruments as nothing short of magic.

But these instruments aren't cheap, which means there can't be a full selection of instruments on every forensic scientist's lab bench. Also, instrumental analyses may be time-consuming—both in terms of preparing specimens for testing and in time needed to run the test—and therefore impractical for analyzing many questioned specimens in a short time. For these reasons, most preliminary screening is done with fast, cheap, low-tech procedures such as color tests and optical microscopes, with the slower, more expensive, instrumental methods reserved for confirmatory tests.

And that's all to the good for anyone who's interested in *doing* real forensic science, instead of just reading about it. Presumably, if you've read this far, that includes you. You don't need a multi-million dollar lab to do real, useful forensic investigations. All you need are some chemicals and basic equipment, much of which can be found around the home, improvised, or purchased inexpensively. There are exceptions, of course. You'll need a decent microscope—the fundamental tool of the forensic scientist—but even an inexpensive student model will serve. You'll need some basic lab equipment and some specialty chemicals, all of which can be purchased from specialty lab supply vendors and law-enforcement forensics supply vendors.

In fact, to make it as easy and inexpensive as possible to acquire the special equipment and chemicals needed for many of the procedures in this book, we sell a customized kit through our company, The Home Scientist, LLC (*www.thehomescientist.com*). You don't need to buy the kit to do the procedures; we provide complete details about what you'll need, and how to make up special reagents yourself. All of the equipment and reagents are readily available from numerous online sources. If you intend to perform only a few of the procedures in this book, it may be less expensive to buy what you need piecemeal. On the other hand, if you plan to do many (or even several) of the procedures, it'll probably be less expensive to buy the kit.

With such minimal equipment, you'll be prepared to delve deeply into real forensics work. You'll analyze soil, hair, and fibers, individualize plastic and glass specimens, develop latent fingerprints and reveal hidden bloodstains, analyze tool marks and other impressions, test for illegal drugs and poisons, analyze gunshot and explosives residues, detect forgeries and fakes, individualize questioned pollen and diatom samples, and extract DNA samples and separate them by gel electrophoresis.

And you'll learn an important lesson as you do the laboratory sessions in this book. On television, the forensics expert always succeeds. Fingerprints are invariably crisp and clear, and technicians always find a hair or fiber on the bad guy's clothes that links him to the victim. There's never any question about the test results. Real life isn't like that. Forensic test results are often ambiguous, and sometimes fail completely to establish any link between questioned and known specimens. Good forensic work is painstaking and difficult. There are seldom any easy answers, but hard work and persistence usually pay off. In doing these lab sessions, you'll gain a real appreciation for just how good real forensic scientists are at what they do, how persistent and inventive they have to be, and just how hard they work to get the job done. Welcome to the world of real forensics.

INDIVIDUAL VERSUS CLASS EVIDENCE

Throughout this book, we refer to the two categories of forensic evidence. *Individual evidence* is evidence—such as a fingerprint or a DNA specimen—that can be identified unambiguously as having originated from a specific, particular source. *Class evidence* is evidence—such as glass or paint specimens—that can at most be identified as being consistent with a particular source, but not necessarily as having originated from that specific source.

The steady improvement in testing methodologies means that some types of evidence that were formerly class evidence can now be individualized. For example, prior to the advent of DNA testing, a blood specimen was inherently class evidence. It could be tested for blood type and other factors—which large numbers of people share—but the blood specimen could not be individualized to a particular person. With DNA testing, that blood specimen becomes individualized evidence, because it can now be identified unambiguously as having originated from one specific individual.

In forensics analyses, we are always comparing the physical, chemical, and other properties of an unknown (or questioned) specimen to those of similar specimens from known sources. If the questioned and known specimens share identical individualizable characteristics, a forensic scientist may categorize them as "matching" specimens. If only class characteristics are present, forensic scientists avoid using the word "match," because it implies a greater degree of certainty than actually exists. Instead, the forensic scientist may describe one specimen as being "consistent with" the other.

Comparing multiple types of class evidence may narrow the possible sources considerably. For example, before DNA testing was available, blood and other body fluids were often analyzed in great detail. A simple ABO blood type test could rule out a significant percentage of the population as possible sources, and testing for the presence or absence of the Rhesus factor and other blood factors could greatly narrow the possible range of sources, sometimes to a small fraction of 1% of the population. As useful as such results are, particularly as exculpatory evidence, they remain class evidence, because they cannot point unambiguously to one individual as the source.

Forensic scientists constantly strive to develop new methods to individualize class evidence, but analyzing class evidence will remain a major part of the work of any forensic lab for the foreseeable future. In that respect, much of forensics work can be considered an attempt to reduce uncertainty, which is often the most that can be hoped for.

WHO THIS BOOK IS FOR

This book is for anyone, from responsible teenagers to adults, who wants to learn about forensic science by doing real, hands-on laboratory work. DIY hobbyists and forensics enthusiasts can use this book to learn and master the essential practical skills and fundamental knowledge needed to pursue forensics as a lifelong hobby. Home school parents and public school teachers can use this book as the basis of a year-long, lab-based course in forensic science.

For a textbook, we recommend *Criminalistics: An Introduction to Forensic Science* by Richard Saferstein (Prentice Hall). As is generally true of textbooks, the current (10th) edition is very expensive. The 9th edition is available used for only a few dollars and is perfectly suitable for a high school or even college-level first-year forensics course. Forensic science has advanced between the 2006 9th edition and the 2010 10th edition, but the changes are not significant for our purposes.

We consider forensics to be the ideal introductory lab-based science course for freshman or sophomore high school students as well as an ideal supplemental science course for 11th or 12th grade students. Even students who dread biology, chemistry, and physics are often excited about doing forensics lab work, and such work is an ideal introduction for later science courses. Although forensic science teaches students about the scientific method and incorporates elements of biology, chemistry, earth science, and the other sciences, detailed knowledge of these subjects is not a prerequisite for an introductory forensics course.

A forensics course is also cost-effective. Most high school science labs and many home-schoolers already possess microscopes, basic chemistry labware, and most of the other equipment and chemicals needed to complete the lab sessions in this book. Home school parents can add a forensic science course to the curriculum at little incremental cost beyond what they'll spend anyway for the equipment and materials required to teach later courses in biology, chemistry, and physics.

> With very few exceptions, included for learning purposes, the forensic science procedures in this book are not merely educational; they're the real deal. Real forensic scientists and technicians actually use these procedures—or ones very like them—every day to analyze real evidence in real criminal cases. In fact, we're honored that major metropolitan law-enforcement organizations have used our materials and videos to train their own CSI staffs.

HOW THIS BOOK IS ORGANIZED

The first part of this book is made up of narrative chapters that cover the essential "book learning" you need to equip your forensics lab and work safely in your lab.

1. Laboratory Safety
2. Equipping a Forensics Lab

The bulk of the book is made up of the following 11 hands-on laboratory chapters, each devoted to a particular topic. Each of the laboratory chapters is self-contained, so you can pick and choose the topics that are most interesting to you, and complete any or all of the chapters in any order you wish. Within a chapter, it's a good idea to do the lab sessions in order, because some sessions use the materials or results from earlier sessions in that chapter.

I. Laboratory: Soil Analysis

II. Laboratory: Hair and Fiber Analysis

III. Laboratory: Glass and Plastic Analysis

IV. Laboratory: Revealing Latent Fingerprints

V. Laboratory: Blood Detection

VI. Laboratory: Impression Analysis

VII. Laboratory: Forensic Drug Testing

VIII. Laboratory: Forensic Toxicology

IX. Laboratory: Gunshot and Explosives Residues Analysis

X. Laboratory: Detecting Forgeries and Fakes

XI. Laboratory: Forensic Biology

ACKNOWLEDGMENTS

Although only our names appear on the cover, this book is very much a collaborative effort. It could not have been written without the help and advice of our editor, Brian Jepson, who contributed numerous helpful suggestions. As always, the O'Reilly design and production staff, who are listed individually in the front matter, worked miracles in converting our draft manuscript into an attractive finished book.

Finally, special thanks are due to our technical reviewers.

Dennis Hilliard is the director of the Rhode Island State Crime Laboratory. In addition to the administration of the State Crime Laboratory, his work includes analysis of evidence and court testimony in the areas of fire debris analysis, hair and fiber analysis, DNA analysis, and breath and blood alcohol analysis. He has worked in the forensic field since 1980. He was appointed acting director of the State Crime Laboratory in 1992, appointed to the director's position in 1995, and has held a position in the University of Rhode Island College of Pharmacy as an adjunct assistant professor of biomedical sciences since 1994. He is a member of several professional forensic organizations and is a past president of the NorthEastern Association of Forensic Scientists (NEAFS).

Mary Chervenak holds a Ph.D. in organic chemistry from Duke University and is a research chemist for Arkema. Mary has long been interested in forensic science in general and forensic chemistry in particular, and jumped at the opportunity to contribute her thoughts to this book.

Paul Jones holds a Ph.D. in organic chemistry from Duke University and is a professor of organic chemistry at Wake Forest University. Our thanks to Paul for his great patience in answering a lot of dumb questions without making us feel stupid.

Dennis, Mary, and Paul outdid themselves as technical reviewers, flagging our mistakes and contributing innumerable useful suggestions and comments. With their help, this is a much better book than it might otherwise have been. Thanks, guys.

HOW TO CONTACT US

We have verified the information in this book to the best of our ability, but you may find things that have changed (or even that we made mistakes!). As a reader of this book, you can help us to improve future editions by sending us your feedback. Please let us know about any errors, inaccuracies, misleading or confusing statements, and typos that you find anywhere in this book.

Please also let us know what we can do to make this book more useful to you. We take your comments seriously and will try to incorporate reasonable suggestions into future editions. You can write to us at:

Maker Media, Inc.
1160 Battery Street East, Suite 125
San Francisco, CA 94111
877-306-6253 (in the United States or Canada)
707-639-1355 (international or local)

Maker Media unites, inspires, informs, and entertains a growing community of resourceful people who undertake amazing projects in their backyards, basements, and garages. Maker Media celebrates your right to tweak, hack, and bend any technology to your will. The Maker Media audience continues to be a growing culture and community that believes in bettering ourselves, our environment, our educational system—our entire world. This is much more than an audience, it's a worldwide movement that Maker Media is leading—we call it the Maker Movement.

For more information about Maker Media, visit us online:

Make: and Makezine.com magazine: *http://makezine.com*

Maker Faire: *http://makerfaire.com*

Maker Shed: *http://makershed.com*

To comment on the book, send email to:

bookquestions@oreilly.com

The O'Reilly website for *Illustrated Guide to Home Forensic Science Experiments* lists examples, errata, and plans for future editions. You can find this page at:

http://oreil.ly/home_forensic_science_exp

For more information about this book and others, see the O'Reilly web site:

http://www.oreilly.com

To contact the authors directly, send mail to:

robert@thehomescientist.com
barbara@thehomescientist.com

We read all mail we receive from readers, but we cannot respond individually. If we did, we'd have no time to do anything else. But we do like to hear from readers.

We also maintain a dedicated landing page on our main website to support *Illustrated Guide to Home Forensic Science Experiments*. This page contains links to equipment kits customized for this book, corrections and errata, supplemental material that didn't make it into the book, and so on. Visit this page before you buy any equipment or chemicals and before you do any of the experiments. Revisit it periodically as you use the book.

www.thehomescientist.com/forensics

THANK YOU

Thank you for buying *Illustrated Guide to Home Forensic Science Experiments*. We hope you enjoy reading and using it as much as we enjoyed writing it.

AUTHOR BIOS

Robert Bruce Thompson is the author of numerous articles, training courses, and books about computers, science, and technology, including many co-authored with his wife, Barbara. He built his first home lab as a teenager and went on to major in chemistry in college and graduate school. Robert maintains a home laboratory equipped for doing real chemistry, forensics, biology, earth science, and physics.

Barbara Fritchman Thompson is, with her husband Robert, the co-author of numerous books about computers, science, and technology. With her masters in library science and 20 years' experience as a public librarian, Barbara is the research half of our writing team.

Laboratory Safety

1

First things first. This is a short chapter, but a very important one. Many of the lab sessions described in this book use chemicals, such as strong acids and bases, that are dangerous if handled improperly. Some lab sessions use open flame or other heat sources, and many use glassware. To state the obvious, you can get hurt working in a lab. Fortunately, there are steps you can take to minimize or eliminate hazards.

If you remember one thing from this chapter, remember this: **If there is even the slightest chance that you will be exposed to any hazardous chemical, always wear chemical splash goggles, gloves, and protective clothing.** We follow this advice ourselves, without exception.

DENNIS HILLIARD COMMENTS

Whenever you are working with any chemicals, glassware, and/or biological material, always wear chemical splash goggles, gloves, and protective clothing. Protective clothing works two ways: it protects the analyst from chemical, sharps, and biological hazards associated with evidence collection and processing; and it protects from contamination of the evidence by the collector/analyst.

Although working in any lab has its dangers, so does driving a car. And, just as you must remain constantly alert while driving, you must remain constantly alert while working in a lab. But it's also important to keep things in perspective. More serious injuries occur every year among a few hundred thousand high school football players than have ever occurred in total among millions upon millions of student scientists in the 200-year history of student labs. Statistically, students are much, much safer working in a home or school lab than they are out skateboarding or riding bicycles.

Most injuries that occur in student labs are minor and easily avoidable. Among the most common are nicks from broken or chipped glassware and minor burns. Serious injuries are very rare. When they do occur, it's nearly always because someone did something incredibly stupid, such as using a flammable solvent near an open flame or absentmindedly taking a swig from a beaker full of a toxic liquid. (That's why one of the rules of laboratory safety is never to smoke, drink, or eat in the lab.)

The primary goal of laboratory safety rules is to prevent injuries. Knowing and following the rules minimizes the likelihood of accidents and helps to ensure that any accidents that do occur will be minor ones.

The following are the laboratory safety rules we recommend:

Prepare properly

- All laboratory activities must be supervised by a responsible adult.

Direct adult supervision is mandatory for all of the activities in this book. This adult must review each activity before it is started, understand the potential dangers of that activity and the steps required to minimize or eliminate those dangers, and be present during the activity from start to finish. Although the adult is ultimately responsible for safety, students must also understand the potential dangers and the procedures that should be used to minimize risk.

- Familiarize yourself with safety procedures and equipment.

  Think about how to respond to accidents before they happen. Have a fire extinguisher and first-aid kit readily available and a telephone nearby in case you need to summon assistance. Know and practice first-aid procedures, particularly those required to deal with burns and cuts. If you have a cell phone, keep it with you while you're working in the lab.

  One of the most important safety items in any lab is the cold water faucet. If you burn yourself, immediately (seconds count) flood the burned area with cold tap water for several minutes to minimize the damage done by the burn. If you spill a chemical on yourself, immediately rinse the chemical off with cold tap water, and keep rinsing for several minutes. Ideally, every lab should have an eyewash station, but most home labs do not. If you do not have an eyewash station and you get any chemical in your eyes, immediately turn the cold tap on full and flood your eyes until help arrives.

WARNING

Everyone rightly treats strong acids with great respect, but many students handle strong bases casually. That's a very dangerous practice. Strong bases, such as solutions of sodium hydroxide, can blind you in literally seconds. Treat every chemical as potentially hazardous, and always wear splash goggles.

  Keep a large container of baking soda (sodium bicarbonate) on hand to deal with acid or base spills. Baking soda neutralizes either type of spill. We keep a 12-pound bag from Costco on hand for this purpose.

- Always read the Material Safety Data Sheet (MSDS) for every chemical you will use in a laboratory session.

  The MSDS is a concise document that lists the specific characteristics and hazards of a chemical. Always read the MSDS for every chemical that is to be used in a lab session. If an MSDS was not supplied with the chemical, locate one on the Internet. For example, before you use lead nitrate in an experiment, do a Google search using the search terms "lead nitrate" and "MSDS".

- Organize your work area.

  Keep your lab bench and other work areas clean and uncluttered, before, during, and after laboratory sessions. Every laboratory session should begin and end with your glassware, chemicals, and laboratory equipment clean and stored properly.

Dress properly

- Wear approved eye protection at all times.

  Everyone present in the lab must at all times wear splash goggles that comply with the ANSI Z87.1 standard. Standard eyeglasses or shop goggles do not provide adequate protection, because they are not designed to prevent splashed liquids from getting into your eyes. Eyeglasses may be worn under the goggles, but contact lenses are not permitted in the lab. (Corrosive chemicals can be trapped between a contact lens and your eye, making it difficult to flush the corrosive chemical away.)

- Wear protective gloves and clothing.

  Never allow laboratory chemicals to contact your bare skin. When you handle chemicals, particularly corrosive or toxic chemicals or those that can be absorbed through the skin, wear gloves of latex, nitrile, vinyl, or another chemical-resistant material. We recommend disposable nitrile gloves, which you can purchase at Costco, Walmart, or any drugstore. We are comfortable using disposable nitrile gloves for handling any of the chemicals used in this book. If you want to be extra cautious when handling corrosive and/or toxic chemicals, either double-glove with disposable nitrile gloves or wear heavier gloves, such as the thick "rubber" gloves sold by lab supply vendors and in the supermarket for household use.

  Wear long pants, a long-sleeve shirt, and leather shoes or boots that fully cover your feet (NO sandals). Avoid loose sleeves. To protect yourself and your clothing, wear a lab coat or a lab apron made of vinyl or another resistant material. Wear a disposable respirator mask if you handle chemicals that are toxic by inhalation.

Avoid laboratory hazards

- Avoid chemical hazards.

  Never taste any laboratory chemical or sniff it directly. (Use your hand to waft the odor toward your nose.) Never use your mouth to fill a pipette. When you heat a test tube or flask, make sure the mouth points in a safe direction. Always use a boiling chip or stirring rod to prevent liquids from boiling over and being ejected from the container. Never carry open containers of chemicals around the lab. Always dilute strong acids and bases by adding the concentrated solution or solid chemical to water slowly and with stirring. Doing the converse can cause the liquid to boil violently and be ejected from the container. Use the smallest quantities of chemicals that will accomplish your goal. In particular, the first time you run a reaction, do so on a small scale. If a reaction is unexpectedly vigorous, it's better if it happens with 1 mL of chemicals in a spot plate than 500 mL in a large beaker.

- Avoid fire hazards.

  Never handle flammable liquids or gases in an area where an open flame or sparks might ignite them. Extinguish burners as soon as you finish using them. Do not refuel a burner until it has cooled completely. If you have long hair, tie it back or tuck it up under a cap, particularly if you are working near an open flame.

- Avoid glassware hazards.

  Assume all glassware is hot until you are certain otherwise. Examine all glassware before you use it, and particularly before you heat it. Discard any glassware that is cracked, chipped, or otherwise damaged. Learn the proper technique for cutting and shaping glass tubing, and make sure to fire-polish all sharp ends.

Don't Do Stupid Things

- Never eat, drink, or smoke in the laboratory.

  All laboratory chemicals should be considered toxic by ingestion, and the best way to avoid ingesting chemicals is to keep your mouth closed. Eating or drinking (even water) in the lab is very risky behavior. A moment's inattention can have tragic results. Smoking violates two major lab safety rules: putting anything in your mouth is a major no-no, as is carrying an open flame around the lab.

- Never work alone in the laboratory.

  No one—adult or student—should ever work alone in the laboratory. Even if the experimenter is adult, there must at least be another adult within earshot who is able to respond quickly in an emergency.

- Never horse around.

  A lab isn't the place for practical jokes or acting out, nor for that matter for catching up on gossip or talking about last night's ball game. When you're in the lab, you should have your mind on lab work, period.

- Never combine chemicals arbitrarily.

  Combining chemicals arbitrarily is among the most frequent causes of serious accidents in home labs. Some people seem compelled to mix chemicals more or less randomly, just to see what happens. Sometimes they get more than they bargained for.

Laboratory safety is mainly a matter of common sense. Think about what you're about to do before you do it. Work carefully. Deal with minor problems before they become major problems. Keep safety constantly in mind, and chances are any problems you have will be very minor ones.

Equipping Your Forensics Laboratory

2

To do a serious study of forensic science lab work, you'll need some specialized equipment and chemicals. Fortunately, it needn't be expensive to acquire the items you need. You may already have some of those items, such as a microscope purchased for a biology course or a balance and other lab equipment purchased for a chemistry course.

To make matters as easy and inexpensive as possible, as we wrote this book we designed a custom lab kit to go with it (*http//:www.thehomescientist.com/kits/FK01/fk01-main.html*). With the exception of readily available materials, major items (such as a microscope and balance), and some optional items, this kit includes the specialized equipment and chemicals you need to complete the lab sessions in this book. Of course, you don't have to buy the kit to use this book. We provide full details of what's needed for each lab session, and all of those materials can be obtained locally or purchased individually from numerous online lab supply vendors and law-enforcement forensic supply vendors.

OPTICAL EQUIPMENT

The iconic image of Sherlock Holmes has him smoking a pipe, wearing a deerstalker cap, and examining evidence with his magnifying glass. The pipe and cap are optional but, as Holmes knew well, optical equipment is essential for successful forensic investigations. Like Holmes, we'll use a variety of optical aids in our investigations, all of which are described in the following sections.

IMAGING EQUIPMENT

One of the fundamental principles of crime scene management is that everything possible should be recorded *in situ* photographically. Once an object has been moved, the scene is permanently altered. The primary goal of crime scene photography is therefore to preserve a permanent record of the crime scene as it was first encountered, with both overall images to show the crime scene in context and close-up images to show details of objects of particular evidentiary value.

Photography is also used extensively for recording specimens that have been recovered from the crime scene and transported to the forensics laboratory. Such images provide a permanent record of specimens as they appeared when they were received by the lab, and are particularly useful for situations in which subsequent testing may alter the appearance of the evidence, perhaps permanently.

CAMERA (OPTIONAL)

Forensic technicians have routinely photographed crime scenes since the last quarter of the 19th century. At that time, the best cameras available were large, bulky, and used glass plate emulsions that for anything other than full daylight required long exposure times or flash exposures that scattered unburned flash powder and ashes all over crime scenes. As camera technology improved, so too did the quality and quantity of crime scene images. Today, most crime scene photographers use both standard digital SLRs—film cameras are seldom used any longer, other than in special situations—and specialized cameras that capture images in the infrared or ultraviolet portions of the spectrum.

Because any camera inherently distorts reality by representing three-dimensional scenes in two dimensions and by rendering colors and contrasts imperfectly, crime scene photographers take great care to minimize any such effects their equipment may impose on the final images and to record the pertinent data about each image. For example, because the focal length of a lens affects perspective, back when film cameras were in common use, many crime scene photographers used only fixed-focal length lenses or, alternatively, recorded the focal length used for each image. That's seldom a problem nowadays, because nearly all digital cameras record such data automatically and store it with the image file. Crime scene photographers also use grids to show the distances and angular relationships among objects in the image. Similarly, where image scale is not obvious, crime scene photographers take great care to make image scale clear within the image itself, for example, by including a section of a ruler next to an object in an image.

In the last few years, digital cameras have come to dominate crime scene photography. Digital images are immediately accessible, much less expensive than film images, and much easier to store, copy, search, and transfer. The main early impediments to widespread adoption of digital imaging for forensics were the relatively low resolution of early digital cameras and the perception that digital images could be altered easily and undetectably, making them useless as evidence.

Digital cameras soon achieved resolution nearly as good as the best available film cameras—and in some cases, better—which answered the first objection. And, as it turned out, sophisticated digital image analysis algorithms can detect changes to digital images, answering the second objection, as well.

> **DENNIS HILLIARD COMMENTS**
>
> Despite their relatively low image quality, camcorders were formerly widely used to record crime scene video. With the use of digital still cameras, crime scene video has fallen out of favor. The digital camera is used to document the entire scene and photos can be stitched together in a computer program such as Adobe Photoshop to give panoramic views.

Although you won't be photographing any crime scenes for the lab sessions in this book, a camera is also useful for recording images of specimens for your lab notebook, shooting photomicrographs (images through the microscope), and so on. If you don't have a suitable camera, you can make sketches, instead.

> **DENNIS HILLIARD COMMENTS**
>
> Sketches are often a part of the analyst's notes and are required in accredited laboratories, because notes are subject to review and photos are not always able to be taken of some evidence or are difficult to take through a microscope.

If possible, use a digital camera with a macro feature, ideally one that permits imaging to at least a 1:3 scale (1:3 means the image on the sensor is a third the size of the actual object), and 1:2 or 1:1 is better. A point-and-shoot digital camera is acceptable; a digital SLR is preferable. Although some standard "kit" zoom lenses supplied with digital SLRs allow focusing down to a 1:3 or closer ratio, these lenses do not provide the best image quality for extreme close-ups. A macro lens—one optimized for best image quality at very short distances—is the best choice for macro shots. We use Pentax K100D Super and K-r digital SLRs with a Pentax 50mm macro lens.

Because an on-camera flash unit provides flat lighting other than at very close focusing distances, it's also helpful to have a small slave flash unit to provide low-angle or cross lighting for contrast when photographing small specimens. We use a $25 Vivitar DF120, which is triggered by the main flash.

SCANNER (OPTIONAL)

One relatively recent development has been the use of ordinary PC flat-bed scanners for recording high-resolution images of flat specimens, usually in the lab but sometimes at the crime scene. One major advantage of using a scanner is that the images are automatically calibrated, with the resolution stored in the image file header. You can open an image file with Photoshop or a similar graphics program and perform direct measurements on the image. Alternately, you can count pixels and simply convert pixels to millimeters or inches using a spreadsheet. Because typical scanners record images at 2,400 to 9,600 dpi (dots per inch) or higher, such measurements are accurate and precise. Figure 2-1 shows a small fabric tear imaged at 3,200 dpi using an inexpensive Epson scanner.

Figure 2-1: *A fabric tear imaged at 3,200 dpi using a scanner*

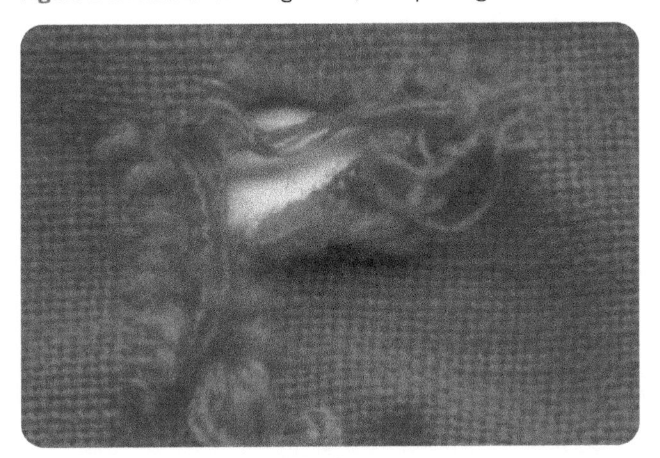

With some scanners, you can remove the lid, invert the scanner over (for example) a stain on the floor, and record an image directly. Results are often better than camera images because the scanner provides a flat, even light, eliminating the problems with hot spots and reflections that are common with an on-camera flash.

> A camera records an image from a fixed sensor (lens) position, which means that different objects within the field of view are imaged at slightly different angles and distances, even if the object is flat. For example, if you use a camera to shoot an image of a document, the camera lens is slightly closer to the center of the document than to the corners, introducing some distortion. We think nothing of this, because that's also how our eyes work.
>
> A scanner, conversely, images objects with a moving sensor. Each tiny part of the image is made with the sensor effectively in contact with the subject, so there is essentially no distortion of flat objects, such as documents, surfaces, or clothing specimens. It is because scanner images have nearly zero distortion that accurate direct measurements may be made from them.

MAGNIFIER (REQUIRED)

One of the most useful items in any forensic kit is a magnifier that provides moderate magnification, something in the range of 5X to 15X. A magnifier provides a close-up view of small objects, fingerprints, and other small but significant evidence.

> The FK01 Forensic Science Kit includes an inexpensive folding magnifier, but if you have a better magnifier or loupe, by all means use it.

Sherlock Holmes used a simple magnifying glass, but there are better tools available today. The best choice is a photographer's loupe with a transparent base, such as the 5X Viewcraft Lupe shown in Figure 2-2.

Figure 2-2: *A 5X Viewcraft Lupe*

A loupe with a built-in scale or reticle allows you to determine the size of objects directly, as shown in Figure 2-3. (A scale appears alongside the object in the field of view, while a reticle is superimposed optically on the image.) A basic loupe costs $5 to $15 and can be purchased from any vendor of photography supplies. High-quality loupes from German and Japanese makers are considerably more expensive. Some loupes have built-in illuminators, but most do not. Even if your loupe is illuminated, but particularly if it is not, add a small flashlight to your portable forensics kit. We use a two-AA model with a white LED bulb, although many forensics technicians prefer to use a flashlight with an incandescent bulb because it renders colors more accurately.

Figure 2-3: *A loupe with a scale or reticle allows you to measure objects directly*

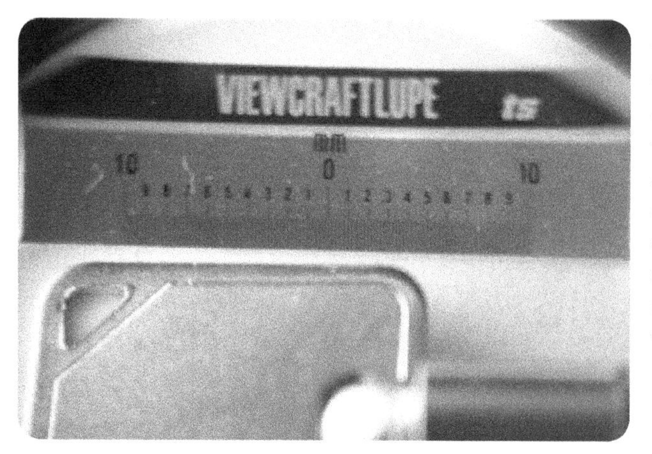

A stand magnifier, preferably illuminated, is useful for examining objects while keeping both hands free. The model shown in Figure 2-4 provides 2X magnification overall and has a smaller embedded lens that provides 4X magnification of small parts of the object. Magnifiers like this one are available from stores and catalogs that serve the needs of the elderly.

Figure 2-4: *A stand magnifier allows you to work with both hands free*

MICROSCOPE (REQUIRED)

Microscopes take over where magnifiers leave off. Typical magnifiers enlarge an object by 5X to 15X, while microscopes enlarge objects from about 10X to 1000X or more. Modern forensics labs use many different kinds of microscopes, including electron microscopes, microscopes designed to work with polarized or UV light, and comparison microscopes that allow viewing two samples side by side in the same field of view. Such specialized microscopes are much too expensive for a basic forensics lab, but all of the lab sessions in this book that require a microscope can be completed with a standard compound microscope. A 40/100/400X model will suffice for all but the lab sessions that require observing diatoms and pollen; for those, a 40/100/400/1000X model is a better choice. If at all possible, use a microscope with a mechanical stage. Many microscopes include an ocular micrometer/reticle as a standard or optional feature. This is worth having, because it allows you to determine the actual sizes of objects you're viewing through the microscope.

For detailed information about choosing and buying a microscope, see *Illustrated Guide to Home Biology Experiments*.

MICROSCOPE ACCESSORIES

In addition to the compound microscope itself, you'll need a modest selection of supplies and accessories. Most laboratory supply vendors offer starter kits that contain many of these items.

Slides

You'll need a reasonable number of standard (25×75 mm or 1×3") microscope slides. Buy a box of 72; you'll use a lot of slides. Avoid the cheapest slides, which may be fragile and have sharp edges. You'll also need a few concavity slides (also called well slides). These are useful for containing small specimens and performing micro-procedures on them. Buy deep-cavity slides, which are about three times the thickness of standard slides (~3 mm versus ~1 mm). The deep-cavity models are harder to find and more expensive than standard well slides, but they're also much more useful. The FK01 Forensic Science Kit includes a box of 72 flat slides and a box of 12 deep-cavity slides.

Coverslips

Buy a half-ounce or one-ounce box of standard 18×18 mm to 24×24 mm square glass coverslips. #1.5 coverslips are the ideal thickness for most standard microscopes, but they're relatively difficult to find. #2 coverslips are suitable for most microscopes. Buy the very thin #1 coverslips only if your microscope requires them. The FK01 Forensic Science Kit includes a box of standard coverslips.

Avoid all plastic slides and coverslips, which are grossly inferior optically to glass models. The one exception is the lab session on doing cross sections of fibers, for which you'll need at least one plastic slide.

Mounting fluids

For routine use, you can simply make a temporary wet mount by placing a drop of water, glycerol, or vegetable oil on the specimen and then placing a coverslip on top of the specimen, but if you want to mount specimens permanently, you'll need mounting fluid. You can buy special permanent mounting fluids such as Permount or Melt-Mount, but a drop of colorless nail polish (we use Sally Hansen Hard as Nails) works about as well. The FK01 Forensic Science Kit includes glycerol and olive oil, both of which are suitable for temporary wet mounts.

Slide storage

Most science supply vendors carry a variety of slide storage boxes in plastic and wood as well as slide folders. If you intend to permanently mount specimens, buy boxes or folders as necessary to store your slides.

Immersion oil

If your compound microscope has an oil-immersion objective (usually the 100X objective), you'll need immersion oil. If your microscope doesn't have an oil-immersion objective or if you don't plan to use high magnification, you can do without immersion oil.

Cleaning kit

You'll need to clean the ocular and objective lenses of your microscope periodically. If you use immersion oil, you should clean the immersion objective each time you use it. Science supply vendors carry cleaning kits that include a blower, brush, lens-cleaning tissue and fluid, and so on.

Dust cover or storage case

One of the most important and frequently overlooked accessories is a dust cover or storage case for the microscope. Always keep your microscope covered when you are not using it to protect it from dust and damage. If you don't have a cover or case, a kitchen trash bag makes an adequate substitute

PHOTOMICROGRAPHY EQUIPMENT (OPTIONAL)

Photomicrography is the process of recording images through a microscope. (Microphotography, conversely, is the process of making very tiny photographs, such as the microdots formerly used by spies.) Using a camera to record the specimens you observe with your microscope is a very useful adjunct to narrative descriptions of your observations. Expert forensic witnesses often use photomicrographs in court to supplement their verbal testimony.

The ideal setup for photomicrography in a basic forensics lab is a dual-head microscope with a digital SLR mounted to the vertical head by means of a T-ring and microscope adapter (see Figure 2-5). Such adapters are available from camera stores, Edmund Optics (*http//:www.edmundoptics.com*), and other vendors. We used the Edmund Microscope Adapter (#41100) and Pentax K100D Super and K-r digital SLRs with a K-mount T-ring adapter to shoot many of the photomicrographs in this book. Similar microscope adapters are available for specific point-and-shoot digital camera models.

Figure 2-5: *A DSLR coupled to a microscope with a T-ring and microscope adapter*

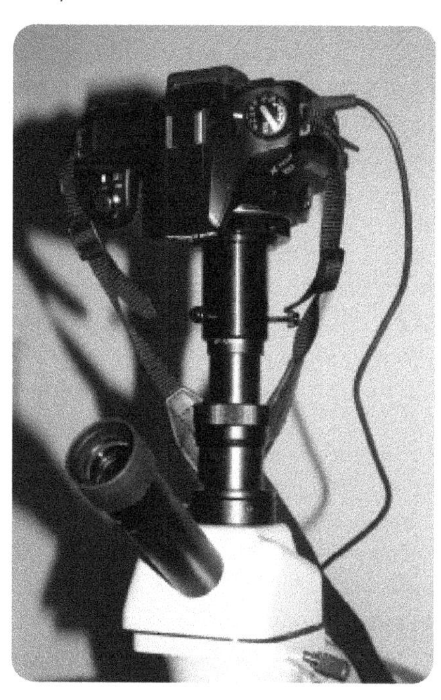

Even if you don't have a dual-head microscope or a camera adapter, it's possible with patience and trial-and-error to shoot usable photomicrographs merely by setting a point-and-shoot digital camera to macro mode and holding it up to the microscope eyepiece. If you attempt this method, you may get better results if you use a short length of cardboard or plastic tube between the eyepiece and the camera lens to block extraneous light and make alignment easier.

SPECTROMETER (OPTIONAL)

Professional forensics labs use a variety of *spectrometers* and *spectrophotometers* to analyze unknown substances by examining the light they emit or absorb at specific wavelengths. Technically, a spectrometer may cover any range of wavelengths, visible and/or invisible, while a spectrophotometer is limited to visible wavelengths, but spectrophotometers are often referred to by the more inclusive (and shorter) term as spectrometers.

Broadly speaking, there are two classes of spectrometers.

An *absorption spectrometer* measures how much light at specific wavelengths is absorbed by a sample in liquid or gaseous form. For example, a copper sulfate solution appears blue because it selectively absorbs yellow wavelengths strongly but is nearly transparent to blue wavelengths. Observing the absorption spectrogram allows a scientist to identify the compound as copper sulfate and determine its concentration. Similarly, an *atomic absorption spectrometer* can be used to identify the specific chemical elements present in an unknown gaseous sample because each element absorbs light at specific wavelengths characteristic to that element.

An *emission spectrometer* measures the light emitted by an unknown sample that has been vaporized and heated to incandescence. Just as each element when in an unexcited state absorbs light at specific wavelengths, each chemical element when in an excited state emits light at those same wavelengths. For example, Figure 2-6 shows the line emission spectrum of a compact fluorescent lamp, showing strong emission lines for mercury.

Figure 2-6: *The spectrum of a compact fluorescent lamp, showing strong mercury lines (image courtesy Rob Brown)*

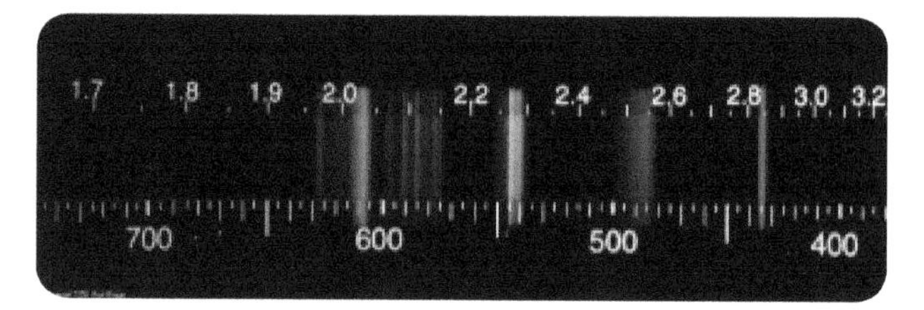

ABOUT FIGURE 2-6

We had little luck in shooting our own images of emission spectra, so we asked Rob Brown (*http//:home.comcast.net/~emcculloch-brown/astro/spectrostar.html*), who created the image shown, for permission to use one of his images, which he kindly granted. In fact, Rob shot a new image for us, and included these comments when he emailed the image to me.

"The calibration is a little off. The strong green line should line up with 546 nm, but it's a little short of 545 nm.

The camera does not do the image justice. The colors are very compressed due to the color filters in the camera detector. Notice the lack of yellow (575 nm), cyan (490 nm), and violet (400 nm). Even the orange looks muddy. I can see visually lines at 710, 705, 685, 656, 645, and 400 nanometers. The camera can't."

Figure 2-6 was made by using the Project Star Spectrometer shown in Figure 2-7. This assembled, calibrated plastic version sells for $40.

Figure 2-7: *The Project Star Spectrometer, an inexpensive spectrophotometer*

Although they look like toys and are priced accordingly, these inexpensive spectrophotometers are serious scientific instruments. Obviously, they are not as sensitive or as accurate as professional models that sell for a thousand to ten thousand times the price, but they can be used to do real science nonetheless. We'll use one of these spectrophotometers in toxicology and soil analysis labs to detect barium and other heavy metals.

MASS SPEC

Despite its name, the *mass spectrometer*, another instrument used in forensics labs, is used to separate ions by mass, and so is not a spectrometer at all in the usual sense of the word.

ULTRAVIOLET LIGHT SOURCE (OPTIONAL)

Ultraviolet (*UV*) is electromagnetic radiation at wavelengths too short to be visible to the human eye. UV wavelengths range from 400 nanometers (nm), just beyond visible violet, down to 1 nm, beyond which lies the class of light called X-rays. Forensic scientists use UV light in the *near UV* range—400 nm down to 200 nm—for many purposes, including detecting blood at crime scenes, revealing alterations in questioned documents, and examining chromatograms of colorless compounds such as drugs and poisons.

The near UV range is further divided into the *UVA* (400 nm to 320 nm), *UVB* (320 nm to 280 nm), and *UVC* (280 nm to 200 nm) ranges. UVA light, also called *long-wave UV* or *blacklight*, has relatively low energy and is therefore safe to work with. We'll use UVA light in several lab sessions in this book. UVB and particularly UVC light has higher energy and is therefore dangerous to work with, requiring special protective goggles and clothing to prevent damage to eyes and skin. Because UVB and UVC light sources are also considerably more expensive than UVA light sources, we decided to limit the use of UV light in these lab sessions to UVA.

UVA light sources are sold in toy and novelty shops as "black light" lamps. Battery-powered portable UVA lights, often described as "long-wave UV" lights, are sold by laboratory supply vendors for as little as $10. If you already have a portable fluorescent light such as a camping lantern, you can simply replace the standard fluorescent tube with a BLB (blacklight blue) fluorescent tube of the correct size and wattage. Figure 2-8 shows a typical $10 battery-powered "black light" that is suitable for the lab sessions in this book.

Figure 2-8: *A typical $10 battery-powered UVA "black light"*

Another option is an ultraviolet LED flashlight. Typical pocket models use three AAA cells to drive an array of six to nine UV LEDs, providing bright illumination at 395 nm. These flashlights are widely available from online vendors for $5 to $10. We've come to prefer UV LED flashlights because they provide a relatively tight, intense beam of UV and because battery life is much better than the tube-based units. Although none of the lab sessions in this book requires a UV light source, having one available is useful for several lab sessions.

Table 2-1 lists the optical equipment we recommend. Items flagged in the FK01 column are included in the FK01 Forensic Science Kit.

Table 2-1: *Recommended optical equipment*

Item	FK01	Sources
Camera with macro capability (optional)	○	Local/on-line vendors
Magnifier	●	Local/on-line vendors
Microscope (ocular micrometer/reticle recommended)	○	Lab supply vendors
Microscope adapter for camera (optional)	○	Camera/microscope vendors
Microscope cleaning kit	○	Lab supply vendors
Microscope cover	○	Microscope vendor
Microscope coverslips, box	●	Lab supply vendors
Microscope immersion oil (optional)	○	Lab supply vendors
Microscope slides, flat, box of 72	●	Lab supply vendors
Microscope slides, deep-cavity, box of 12	●	Lab supply vendors
Mounting fluid for permanent mounts (optional)	○	Lab supply vendors, drugstores
Mounting fluids for temporary mounts	●	drugstores, supermarkets
Scanner with software (optional)	○	Local/online vendors
Spectrometer (optional)	○	See text
Ultraviolet light source	○	See text

Laboratory Equipment

In designing the lab sessions for this book, we made every effort to keep equipment requirements as modest as possible. With few exceptions—mainly optional items and expensive items such as a microscope—the FK01 Forensic Science Kit includes the specialty equipment and chemicals needed to do the lab sessions.

Most of the lab sessions in this book use microscale techniques. These techniques are in accord with real forensic practices, which usually (not always!) operate on very small samples, often as small as a single hair or a few milligrams of a questioned specimen. Microscale experiments also have the advantage of using smaller and less expensive equipment and requiring smaller quantities of expensive chemicals. That makes setup, teardown, cleanup, and storage faster and easier, and also minimizes disposal issues.

The following sections list and describe the equipment needed to complete the laboratory sessions in this book. Many of the items you'll need are common household items or things that are readily available locally, such as the following:

- Aluminum foil
- Bags, plastic zip (several quart/liter and one or two gallon)
- Bottles, storage (250 and 500 mL, 1 and 2 L soft drink bottles)
- Butane lighter
- Cotton balls (real cotton)
- Cotton swabs
- Cups, foam, 500 mL/pint
- Hair dryer (optional)
- Hammer
- Nickel (US coin)
- Newspapers or brown paper bags
- Ovens (standard and microwave)
- Paper (white copy and bond; black)
- Paper towels
- Pen, permanent marking, ultra-fine point (Sharpie or similar)
- Pencil
- Refrigerator/freezer
- Scissors
- Tape, masking
- Tape, transparent, wide (packing tape or similar)
- Toothpicks, plastic

In addition, you'll need the following items:

Goggles

All lab sessions that involve handling chemicals require goggles that are designed to prevent liquids from penetrating the goggles and getting into your eyes. You can purchase chemical splash goggles from any lab supply vendor.

WARNING

Safety glasses or impact goggles are **not** sufficient to protect your eyes against chemical splashes. You need chemical splash goggles, which are vented by cap vents rather than holes. Figure 2-9 shows a set of impact goggles on the left and proper chemical splash goggles on the right.

Figure 2-9: *Impact goggles (left) and chemical splash goggles*

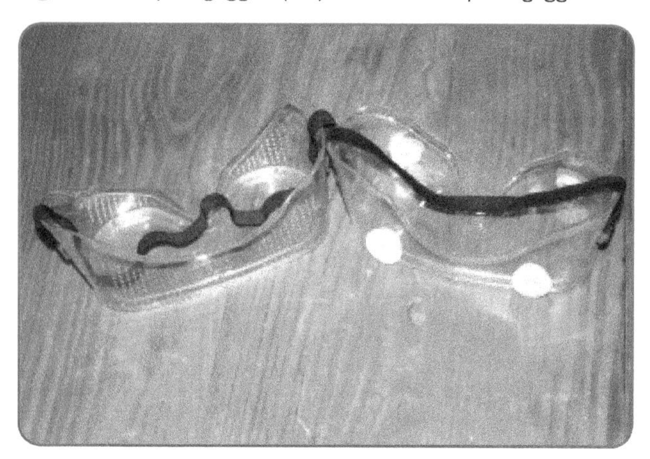

Balance

A balance is required for Lab Sessions I-2, Examine the Physical Characteristics of Soil, and III-1, Determine Densities of Glass and Plastic Specimens, and optional but recommended for several other lab sessions. Suitable pocket electronic balances are available from lab supplies vendors for as little as $25. Look for a model with centigram (0.01 gram) resolution and a capacity of 100 to 200 grams. Such a balance is also useful for the other sciences, including biology, chemistry, and physics.

Beakers, 100 mL and 250 mL

Many lab sessions require making up solutions and other activities for which 100 mL and 250 mL polypropylene beakers are useful. You can can purchase suitable beakers from any lab supplies vendor, or substitute measuring cups or similar containers.

Bottle, sprayer

Some lab sessions require misting a surface with a particular reagent. The FK01 Forensic Science Kit includes a suitable sprayer bottle. If you don't have the kit, you can buy small "fingertip sprayer bottles" at most drugstores for a dollar or so.

Burner, gas

Lab I-5, Examine the Spectroscopic Characteristics of Soil, requires a gas burner of some sort. Hardware store propane torches are suitable, as are the small gas torches sold for soldering and other hobby purposes. In a pinch, you can substitute a natural gas kitchen stove burner.

Centrifuge tubes, 1.5 mL, 15 mL, and 50 mL

Polypropylene centrifuge tubes have numerous uses: everything from storing specimens to holding solutions to acting as small reaction vessels to developing chromatography strips. The FK01 Forensic Science Kit includes fifteen 1.5 mL tubes with snap caps and six each of the 15 mL and 50 mL tubes with screw-on caps. If you don't have the kit, you can purchase these tubes from most lab supply vendors, or simply substitute suitable small glass or plastic containers.

Chromatography paper

Chromatography paper is used in several lab sessions. You can purchase chromatography paper from lab supply vendors in letter-size or larger sheets or as pre-cut strips. If you don't have the FK01 kit, or if you need additional chromatography paper, you can substitute strips cut from filter paper or bleached white coffee filters.

Dissection tools

Some of the lab sessions involve manipulating tiny specimens such as a single hair or fiber. The FK01 Forensic Science Kit includes bent and straight dissecting needles and forceps. If you don't have the kit, you can buy these items individually or as part of a dissecting kit from any lab supply vendor, or you can substitute household tweezers and large sewing needles.

Digital voice recorder

Although it is optional, a digital voice recorder is extremely useful for taking voice notes hands-free during procedures. We use an old Olympus WS-100 digital voice recorder, which hangs around Robert's neck on a lanyard, but you can substitute a cell phone, MP3 player, or other device that has a digital voice recorder function.

Dishes, drying

Two of the lab sessions in the soil analysis group require drying soil specimens in the oven. You can use any oven-safe flat containers—saucers, oven dishes, and so on—or even boats made from aluminum foil. Disposable aluminum pie plates, which are used in another lab session, are also an excellent choice.

Filters, plane-polarizing

Two of the lab sessions involved examining specimens by polarized light. For those sessions, you'll need a pair of plane-polarizing filters, which are included in the FK01

Forensic Science Kit. If you don't have the kit, you can purchase these filters from most lab supply vendors, or you can substitute a pair of plane-polarizing camera filters. (Although we haven't tested it, you may also be able to substitute polarizing sun glasses.)

Fingerprint brush

The lab session on dusting for fingerprints requires a suitable brush, which is included in the FK01 Forensic Science Kit. If you don't have the kit, you can substitute a small artist's paintbrush, a makeup brush, or a similar soft-bristled brush, ideally camelhair. In a pinch, you can substitute a feather. (Actually, a few professional fingerprint technicians prefer a feather to any brush.) The size of the brush is a matter of personal preference. Some professional fingerprint technicians prefer a very small brush, while others prefer a brush with bristles 2 cm or more in diameter.

Fuming chambers

Two lab sessions in the fingerprinting group require fuming chambers, one for iodine and the other for superglue. The required sizes of both depend on the sizes of your specimens. For the superglue fuming chamber, we used a disposable plastic one-quart Gladware kitchen container with a snap lid, which is also suitable for iodine-fuming small specimens. If you are iodine-fuming larger specimens, use a one-gallon zip lock plastic bag.

> Any chamber you use for superglue fuming will have a cloudy layer of superglue covering all its interior surfaces, so use something disposable. Iodine stains can be removed with a solution made by dissolving a vitamin C tablet in a few milliliters of tap water.

Gloves

Wear chemical-resistant gloves at all times while working with chemicals and/or specimens. The purpose of the gloves is two-fold: to protect you from chemicals, and to protect the specimens from being contaminated by oils and other substances present on your skin. The best choice is disposable latex or nitrile exam gloves, which are inexpensive and sufficient to protect your hands from the chemicals you'll be using. It's important to use gloves that fit your hands properly. If the gloves are too small, they stretch excessively and are more likely to develop pinholes or even tear. If they're too large, it's more difficult to manipulate small items while wearing them.

> When you're working with particularly corrosive chemicals—such as diphenylamine reagent, Mandelin reagent, and Marquis reagent—you may wish to use more protection than a single thin layer of latex or nitrile. In that case, either double-glove with exam gloves or wear heavier gloves, such as the heavy "rubber" gloves sold in supermarkets for household use.

Graduated cylinders, 10 mL and 100 mL

Many lab sessions require measuring solutions accurately. You'll need both 10 mL and 100 mL graduated cylinders, either glass or polypropylene. Glass cylinders are transparent, but are easily broken. Polypropylene cylinders are translucent, but are unbreakable and have no meniscus. Which you use is personal preference. The FK01 Forensic Science Kit includes 10 mL and 100 mL polypropylene cylinders. If you don't have the kit, you can purchase graduated cylinders from any lab supply vendor.

Inoculating loop

The lab session on spectroscopic analysis of soil requires an inoculating loop, which the FK01 Forensic Science Kit includes. If you don't have the kit, you can purchase an inoculating loop from any lab supply vendor, or you can substitute a large sewing needle with the tip embedded in a wooden dowel or pencil. If you don't intend to do that lab session, you don't need the loop. If you don't have the kit, inoculating loops are available from any lab supply vendor.

Leads, alligator clip

In Lab Session XI-4, DNA Analysis by Gel Electrophoresis, we construct and use a DNA gel electrophoresis apparatus. To construct that apparatus, you'll need a pair of alligator clip leads, which are included in the FK01 Forensic Science Kit. If you don't have the kit, you can purchase these leads at a local Radio Shack or other electronics supply store.

Light sources

Several lab sessions require a bright light source, such as a small desk lamp. (If you have a work surface that is illuminated by bright natural light, the lamp is unnecessary.) Examining opaque objects under the microscope requires a source of incident (top) illumination. We use a white LED book light with a flexible neck, which allows positioning the angle and distance of the light. You can also use a bright desk lamp or other light source.

Mesh, fiberglass

Lab Session I-2, Examine the Physical Characteristics of Soil, involves sifting soil specimens through a mesh to separate particles by size. The FK01 Forensic Science Kit includes a suitable piece of mesh, but you can substitute any similar mesh, such as a piece of window screen or a kitchen flour sifter.

Microscope slide, cross-sectioning

A cross-sectioning slide is a standard plastic or metal microscope slide drilled with a small hole or holes. Fibers are drawn through the hole and then cut flush with the top and bottom of the slide using a scalpel or razor blade to allow viewing the fibers in cross section through the microscope. The FK01 Forensic Science Kit includes a drilled plastic cross-sectioning slide. You can purchase metal versions of these slides from forensic supply vendors or make your own by drilling a plastic slide or using a heated needle to melt a hole in it.

Modeling clay

Modeling clay is included in the FK01 Forensic Science Kit, but can also be purchased from toy, hobby, and craft stores. It's used to mount small opaque specimens for viewing under the microscope.

Paint chips

Lab I-2, Examine the Physical Characteristics of Soil, requires a selection of soil-colored paint chips, which you can obtain at a hardware store or paint store.

Pie pans, disposable aluminum

Lab VI-1, Tool Mark Analysis, requires thin aluminum sheets, thicker than foil, for making tool marks. One convenient and inexpensive source is the disposable aluminum pie pans sold in supermarkets. Buy a pack. These can also be used as drying dishes.

Plastic sheet, transparent

Lab IV-5, Revealing Latent Fingerprints On Sticky Surfaces, optionally uses a plastic sheet as a fingerprint transfer sheet. The transparent plastic sheets sold for overhead transparencies or as notebook sheet protectors are suitable.

Pipettes, plastic graduated

Plastic graduated pipettes are used for measuring and transferring small amounts of liquids (see Figure 2-10). The pipettes supplied with the FK01 Forensic Science Kit are graduated with four lines on the stem of the pipette at 0.25, 0.5, 0.75, and 1 mL. They can also be used to measure very small amounts of dilute aqueous solutions because they deliver about 36 drops per mL, which translates to about 27.8 microliters per drop. (This value may differ for solutions that are more or less viscous than water.) If you don't have the kit, these pipettes can be purchased from any lab supply vendor.

Figure 2-10: *Graduated plastic pipettes*

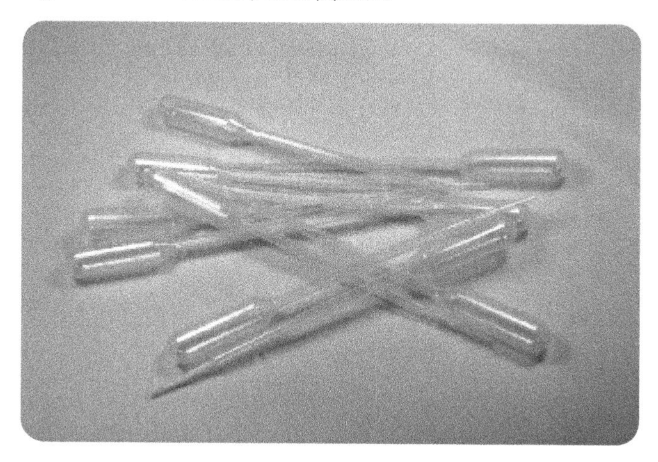

Plates, reaction and spot

The FK01 Forensic Science Kit includes a 24-well polystyrene deep-well reaction plate with a lid, shown in

Figure 2-11. This reaction plate is used in many lab sessions for running chemical tests, mixing or diluting solutions, and so on. The FK01 kit also includes a 12-well polypropylene shallow-well spot plate. The 12-well plate is used for liquids—strong acids, solvents, and so on—that may damage the polystyrene plate. If you don't have the kit, you can purchase these plates from any lab supply vendor.

Figure 2-11: *A reaction plate*

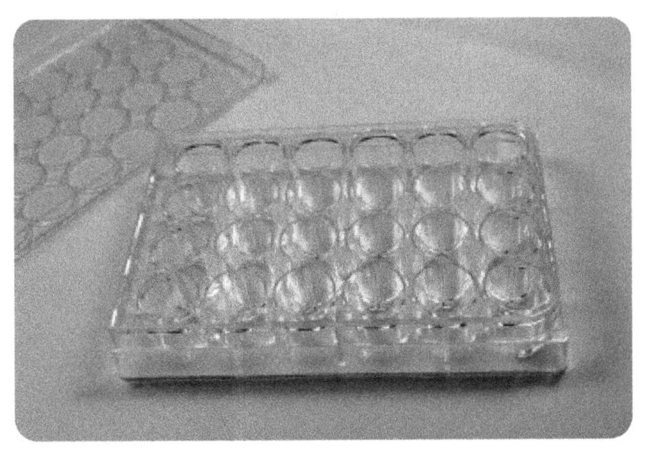

Ruler

The FK01 Forensic Science Kit includes a standard 6"/150 mm ruler. If you don't have the kit, you can substitute any millimeter-graduated ruler.

Scalpel

The FK01 Forensic Science Kit includes a standard disposable scalpel. If you don't have the kit, you can purchase a scalpel from any lab supplies vendor or substitute a single-edge razor blade.

Sieves

Lab I-2, Examine the Physical Characteristics of Soil, involves separating soil specimens into fractions based on particle size. The FK01 Forensic Science Kit includes one piece of mesh, which allows separating the specimens into two fractions. Optionally, you can use other mesh sizes to separate the specimens into three or more fractions. Household items such as a flour sifter or a fine mesh metal coffee filter can be used to do this.

Small items

The FK01 Forensic Science Kit includes a flat/spoon microspatula, a ruler, and a stirring rod. If you don't have the kit, you can purchase these items from any lab supply vendor, or substitute similar household items.

Steam iron (or oven)

Lab session IV-3, Revealing Latent Fingerprints Using Ninhydrin, requires a steam iron (by preference) or a kitchen oven.

Test tubes and accessories

The FK01 Forensic Science Kit includes six test tubes, a test tube clamp, and a test tube rack. If you don't have the kit, you can purchase these items from any lab supplies vendor.

Timer

Several lab sessions require timing with more or less accuracy. Any watch or clock with a second hand is sufficient for these sessions.

Transfer sheets

Lab session II-1, Gathering Hair Specimens, requires transfer sheets to contain the hair specimens you obtain. You can use something as simple as index cards or even sheets of paper.

Table 2-2 summarizes the laboratory equipment we recommend.

Table 2-2: *Recommended laboratory equipment*

Item	FK01	Sources
Balance (optional)	○	Lab supply vendors
Beaker, 100 mL (glass or polypropylene)	●	Lab supply vendors
Beaker, 250 mL (glass or polypropylene)	●	Lab supply vendors
Bottle, sprayer	●	Drugstores
Burner, gas	○	See text
Centrifuge tubes, 1.5 mL	●	Lab supply vendors
Centrifuge tubes, 15 mL	●	Lab supply vendors
Centrifuge tubes, 50 mL	●	Lab supply vendors
Chromatography paper	●	Lab supply vendors
Digital voice recorder (optional)	○	See text
Dishes, drying	○	See text
Filters, plane-polarizing (2)	●	Lab supply vendors
Fingerprint brush	●	Forensic supply vendors
Forceps	●	Lab supply vendors
Fuming chambers	○	See Text
Gloves	○	Drugstores, supermarkets
Goggles, chemical splash	●	Lab supply vendors
Graduated cylinder, 10 mL (glass or polypropylene)	●	Lab supply vendors
Graduated cylinder, 100 mL (glass or polypropylene)	●	Lab supply vendors
Inoculating loop	●	Lab supply vendors
Lab notebook (bound composition book or similar)	○	Office supply vendors
Leads, alligator clip (2; red and black)	●	Electronics supply vendors
Light sources	○	See text
Mesh, fiberglass	●	Hardware stores
Modeling clay	●	Toy/hobby/craft stores
Needle, dissecting (teasing), bent	●	Lab supply vendors
Needle, dissecting (teasing), straight	●	Lab supply vendors
Paint chips	○	See text

Item	FK01	Sources
Pie pans, disposable aluminum	○	Supermarkets
Pipettes, plastic, graduated	●	Lab supply vendors
Plastic sheet, transparent (optional)	○	See text
Plate, reaction, 24-well polystyrene with lid	●	Lab supply vendors
Plate, spot, 12-well polypropylene	●	Lab supply vendors
Ruler, millimeter scale	●	Office supply vendors
Scalpel	●	Lab supply vendors
Sieves (optional)	○	See text
Slide, microscope, cross-sectioning	●	Forensic supply vendors
Spatula	●	Lab supply vendors
Steam iron (or oven)	○	See text
Stirring rod	●	Lab supply vendors
Test tube clamp	●	Lab supply vendors
Test tube rack	●	Lab supply vendors
Test tubes	●	Lab supply vendors
Timer (watch or clock with second hand)	○	Obtain locally
Toothpicks, plastic	○	Supermarkets
Transfer sheets	○	See text

Chemicals and Reagents

In addition to the equipment described in the previous section, you'll need an assortment of raw chemicals and reagents to complete the lab sessions. (Broadly speaking, raw chemicals are materials that are supplied and used as-is in pure form or as simple solutions, while reagents are specific mixtures of chemicals, usually in liquid form.)

The word reagent has two meanings in lab parlance: a reagent may be a mixture of chemicals, and often bears a person's name. For example, Marquis reagent is a mixture of the raw chemicals sulfuric acid and formaldehyde. But "reagent" is also used to specify a purity grade for raw chemicals. Chemicals that are specified as "reagent," "reagent grade," or "ACS reagent grade" meet specific purity standards, are assayed to determine their exact contents, and are extremely pure. Reagent-grade chemicals are suitable for the lab sessions in this book but are often purer (and more expensive) than actually necessary. For most of the lab sessions in this book, laboratory-grade, USP-grade, or another relatively pure grade of chemicals is sufficient.

All of the lab sessions that require special reagents include instructions for preparation of those reagents from raw chemicals. If you make up these reagents yourself, be sure to follow all safety procedures, including reading the MSDSs (Material Safety Data Sheets) for the chemicals you use. Always wear goggles, gloves, and protective clothing while making up and using these reagents.

Acetone

Several lab sessions require acetone, which is not included in the FK01 Forensic Science Kit. You can purchase pure acetone in pint (500 mL) or quart (1 L) containers at any hardware store or paint store.

Agar powder

The FK01 Forensic Science Kit includes 10 grams of agar powder, which is used in the DNA gel electrophoresis lab session. If you don't have the kit, you can purchase agar from a lab supply vendor. You can also substitute food-grade agar, which sells for a few dollars per ounce in some supermarkets and most stores that sell Chinese and Japanese specialty foods. The gelling quality of food-grade agar is variable, so we recommend using twice the amounts specified in the lab session.

Alcohol, ethyl (ethanol), 95%

Several lab sessions require ethanol, which is not included in the FK01 Forensic Science Kit. You can purchase 95%/96% ethanol under that name or as ethyl alcohol in hardware and paint stores. Most drugstores carry 70% ethanol, which is not ideal but is generally usable.

Alcohol, isopropyl (isopropanol), 99%

Several lab sessions require isopropanol, which is not included in the FK01 Forensic Science Kit. You can purchase 99% isopropanol under that name or as isopropyl alcohol in some hardware stores. Most drugstores carry 70% isopropanol, which is not ideal but is generally usable. Some drugstores also carry 91% isopropanol, which is superior to the 70% concentration for our purposes.

> In general, ethanol and isopropanol can be used interchangeably for the lab sessions in this book. Sometimes one or the other yields superior results. If we specify one and you don't have it, the other will almost certainly work.

Ammonia, household (clear, non-sudsy)

Two lab sessions require ammonia, which is not included in the FK01 Forensic Science Kit. Household ammonia is generally a 5% to 10% concentration, which is fine for our purposes. Get the clear, non-sudsy version.

Ascorbic acid (vitamin C) tablets, 500 mg

The FK01 Forensic Science Kit includes 500 mg ascorbic acid (vitamin C) tablets. If you don't have the kit or if you require additional tablets, simply use standard 500 mg vitamin C tablets.

Baking soda (sodium bicarbonate)

One lab session requires baking soda, which is not included in the FK01 Forensic Science Kit. Use ordinary baking soda from the supermarket.

Bleach, chlorine laundry

Two lab sessions require chlorine laundry bleach, which is not included in the FK01 Forensic Science Kit. Use ordinary 5.25% sodium hypochlorite bleach. The cheap, no-name stuff is fine.

Blood, synthetic

The FK01 Forensic Science Kit includes synthetic blood, which mimics the chemical behavior of actual blood. If you don't have the kit, you can purchase synthetic blood from a forensic supply vendor, or simply substitute actual blood from raw meat or another source. (Note that there are several types of synthetic blood available. Theater blood is designed to look like real blood, but is unsuitable for our purposes. Synthetic blood designed for use in spatter-pattern analysis is also unsuitable. If you buy synthetic blood, you want the type used for training analysts on blood detection activities.)

Buffers, DNA loading and running

The FK01 Forensic Science Kit includes a 6X concentrate of DNA loading buffer and a 20X concentrate of DNA running buffer. If you don't have the kit, you can make up these buffers yourself or purchase them from forensic supply vendors or general lab supply vendors.

Copper(II) sulfate solution

One of the forensic drug testing sessions requires a copper(II) sulfate solution, which is included in the FK01 Forensic Science Kit. If you don't have the kit, you can make up the solution from copper(II) sulfate crystals purchased from a lab supply vendor or another source.

Diphenylamine reagent

The lab session on testing for explosives residues requires diphenylamine reagent, which is included in the FK01 Forensic Science Kit. If you don't have the kit, you can make up this reagent from chemicals purchased from a lab supply vendor.

Dishwashing detergent (Dawn or similar)

Three lab sessions require a liquid detergent, which is not included in the FK01 Forensic Science Kit. Use Dawn or a similar dishwashing liquid.

Dragendorff's reagent

One forensic toxicology lab session requires Dragendorff's reagent, which is included in the FK01 Forensic Science Kit. If you don't have the kit, you can purchase Dragendorff's reagent from most forensic supply vendors and many lab supply vendors, or you can make it up yourself as described in the lab session.

Fingerprint powders, black and white

The lab session on dusting for fingerprints requires these fingerprint powders, which are included in the FK01 Forensic Science Kit. If you don't have the kit, you can purchase fingerprint powders from a forensic supply vendor, or you can use substitutes such as soot, lampblack, or charcoal powder as the black powder, and talcum powder or cornstarch as the white powder.

Gentian violet solution

The lab session on detecting fingerprints on sticky tape requires a 0.1% aqueous solution of gentian violet, which is included in the FK01 Forensic Science Kit. If you don't have the kit, you can substitute drugstore gentian violet (also called crystal violet) solution. These drugstore solutions are usually 1%, so they need to be diluted one volume of 1% gentian/crystal violet solution to nine volumes of distilled or deionized water.

Glycerol

Several lab sessions require glycerol, which is included in the FK01 Forensic Science Kit. If you don't have the kit, you can purchase glycerol (also called glycerin) in some drugstores and health food stores.

Herzberg's stain

The lab session on paper analysis requires Herzberg's stain, which is included in the FK01 Forensic Science Kit. If you don't have the kit, you can make up this stain yourself from raw chemicals. We know of no commercial source for the prepared stain.

Hydrochloric acid

Several lab sessions require hydrochloric acid, a 6 molar solution of which is included in the FK01 Forensic Science Kit. If you don't have the kit, you can make up 6 M hydrochloric acid by diluting concentrated (12 M, 37%) hydrochloric acid from a lab supply vendor 1:1 with distilled or deionized water.

Hydrogen peroxide, 30%

The lab session on diatom analysis requires 30% hydrogen peroxide, which is included in the FK01 Forensic Science Kit. If you don't have the kit, you can purchase 30% hydrogen peroxide from any lab supply vendor.

Hydrogen peroxide, 3%

Two lab sessions require 3% hydrogen peroxide, which is not included in the FK01 Forensic Science Kit. Although you can dilute one volume of the 30% hydrogen peroxide supplied with the kit with nine volumes of water to make 3% hydrogen peroxide, 3% hydrogen peroxide is cheaply available in any drugstore.

Iodide/glycerol reagent

The lab session on revealing alterations in documents requires an iodide/glyerol reagent, which is included in the FK01 Forensic Science Kit. If you don't have the kit, you can make up the solution yourself using glycerol and potassium iodide from a lab supply vendor. We know of no commercial source for this reagent.

Iodine crystals

Three lab sessions require a few small crystals of solid iodine, which is included in the FK01 Forensic Science Kit. If you don't have the kit, you can purchase iodine crystals from a lab supply vendor, but because solid iodine is used illegally to manufacture methamphetamine, most vendors refuse to sell solid iodine to anyone other than established customers. Fortunately, it's easy enough to make iodine yourself. Robert describes how to do it in this video: *http://www.youtube.com/watch?v=CLhwkFKLdPA.*

Iodine solution

One of the forensic toxicology lab sessions requires an iodine solution, which is included in the FK01 Forensic Science Kit. If you don't have the kit, you can make up the solution using the instructions in the lab session.

Jenk's stain

The lab session on paper analysis requires Jenk's stain, which is included in the FK01 Forensic Science Kit. If you don't have the kit, you can make up this stain yourself from raw chemicals. We know of no commercial source for the prepared stain.

Kastle-Meyer reagent

The lab session on blood detection requires Kastle-Meyer reagent, which is included in the FK01 Forensic Science Kit. If you don't have the kit, you can purchase this reagent from a forensic supply vendors or make it up yourself. We don't recommend making it up yourself unless you have a reasonably complete home lab, including a ground-glass distillation apparatus, and are comfortable working with boiling solutions of concentrated potassium hydroxide.

Lead nitrate

Two lab sessions, on gunshot residue analysis and explosives residue analysis, use a dilute solution of lead nitrate, which is included in the FK01 Forensic Science Kit. If you don't have the kit, you can make up a 0.1% solution of lead nitrate by dissolving 0.1 gram of lead nitrate in 100 mL of distilled or deionized water.

Mandelin reagent

The lab session on presumptive drug testing requires Mandelin reagent, which is included in the FK01 Forensic Science Kit. If you don't have the kit, you can make up this reagent yourself from the raw chemicals, or you can purchase it from a forensic supply vendor.

Marquis reagent

The lab sessions on presumptive drug testing and detecting cocaine and methamphetamine on paper currency require Marquis reagent, which is included in the FK01 Forensic Science Kit. If you don't have the kit, you can make up this reagent yourself from the raw chemicals, or you can purchase it from a forensics supply vendor.

Methylene blue reagent, parts A and B

The lab session on explosives residues analysis requires methylene blue reagent, a two-part reagent that is included in the FK01 Forensic Science Kit. If you don't have the kit, you can make up this reagent yourself from the raw chemicals, or you can purchase it from a forensic supply vendor.

Methylene blue stain

Two lab sessions on DNA analysis require methylene blue stain, which is included in the FK01 Forensic Science Kit. If you don't have the kit, you can purchase methylene blue stain from any lab supply vendor or use the methylene blue solution sold in pet stores for treating fish.

Modified Griess reagent, Parts A and B

The lab sessions on gunshot residues analysis and explosives residues analysis require modified Griess reagent, a two-part reagent that is included in the FK01 Forensic Science Kit. If you don't have the kit, you can purchase modified Griess reagent from a forensic supply vendor or make it up yourself.

Molybdate reagent

The lab session on chemical analysis of soils requires molybdate reagent, which is included in the FK01 Forensic Science Kit. If you don't have the kit, you can purchase molybdate reagent from Hach (*http://www.hach.com/*) in either powder or liquid form or make it up yourself.

Nail polish

In addition to being useful as a permanent slide-mounting fluid, a clear, colorless nail polish is used in lab sessions on hair analysis. The FK01 Forensic Science Kit does not include this chemical. We recommend purchasing a small bottle of clear, colorless Sally Hansen Hard as Nails at the drugstore.

Ninhydrin powder

One of the lab sessions on fingerprinting requires ninhydrin, which is included as a powder in the FK01 Forensic Science Kit. If you don't have the kit, you can purchase ninhydrin in powder or solution form from a forensic supply vendor.

Oils

Some of the lab sessions use cassia oil and olive oil, which are included in the FK01 Forensic Science Kit, and clove oil (optional), which is not. If you don't have the kit, you can purchase olive oil at the supermarket. Cassia (cinnamon) oil and clove oil are expensive and harder to find, but many specialty food stores carry them.

Papain powder

One of the DNA labs uses papain powder, which is included in the FK01 Forensic Science Kit. If you don't have the kit, you can substitute Adolph's Meat Tenderizer or another papain-based meat tenderizer.

Phosphate extraction reagent, concentrate

The lab session on chemical analysis of soil requires this reagent, which is included in the FK01 Forensic Science Kit. If you don't have the kit, you can make up this reagent yourself from the raw chemicals.

Phosphate standard, 1,000 ppm phosphate

The lab session on chemical analysis of soil requires a standardized phosphate solution. The FK01 Forensic Science Kit includes a solution that is 1,000 ppm with respect to phosphate ion. If you don't have the kit, you can purchase a standardized phosphate solution from a specialty supplier or make up the solution yourself.

Salicylate reagent

One of the forensic toxicology lab sessions requires salicylate reagent, which is included in the FK01 Forensic Science Kit. If you don't have the kit, you can make up this solution simply by dissolving iron(III) nitrate (ferric nitrate) in distilled or deionized water. The concentration is not critical; anything close to 1% w/v is adequate.

Salicylate standard solution

That same forensic toxicology lab session requires a standardized salicylate solution, which is included in the FK01 Forensic Science Kit. If you don't have the kit, you can purchase a standardized salicylate solution or make up the solution yourself.

Scott reagent

Two of the lab sessions on forensic drug testing require Scott reagent, which is included in the FK01 Forensic Science Kit. If you don't have the kit, you can purchase Scott reagent from a forensic supply vendor, or make it up yourself.

Sodium carbonate solution

One of the lab sessions on forensic drug testing requires a solution of sodium carbonate, which is included in the FK01 Forensic Science Kit. If you don't have the kit, you can make it up yourself with sodium carbonate purchased from a lab supply vendor or from washing soda. Concentration is not critical.

Sodium dithionite

One of the lab sessions on fiber testing requires sodium dithionite (also called sodium hydrosulfite), which is included in the FK01 Forensic Science Kit. If you don't have the kit, you can purchase it from a lab supply vendor or substitute RIT Color Remover.

Sodium nitrite, 1% w/v

Two of the lab sessions on gunshot residues analysis and explosives residues analysis use a sodium nitrite solution. The FK01 Forensic Science Kit includes a 1% w/v sodium nitrite solution. If you don't have the kit, you can make up a solution by dissolving sodium nitrite in distilled or deionized water. Concentration is not critical.

Starch indicator solution

One of the forensic toxicology lab sessions requires starch indicator solution, which is included in the FK01 Forensic Science Kit. If you don't have the kit, you can purchase it from a lab supply vendor or simply use water in which potatoes, rice, or pasta has been boiled.

Superglue, cyanoacrylate

Lab session IV-4, Revealing Latent Fingerprints Using Superglue Fuming, requires cyanoacrylate-based superglue. Most common superglues, sold under brand names like Super Glue and Krazy glue, are based on ethyl-2-cyanoacrylate, but some are based on methyl 2-cyanoacrylate. Either is suitable, as are medical and veterinary glues based on n-butyl cyanoacrylate or 2-octyl cyanoacrylate. Only a few drops are required, so a small bottle or tube is sufficient. You can retain any unused superglue for use around the house.

Testfabrics dyes

One of the fiber testing lab sessions requires Testfabrics, Inc. TIS #1 and TIS #3A dyes, which are included in the FK01 Forensic Science Kit. If you don't have the kit, you can purchase these dyes directly from Testfabrics, Inc. (*http://testfabrics.com/*) or from a forensic supply vendor.

Vinegar, distilled white

Several lab sessions require distilled white vinegar, which is not included in the FK01 Forensic Science Kit. Purchase a bottle at the supermarket.

Water, distilled or deionized

Many lab sessions require distilled or deionized water, which is not included in the FK01 Forensic Science Kit. Purchase a gallon bottle at the supermarket.

Zinc chloride solution

One of the lab sessions on fingerprinting requires a zinc chloride solution, which is included in the FK01 Forensic Science Kit. If you don't have the kit, you can make up the solution from solid zinc chloride, sold by any lab supply vendor, and distilled or deionized water.

Table 2-3 summarizes our recommendations for chemicals and reagents.

Table 2-3: *Recommended chemicals and reagents*

Item	FK01	Sources
Acetone	○	Hardware stores or paint stores
Agar powder	●	Lab supply vendors
Alcohol, ethyl (ethanol), 95%	○	Drugstores or hardwares store
Alcohol, isopropyl (isopropanol), 95%	○	Drugstores or hardwares store
Ammonia, household (clear, non-sudsy)	○	Supermarkets
Ascorbic acid (vitamin C) tablets, 500 mg	●	Drugstores, supermarkets
Baking soda (sodium bicarbonate)	○	Supermarkets
Bleach, chlorine laundry	○	Supermarkets
Blood, synthetic	●	Forensic supply vendors
Buffer, DNA loading, 6X concentrate	●	Lab supply vendors
Buffer, DNA running, 20X concentrate	●	Lab supply vendors
Copper(II) sulfate solution	●	Lab supply vendors
Diphenylamine reagent	●	Forensic supply vendors
Dishwashing detergent (Dawn or similar)	○	Supermarkets
Dragendorff's reagent	●	Forensic supply vendors
Fingerprint powder, black	●	Forensic supply vendors
Fingerprint powder, white	●	Forensic supply vendors
Gentian violet solution, 0.1% aqueous	●	Drugstores, lab supply vendors
Glycerol	●	Drugstores, supermarkets
Herzberg's stain	●	Forensic supply vendors
Hydrochloric acid	●	Lab supply vendors
Hydrogen peroxide, 3%	○	Drugstores, supermarkets

Item	FK01	Sources
Hydrogen peroxide, 30%	●	Lab supply vendors
Iodide/glycerol reagent	●	Forensic supply vendors
Iodine crystals	●	Lab supply vendors
Iodine solution	●	Lab supply vendors
Jenk's stain	●	Forensic supply vendors
Kastle-Meyer reagent	●	Forensic supply vendors
Lead nitrate, 0.1%	●	Lab supply vendors
Mandelin reagent	●	Forensic supply vendors
Marquis reagent	●	Forensic supply vendors
Methylene blue reagent, Part A	●	Forensic supply vendors
Methylene blue reagent, Part B	●	Forensic supply vendors
Methylene blue stain	●	Lab supply vendors
Modified Griess reagent, Part A	●	Forensic supply vendors
Modified Griess reagent, Part B	●	Forensic supply vendors
Molybdate reagent	●	Forensic supply vendors
Nail polish, clear colorless (Sally Hansen Hard as Nails)	○	Drugstores
Ninhydrin powder	●	Forensic supply vendors
Oil, cassia	●	Health food stores
Oil, clove (optional)	○	Health food stores
Oil, olive	●	Supermarkets
Papain powder	●	Health food stores
Phosphate extraction reagent, concentrate	●	Lab supply vendors
Phosphate standard, 1000 ppm phosphate	●	Lab supply vendors
Salicylate reagent (1% ferric nitrate)	●	Lab supply vendors
Salicylate standard solution	●	Lab supply vendors
Scott reagent	●	Forensic supply vendors
Sodium carbonate solution	●	Lab supply vendors
Sodium dithionite	●	Lab supply vendors
Sodium nitrite, 1% w/v	●	Lab supply vendors
Starch indicator solution	●	Lab supply vendors

Continued

Table 2-3: *Recommended Chemicals and Reagents (continued)*

Item	FK01	Sources
Superglue, cyanoacrylate	○	Hardware stores, supermarkets
Testfabrics TIS #1 dye	●	Forensic supply vendors
Testfabrics TIS #3A dye	●	Forensic supply vendors
Vinegar, distilled white	○	Supermarkets
Water, distilled or deionized	○	Supermarkets
Zinc chloride solution	●	Lab supply vendors

Specimens

Forensic science focuses on the comparison of known and questioned specimens. We'll need a lot of specimens, some of which are included in the FK01 Forensic Science Kit, and others you'll need to obtain locally yourself. Here are the specimens we recommend.

Acetaminophen

Three of the lab sessions on forensic drug testing require acetaminophen, which is included in the FK01 Forensic Science Kit. If you don't have the kit, use standard acetaminophen tablets from the drugstore.

Aspirin

Three of the lab sessions on forensic drug testing require aspirin, which is included in the FK01 Forensic Science Kit. If you don't have the kit, use standard aspirin tablets from the drugstore.

Caffeine

Two of the lab sessions on forensic drug testing and one of the lab sessions on forensic toxicology require caffeine, which is included in the FK01 Forensic Science Kit. If you don't have the kit, use standard caffeine tablets from the drugstore, which are sold as generics and under brand names like NoDoz and Vivarin.

Cartridge case, brass, fired

One of the fingerprinting lab sessions requires a brass cartridge case, which is included in the FK01 Forensic Science Kit. If you don't have the kit, you may be able to obtain fired brass at a local shooting range. Otherwise, you can substitute any uncoated brass object.

Chlorate known

Lab Session IX-2, Presumptive Color Tests for Explosives Residues, requires a known specimen that contains chlorate ion, which is included in the FK01 Forensic Science Kit. If you don't have the kit, you can use scrapings from the head of a safety match, which typically contains about 50% potassium chlorate.

Chlorpheniramine

Lab Session VII-1, Presumptive Drug Testing, requires a specimen of chlorpheniramine, which is included in the FK01 Forensic Science Kit. If you don't have the kit, you can use chlorpheniramine maleate tablets sold as allergy medication in any drugstore.

Chromatography questioned

Two of the lab sessions on forensic drug testing require a questioned (unknown) specimen for analysis, which is included in the FK01 Forensic Science Kit. If you don't have the kit, have someone else make up a questioned specimen for you that contains one, two, or all three of the following materials: acetaminophen and/or aspirin and/or caffeine.

Clothing, unlaundered

Lab Session II-1, Gathering Hair Specimens, requires one or more unlaundered clothing specimens, from which you will extract hairs and other trace evidence. Any clothing from the dirty clothes basket suffices. It can be interesting to compare the specimens obtained from underwear against those from outer clothing.

Digital image files

Lab Session VI-2, Matching Images to Cameras, requires a selection of digital image files, which you can obtain from your own collection as well as those of family members and friends.

Diphenhydramine

Two of the lab sessions on forensic drug testing require diphenhydramine, which is included in the FK01 Forensic Science Kit. If you don't have the kit, use standard diphenhydramine tablets from the drugstore, which are sold as generics and under brand names such as Benedryl. (Note that many brand-name and generic OTC medications contain diphenhydramine in combination with other drugs.)

Drug testing specimens, questioned A and B

Lab Session VII-1, Presumptive Drug Testing, requires two questioned drug specimens. The FK01 Forensic Science Kit includes two questioned specimens, which may be any one of the known materials tested in that lab session or any combination of two of them. If you don't have the kit, have someone make up two questioned specimens for you using those guidelines.

Fabric/fiber A, B, C, D, and E

Several of the fiber testing lab sessions require five known fabric specimens, which are included in the FK01 Forensic Science Kit. If you don't have the kit, you can substitute any five fabric specimens that all differ from each other, are made of various single or mixed fibers, and appear similar. (The kit includes five navy blue fabric specimens.)

Film negatives and/or slides

Lab Session VI-2, Matching Images to Cameras, requires a selection of black-and-white and/or color negatives and/or color slides, which you can obtain from your own collection as well as those of family members and friends.

Foods, assorted

Lab Session V-1, Testing the Sensitivity and Selectivity of Kastle-Meyer Reagent, requires a selection of food items to test for false-positives with Kastle-Meyer reagent. See the lab session for details.

Glass and plastic

Several of the lab sessions on glass and plastic analysis require known specimens of glass and various types of plastics. Your recycling bin is a good source of these.

Hair, animal and human

Several of the lab sessions on hair and fiber analysis require specimens of human and animal hair. You, your family, and your friends are obvious sources of human hair specimens and pet hair specimens.

Ibuprofen

Lab session VII-1, Presumptive Drug Testing, requires an ibuprofen specimen, which is included in the FK01 Forensic Science Kit. If you don't have the kit use ibuprofen tablets sold in any drugstore as generics or under brand names such as Advil, Motrin, and Nuprin.

Perchlorate known

Lab Session IX-2, Presumptive Color Tests for Explosives Residues, requires a known specimen that contains perchlorate, which is included in the FK01 Forensic Science Kit. If you don't have the kit, you can purchase ammonium, potassium, or sodium perchlorate from a lab supply vendor.

Poppy seeds

Two lab sessions in the forensic drug testing and forensic toxicology groups require a poppy seed specimen as a source of known opiates. The FK01 Forensic Science Kit includes poppy seeds. If you don't have the kit, you can buy poppy seeds at the supermarket.

Soil

Several of the lab sessions in the soil analysis group require various soil specimens, which you can obtain yourself. See the lab session for details.

Spectroscopy questioned soil

Lab Session I-5, Examine the Spectroscopic Characteristics of Soil, requires a soil specimen that contains one, some, or all of the following elements at detectable levels: aluminum, barium, boron, calcium, cerium, iron, mercury, potassium, sodium, and strontium. The FK01 Forensic Science Kit includes such a specimen. If you don't have the kit, have someone make up a questioned soil specimen for you using those guidelines.

Tape dispensers

Lab Session VI-3, Perforation and Tear Analysis, requires a selection of at least two or three different standard tape dispensers. See the lab session for details.

Testfabrics Multifiber Fabric #43

Several of the lab sessions in the hair and fiber analysis group require known specimens of various types of fibers. The FK01 Forensic Science Kit includes a specimen of Testfabrics, Inc. Multifiber Fabric #43 (MFF #43), which includes strips of the following fibers: spun diacetate, SEF (modacrylic), filament triacetate, bleached cotton, Creslan 61 (acrylic), Dacron 54 (polyester), Dacron 64 (polyester), Nylon 66 (polyamide), Orlon 75 (acrylic), spun silk, polypropylene (polyolefin), viscose (rayon), and worsted wool. If you don't have the kit, you can purchase MFF #43 directly from Testfabrics, Inc. (*http://www.testfabrics.com/*) or substitute fabrics from a piece-goods shop or similar store.

Tools: cutting pliers and flat screwdrivers

Lab Session VI-1, Tool Mark Analysis, requires specimens of cutting pliers and flat screwdrivers for making the impressions to be analyzed. See the lab session for details.

Table 2-4 summarizes the specimens we recommend.

Table 2-4: *Recommended Specimens*

Item	FK01	Sources
Acetaminophen	●	Drugstores
Aspirin	●	Drugstores
Caffeine	●	Drugstores
Cartridge case, brass, fired	●	Shooting ranges
Chlorate known	●	See text
Chlorpheniramine	●	Drugstores
Chromatography questioned	●	See text
Clothing, unlaundered	○	Laundry baskets
Digital image files	○	See text
Diphenhydramine	●	Drugstores
Drug testing specimens, questioned A & B	●	See text
Fabric/fiber A, B, C, D, and E	●	See text
Film negatives and/or slides	○	See text
Foods, assorted	○	Supermarkets
Glass and plastic	○	Recycling bins
Hair, animal	○	Pets
Hair, human	○	Self, family, and friends
Ibuprofen	●	Drugstores
Perchlorate known	●	See text

Item	FK01	Sources
Poppy seeds	●	Supermarkets
Soil	○	Yards, parks, and so on
Spectroscopy questioned soil	●	See text
Tape dispensers	○	Obtain locally
Testfabrics Multifiber Fabric #43	●	See text
Tools: cutting pliers	○	Self, family, and friends
Tools: screwdrivers, flat	○	Self, family, and friends

Soil Analysis

Group I

Soil is one of the most common forms of physical evidence found at crime scenes. For example, a vehicle suspected of having been used in an armed robbery may later be found to have soil from the crime scene adhering to its tires or wheel wells, thereby establishing that that vehicle was present at the crime scene. Similarly, a suspected rapist or mugger may have soil adhering to his clothes or shoes. (Soil specimens are often particularly easy to obtain from shoes or boots with deep tread, as shown in Figure I-0-1.) If soil found on a suspect's shoes or clothing is consistent with soil present at the crime scene, it establishes that the suspect was probably present at that scene.

Figure I-0-1: *Rubber-soled shoes or boots with deep treads are likely sources for soil specimens*

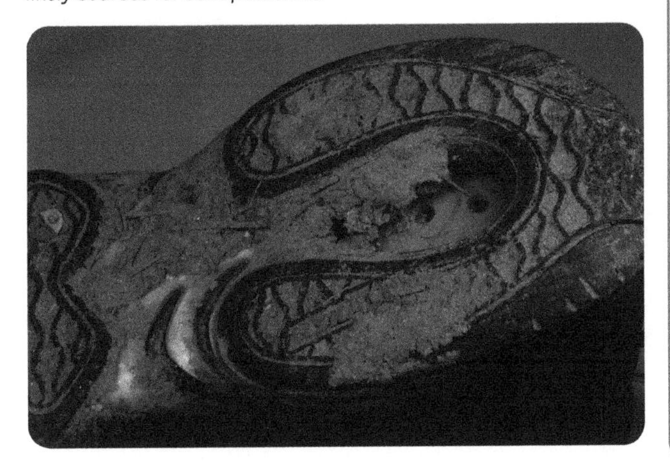

DENNIS HILLIARD COMMENTS

Footwear and tire impressions in soil are an excellent example of examining different types of class evidence to increase certainty. The questioned impressions can be compared against impressions made by known footwear or tires to determine if they are consistent. In addition, the soil found adhering to the footwear or tires can be compared against soil from the scene to determine if they are consistent. If both types of class evidence are consistent, that increases the likelihood that the suspect person or vehicle was in fact present at the scene.

Obviously, impression evidence must be preserved before soil specimens are taken. Depending on the nature of the impressions and the underlying surface, impressions may be preserved by photographing them or by making casts.

Soil evidence by itself is seldom sufficient to secure a conviction because it usually establishes only the likelihood that a suspect person or object was present at the crime scene at some time, but not *when* that visit took place. Still, in combination with other evidence—and particularly if the suspect denies ever having visited the scene—soil evidence may provide an essential building block in the prosecution case. Equally, soil evidence can be exculpatory. For example, the police may suspect someone who has mud stains on her clothing that appear visually to be consistent with the soil at the crime scene. If subsequent forensic tests establish that the mud stains on the clothing are inconsistent with the soil at the crime scene, the police can redirect their efforts elsewhere.

But how can forensic scientists determine whether one soil specimen is consistent with another? After all, dirt is dirt, right? Not at all.

Soil is a complex mixture of mineral, vegetable, and animal material, and may also contain plastic, glass, metal, and other manufactured materials. Soil is anything but uniform. Specimens taken only a few centimeters apart may differ significantly in composition. Two specimens taken at a distance from each other, even if they are of the same general type, are almost certain to have sufficient differences in their composition, physical properties, and chemical properties to make it possible to discriminate between them. These differences mean that soil from any particular location has its own unique fingerprint.

Soil analysis is the process of determining what components make up a soil specimen, and in what proportions. Forensic scientists use microscopic examination and various physical and chemical tests to determine the characteristics of a *questioned soil specimen* (also known as an *associated soil specimen*, a *suspect soil specimen*, or an *unknown soil specimen*). By performing the same tests on a *known soil specimen*, a forensic scientist can, if the specimens are consistent in all respects, state with high confidence that the two specimens in fact originated from the same location.

A *forensic geologist* has primary responsibility for forensic soil analysis. Tests done by the forensic geologist are often sufficient to identify soil specimens as consistent. At times, however, a forensic geologist will call upon other specialists to complete more detailed analyses. For example, a *forensic entomologist* may be needed to identify insect eggs or larvae present in the soil specimen, a *forensic botanist* to identify pollen or other plant material, or a *forensic chemist* to identify trace amounts of organic or inorganic chemicals.

In this group of lab sessions, we'll obtain and dry soil specimens and then use various tests to analyze and characterize those soil specimens, learning as much about them as possible.

Gather and Prepare Soil Samples

EQUIPMENT AND MATERIALS

You'll need the following items to complete this lab session. (The standard kit for this book, available from *http//:www.thehomescientist.com*, includes the items listed in the first group.)

MATERIALS FROM KIT

- Goggles

MATERIALS YOU PROVIDE

- Gloves
- Bags, quart/liter plastic zip
- Drying dishes (saucers, oven dishes, etc.)
- Oven
- Pen, felt-tip marking
- Specimens: soil (see text)

WARNING

Although the only real danger in this lab session is handling hot material, as a matter of good practice, you should always wear splash goggles and gloves when doing lab work.

BACKGROUND

Because the makeup of soil can vary so much over a small area, the first task faced by the forensic analyst is to gather known soil specimens that visually appear as closely similar as possible to the questioned soil specimen, in the hope that at least one of those known specimens will be consistent with the questioned specimen. In practice, the number of specimens needed depends on the particulars of the case. For example, if the questioned specimen was obtained from the tires or wheel well of a vehicle whose tires are consistent with the tracks present at a crime scene, only a few known specimens taken from the vicinity of those tracks may suffice. Conversely, in the absence of any such indication of where the questioned specimen may have originated, the forensic technician may have to obtain literally dozens or even hundreds of known specimens.

Obviously, in the real world, the importance of the case is a major factor in how many specimens are taken, if indeed any are taken at all. For a routine burglary, for example, budget and time constraints often mean that soil specimens are taken only if a questioned specimen is known or likely to be available and only if soil analysis evidence is likely to establish guilt or innocence. Conversely, in an important case such as a high-profile murder or a child kidnapping in which soil evidence appears likely to play a part, the forensics technicians may be directed to obtain numerous soil specimens, even if there is no certainty that a questioned specimen will later become available for analysis.

The questioned soil specimen is usually imperfect, in the sense that it is very unlikely to be consistent in all respects with any known soil specimen, including a known specimen that is collected from exactly the same location where the questioned specimen originated. For example, if a suspected criminal has mud stains on his pants, that mud may consist of only the finest particles from the soil that produced the mud stain. For that reason, the forensic geologist may process the known specimen from that location to remove larger particles and compare only the fine particles from the known specimen against the questioned specimen.

Questioned specimens are gathered as carefully as possible, making sure not to alter them any more than necessary. For example, if the questioned specimen is in the form of chunks of mud that appear to have broken from the underside of a vehicle during a hit-and-run accident, those chunks are, insofar as is possible, preserved as-is rather than crushed or otherwise altered. Similarly, if the questioned specimen is taken from an area where oil has spilled or bits of broken glass are present, every effort is made to obtain a representative specimen that includes those contaminants.

With few exceptions, known specimens are collected from only the top centimeter or so of the soil surface, because soil composition may vary dramatically with depth, and unknown specimens almost always originate at or very near the surface. The major exception to this rule is for known soil specimens collected at a crime scene located at a ditch, road cut, or similar disturbance that exposes the stratigraphy of the soil. In that situation, many known specimens are collected to make sure that at least one known specimen is available to represent each of the exposed layers of soil.

Collected soil specimens are usually stored in plastic jars or ziplock plastic bags, with each specimen labeled with the case number, date, time, and exact location where the specimen was collected; the name of the person who collected the specimen; and so on. After the soil specimens are gathered, they may undergo preliminary processing to prepare them for storage and later analysis. The most common form of preliminary processing is drying, which is the best way to preserve the characteristics of most soil specimens. Storing a moist soil specimen in a sealed container may cause several undesirable changes to occur, including chemical and biological changes. Conversely, the forensic geologist recognizes that drying may also cause changes, including an increase in nitrate concentration, oxidation of some minerals, crystallization of dissolved salts, and significant changes in the numbers and activity of bacteria and other microscopic life forms. Still, on balance, drying usually does a lot less harm than good, so most soil specimens are routinely dried. We'll dry all of our specimens.

If a soil specimen is to be tested for conductivity or if it contains volatile organic compounds, drying the specimen may significantly alter its characteristics and thereby destroy important evidence. In this situation, the soil specimen should be analyzed as soon as possible after it is collected. To minimize changes while awaiting analysis, such soil specimens are stored in sealed containers at just above the freezing point of water.

The size of the questioned specimen is often not within the control of the forensic scientist. In some cases, such as mud from a wheel well, there may be a kilogram or more of the questioned specimen available, while in other cases there may be only a few grams (or even just a few particles) of the questioned specimen available. In the latter case, the small specimen size may limit the number and types of tests that can be done, and the forensic scientist must prioritize those tests according to which are most likely to provide useful information.

Known specimens are a different matter. Often, the only constraint on known specimen size is how much the forensics technician is willing to gather and carry back to the lab. For our series of experiments in this group, we'll obtain relatively large soil specimens, which makes things easier by allowing us to complete as many tests as we wish and to do multiple runs if necessary. Samples of about 500 mL (2 cups) will suffice for our purposes.

In this lab session, we'll obtain five soil specimens and process them in various ways to prepare them for testing in subsequent lab sessions.

WE USED THE FOLLOWING FIVE SOIL SPECIMENS:

A. A specimen of topsoil from our yard, which is fertilized frequently by a commercial lawn service. This specimen was taken from the top centimeter of exposed soil.
B. A second specimen of topsoil from our yard, but taken from a depth between one and two centimeters.
C. A specimen of topsoil taken from the top centimeter of a neighbors' yard, near the edge of a children's sandbox.
D. A second specimen of our neighbors' topsoil, taken from the top centimeter about one meter from the sandbox.
E. Topsoil purchased at a lawn and garden supply store.

Obviously, your own specimens will be different, but try to obtain specimens that are similar in appearance. If your specimens differ greatly in appearance, it becomes trivial to discriminate one from another.

PROCEDURE I-1-1: GATHER SOIL SPECIMENS

For the laboratory sessions in this chapter, you'll need one questioned soil specimen and five known soil specimens. The questioned specimen should be gathered from the same location as one of the known specimens. Obviously, if you're working alone your "questioned" specimen will actually be just another known specimen, so if possible, enlist the aid of a lab partner or friend to gather at least the questioned specimen. If you know someone else who is also doing these labs, gather all six of her specimens for her, and have her gather all six of your specimens for you. We used ordinary quart zip-lock bags as soil specimen containers, but you can substitute wide-mouth jars or any other containers large enough to hold specimens of at least 500 mL.

1. Choose the location where you'll gather your first known soil specimen. As shown in Figure I-1-1, remove any extraneous sticks, leaves, pebbles, or other objects that are lying on top of the soil but aren't a part of it. (For example, as Barbara was removing leaves and pebbles from the surface of the area where we intended to gather our first specimen, our Border Collie dropped a tennis ball right in the middle of our soil specimen area. She removed that, too.)

Figure I-1-1: *Barbara removing extraneous material before gathering a soil specimen*

2. Use the trowel to scrape soil from the surface, as shown in Figure I-1-2, going down no more than one centimeter. Transfer the soil to a clean plastic bag or other specimen container until you have accumulated about 500 mL of specimen (roughly two measuring cups). You should be able to obtain an adequate specimen from an area no more than 25 cm square. As far as possible, transfer only soil, leaving larger embedded objects such as stones, twigs (and tennis balls) in place.

Figure I-1-2: *Barbara using a trowel to gather a soil specimen from the top centimeter of earth*

In Figure I-1-2, note the difference in color between the brown surface topsoil and the reddish subsoil immediately beneath it. If as you gather your soil specimens you scrape down far enough to notice such a color change, you've gone too far.

3. Seal the container, and label it Known #1 (or K1). On the label, write the date, time, and exact location of the specimen.

4. Repeat steps 1 through 3 to gather four more known soil specimens and one questioned (Q1) soil specimen.

PROCEDURE I-1-2: DRY SOIL SPECIMENS

Although you can air dry your soil specimens, it's much faster to dry them in an oven set to its lowest temperature. Don't make the mistake of setting the oven to a higher temperature to dry the specimens faster. Anything higher than the lowest setting may cause undesirable changes to the specimens. You can use any flat heat-resistant containers as drying dishes—old dinner plates, baking dishes, disposable aluminum pie pans, and so on. If you have no suitable containers, make drying boats from aluminum foil.

DENNIS HILLIARD COMMENTS

When drying soil samples in an oven, even at the lowest temperature, it is necessary to have adequate ventilation, preferably with an oven exhaust fan in operation. Added caution: if soil samples are taken in an arson investigation, there may be presence of an ignitable liquid, such as gasoline. Such samples as stated need to be refrigerated and processed for the ignitable liquid prior to further analysis.

1. If you have not already done so, put on your splash goggles and gloves.

2. Transfer the entire Q1 soil specimen to an evaporating dish, plate, or large saucer. Distribute the soil into the thinnest layer possible and place the dish in the oven, as shown in Figure I-1-3.

Depending on the size of your soil specimens, plates, and oven, you may have to dry one specimen on several plates, and you may have to dry different specimens in separate sessions. For best drying, make sure the soil is at most 1 cm thick on the drying plate and stir the specimen occasionally as it dries.

3. Repeat step 2 for all of the other soil specimens, making sure to keep track of which specimen is in which dish.

4. Heat the specimens at the lowest temperature setting, for at least one hour to drive off any water present.

5. Remove the dishes from the oven and allow them to cool completely.

6. Transfer each dried soil specimen back into its labeled storage container and reseal the container.

Figure I-1-3: *Soil specimen drying in the oven*

The visual appearance of the dried specimen may or may not differ from the original specimen, depending upon the amount of moisture present in the original specimen, its composition, and other factors. Many soils have a lighter color when dry, although some darken or appear unchanged.

Figure I-1-4 shows one of our soil specimens before and after drying. Although it may not be obvious in the photograph, the dried soil specimen (on the left) was noticeably lighter and redder in color, versus the darker, browner color of the undried specimen. Also, the dried specimen had a noticeably finer texture. During drying, many of the clumps separated into smaller clumps or individual grains. Although not dramatic, the differences are noticeable at a glance.

Figure I-1-4: *Soil specimens before (right) and after drying*

REVIEW QUESTIONS

Q1: What additional information, if any, might it be useful to record with each soil specimen?

Q2: Why did we not determine the mass of each specimen before and after drying and use that information to calculate moisture content?

Examine the Physical Characteristics of Soil

Lab I-2

EQUIPMENT AND MATERIALS

You'll need the following items to complete this lab session. (The standard kit for this book, available from *http//:www.thehomescientist.com*, includes the items listed in the first group.)

MATERIALS FROM KIT

- Goggles
- Cylinder, graduated, 100 mL
- Mesh, fiberglass
- Pipettes
- Reaction plate, 24-well
- Stirring rod
- Test tubes
- Test tube rack

MATERIALS YOU PROVIDE

- Gloves
- Balance
- Cups, foam
- Dishwashing liquid detergent
- Drying dishes (see text)
- Oven
- Paint chips (see text)
- Paper (copy paper, newspaper, or similar)
- Pen, felt-tip marking
- Sieves (optional; see text)
- Soda bottle, 1 L or 2 L (clean and empty)
- Specimens: soil, known and questioned
- Ultraviolet light source (optional)
- Watch or clock with second hand

BACKGROUND

Many soils can easily be differentiated from each other by observing such physical characteristics as color, density, settling time, and particle size distribution. Physical soil tests have been widely used forensically since the late 19th century. As is true of so many other fields of forensic science, Sherlock Holmes led the way. In the first Holmes novel, Watson has just met Holmes and is trying to figure out just what Holmes' eclectic studies are intended for. Watson lists Holmes' areas of knowledge, including:

"*Knowledge of Geology: Practical, but limited. Tells at a glance different soils from each other. After walks has shown me splashes upon his trousers, and told me by their colour and consistence in what part of London he had received them.*" —Arthur Conan Doyle, *A Study in Scarlet*, 1888

Which makes Sherlock Holmes the world's first forensic geologist. Although it can't be established firmly, we think it likely that Holmes also devised many of the other standard physical soil tests that are still used in forensics laboratories.

In this lab session, we'll examine several physical characteristics of our soil specimens to determine if we can discriminate them based on these physical characteristics.

This lab has four procedures. In the first procedure, we'll observe and categorize soil color. In the second and third procedures, we'll determine the density and settling time for each of our known and questioned soil specimens. In the fourth procedure, we'll determine the particle size distribution of our specimens.

PROCEDURE I-2-1: OBSERVE AND CATEGORIZE SOIL COLOR

The most obvious characteristic of a soil specimen is its color. If the color of the questioned soil specimen is consistent with a known specimen, it's possible (although not certain) that the questioned specimen originated from the same source as the known. Obviously, if the color of the questioned specimen differs significantly from any of the known specimens, it's likely (although not certain) that the questioned specimen did not originate from the same source as any of the known specimens against which it was compared.

It's important to dry the soil specimens completely before doing color comparisons. Some soils have similar color whether they are dry or moist, while the color of other soils may vary significantly. It's also important to use consistent lighting when comparing colors. The standard used by professional forensics labs is open shade (daylight from a clear north sky; not direct sunlight). Soil colors may appear significantly different if viewed under incandescent or fluorescent lighting. One very different form of lighting is useful, however. Two soil specimens that appear identical in daylight may appear very different under ultraviolet lighting. The overall color may be quite different, and some soil specimens contain natural or artificial materials that fluoresce under ultraviolet lighting.

It's also important to use a standard specimen size. If you've ever chosen a paint color from a small sample chip and painted a room in that color, you know that the paint appears very different when it's on the wall than it did when you looked at the small sample chip. (Although, if you hold the chip up to the painted wall, you'll find that the colors are in fact identical.) We placed our specimens in a 24-well reaction plate, as shown in Figure I-2-1, to ensure consistent sample size.

Figure I-2-1: *Questioned and known soil specimens in a reaction plate*

We had plenty of our questioned soil specimen available, so we filled all six cells in the bottom (D) row with it to make it easier to compare against the known specimens in the row above (C). To make sure the lighting or viewing angle wasn't affecting our judgment, we also filled cell C1 (to the far left) with the questioned specimen. Visually, all six D-row cells and cell C1 appear nearly identical. Cells C2 and C5 are much darker than the questioned specimen, and cells C3 and C4 are somewhat darker. Cell C6, which contains Known #5 (K5), is extremely similar visually to the questioned specimen.

Finally, for forensic purposes, it's important to be able to categorize the color of soil specimens unambiguously. In day-to-day life, it's acceptable to describe a color as "a chocolate brown with a hint of purple" or "a reddish sand," but that obviously won't do for forensics. It's necessary to assign a specific color description that can be reproduced by others. To do that, we'll use paint sample chips, available anywhere that paint is sold.

> Professional forensic labs use Munsell Soil Color Charts, but a set of those costs $184.50. Urk. Instead, we visited our local paint store and obtained a full set of paint chips in browns, tans, and other earth tones. (We don't feel too guilty about pillaging their paint chips; we also bought several gallons of paint to repaint our kitchen, dining room, and library.)

With all of that in mind, let's compare the colors of our soil specimens.

1. Place your reaction plate on a sheet of white paper, and fill a row of wells to near the top with portions of your questioned specimen, as shown in Figure I-2-1.

2. Compare the color of the questioned specimen with your paint chip samples, and record the name of the closest paint chip in your lab notebook.

3. Fill an adjacent row of five wells with your known specimens, as shown in Figure I-2-1, noting which well contains which known.

4. Under even daylight lighting, compare Known #1 (K1) against the questioned specimen (Q1) to determine how closely it matches the unknown specimen. Record a short description (for example, "match," "close match," or "no match") in your lab notebook.

5. Repeat step 4 for K2 through K5, and record your observations in your lab notebook.

6. If you have an ultraviolet light source, darken the room as much as possible and then turn on the UV light source. Record the appearance of the questioned specimen (Q1) and the known specimens (K1 through K5) in your lab notebook. Note any fluorescence, including color and intensity, and whether the glow appears uniform or is emitted by only some particles in the specimen.

7. Compare your questioned specimen against your color charts, as shown in Figure I-2-2, and record the best match in your lab notebook.

Figure I-2-2: *Comparing paint chip samples against a soil specimen*

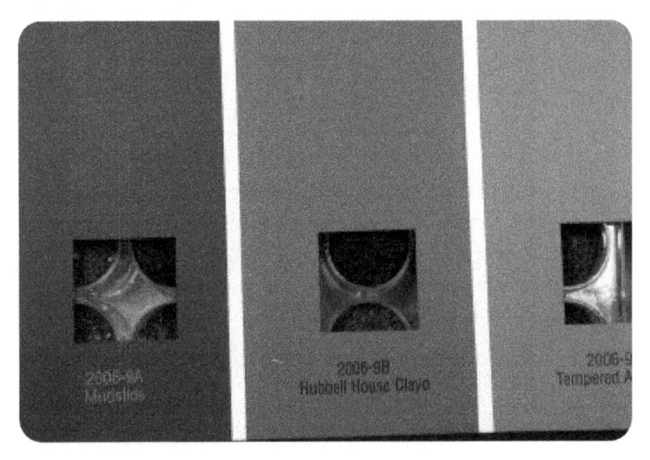

Figure I-2-2 shows the first paint chip sample we chose, based on eyeballing the questioned specimens and the paint chip samples. As you can see, the match is not at all close. We finally settled on a paint chip that was considerably darker than those shown here. It was not an exact match, but came very close.

PROCEDURE I-2-2: DETERMINE SOIL DENSITY

The density of soils varies over a wide range, depending on the composition of the soil, how tightly it is compacted, how moist it is, and other factors. Average soil has a density of about 1.5 g/mL to about 2.7 g/mL. The least-dense soils, often found in forested areas, contain a high percentage of organic matter and may have average densities lower than 1.0 g/mL (less dense than water). The densest soils are made up primarily of very dense minerals and may have densities of 4.0 g/mL or higher. In general, the organic (vegetable, animal, and some plastics) content of soils is of low density, often low enough to float on water. The presence of large amounts of material of low density makes it difficult or impossible to obtain good values for density.

One way to address this problem is to wash the soil specimens to remove as much light material as possible by flotation, leaving only the denser materials. Although washing removes both light and water-soluble materials from the specimen, the results remain valid because you have eliminated the same classes of materials from all specimens and are testing the same subset of the material for each specimen. The drawbacks to washing are that some soil specimens require considerable time (hours to days) to settle completely, and each specimen must be thoroughly dried after washing so that known masses of each specimen can be compared.

Professional forensics labs centrifuge wet soil specimens to speed settling. A specimen that might take days to settle under just the force of gravity can be spun down in a centrifuge in literally seconds.

Another method is to substitute a less-dense liquid for water for determining density by displacement. Acetone, 95% ethanol, and 91% isopropanol all have densities around 0.79 g/mL, which means that some of the light soil components that float in water sink in one of these liquids.

Forensics technicians routinely add a small amount of detergent (typically, dishwashing liquid) to the water or other liquid used for the tests. The detergent allows the liquid to wet the soil particles thoroughly to prevent trapped air bubbles and eliminate surface tension, which reduces the measured density of the soil.

1. Label six foam cups, one for each of your known and questioned soil specimens.

2. Fill each cup roughly a third full with the corresponding specimen.

3. Add tap water until each cup is nearly full. Add a drop of dishwashing liquid to each cup.

4. For each sample, use the stirring rod to break up any clumps of soil. Stir the contents until any vegetable matter and other light material floats to the top.

5. Allow the contents of each cup to settle for a minute or two, and then carefully pour off most of the excess liquid. Avoid pouring off any of the solid soil.

With such a short settling time, some clay and other fine materials may remain suspended in the liquid. That's acceptable because, again, we're treating all of the specimens the same. If we lose a portion of the specimen in one cup, we'll lose the same portion of the specimen in all other cups.

6. Repeat steps 3 through 5 for each cup. After this second wash, all or most of the light material should have been removed from the samples in all of the cups. If not, do a third wash on all of the cups.

7. Pour the damp soil from each cup into an individual drying dishes, transferring as little water as possible. Place the drying dishes in the oven and heat them on low heat until they have dried completely.

You can use any suitable containers as drying dishes. We used old dessert plates, but saucers, oven dishes, disposable pie plates, or similar containers will work as well. If you have nothing else, you can make boats from aluminum foil.

8. While you are waiting for the specimens to dry, fill the soda bottle with tap water and add a few drops of dishwashing liquid. Invert the bottle several times to mix the solution.

9. After the samples have dried, allow them to cool to room temperature.

10. Weigh out about 50 g of the dry questioned specimen and record its mass to the resolution of your balance in your lab notebook.

You can also use the balance to measure water very accurately and verify the accuracy of your graduated cylinder. For our purposes, we can use a value for the density of water of 1.0 g/mL. Place the graduated cylinder on the balance and tare (zero) the balance to read 0.0 g. Transfer water to the graduated cylinder until the balance indicates a mass of 20.0 g. There is now 20.0 mL of water in the cylinder. Check the reading on the cylinder, which should be very close to 20.0 mL. If the reading is off, record the indicated versus actual value in the front of your lab notebook. Continue adding water until the mass is 40.0 g, and again record the indicated reading on the cylinder. Repeat for 60.0 g, 80.0 g, and 100.0 g.

11. Fill the 100 mL graduated cylinder to 50.0 mL with water from the soda bottle, using a disposable pipette to add water dropwise until the cylinder contains as close as possible to 50.0 mL. Record this initial volume as accurately as possible in your lab notebook.

12. Withdraw a few mL of the water from the graduated cylinder with each of two pipettes. Set them aside, inverted to make sure none of the water leaks from the pipettes.

13. Using a folded sheet of paper, carefully transfer the weighed questioned soil specimen to the graduated cylinder. Make sure as little as possible of the soil specimen adheres to the walls of the cylinder above the liquid level. Your goal is to make sure all of the soil is immersed in the liquid. If you get air bubbles under the surface of the liquid, tap the cylinder or use the stirring rod to eliminate them.

14. Use the liquid stored in the disposable pipettes to rinse down any soil that adheres to the inside surface of the graduated cylinder above the liquid line. Make sure to expel all of the liquid from both of the disposable pipettes, restoring the exact amount of liquid to the cylinder that was present at the initial measurement.

15. Determine the new liquid volume as accurately as possible and record it in your lab notebook.

16. Subtract the initial volume from the final volume to determine the volume of liquid displaced by the specimen. For example, if the graduated cylinder initially contained 50.4 mL and the final volume was 84.2 mL, calculate the displaced volume as 84.2 – 50.4 = 33.8 mL. Record the displacement volume in your lab notebook.

17. Divide the mass of the specimen by the volume displaced to determine the density of the specimen in grams per milliliter. For example, if your specimen mass was 50.39 g and the displacement volume was 33.8 mL, calculate the density of the specimen as 50.39 g / 33.8 mL = 1.49 g/mL. Record this value in your lab notebook.

18. Repeat steps 10 through 17 for each of the known specimens. Compare the density value you obtained for the questioned specimen with those you obtained for the known specimens to determine if one or more of the known specimens is similar in density.

PROCEDURE I-2-3: DETERMINE SOIL SETTLING TIME

Soil settling time is a commonly used metric in forensic soil analysis. The amount of time required for a suspension of soil to settle completely out of a liquid depends on many factors, including the liquid used, the average size of the particles, and particle density. Some soils, such as very sandy soils, settle very quickly, sometimes in a minute or less. Other soils, particularly those that contain significant amounts of clay, form suspensions that are almost colloidal and may require anything from hours to literally days or longer to settle completely.

In this procedure, we'll use tap water as the liquid. Professional forensics labs routinely use water for settling tests, but sometimes use other liquids as well. For example, if the settling times in water are similar for a questioned and known specimen, before concluding that the specimens are consistent, the lab may retest both specimens using a more viscous liquid such as a 1:1 mixture of glycerol and water. In that liquid, the two specimens may exhibit noticeably different settling characteristics.

We'll determine by eye when settling is complete for each specimen, which will introduce subjective factors. Professional labs may use spectrometers to quantify settling times accurately by measuring the decrease in density over time as the soil components settle out. The output of the spectrometer in density versus time provides a characteristic "fingerprint" for each specimen with regard to settling time.

1. Label six test tubes Q1 and K1 through K5.

2. Transfer the questioned soil specimen to tube Q1 until the tube is about one quarter full. Tap the tube gently to settle the soil specimen.

3. Fill tube Q1 with tap water (with detergent added) to about 1 cm from the rim.

4. Agitate the contents of the tube to suspend the soil in the liquid, and immediately note the start time in your lab notebook. Replace the tube in the rack and observe it as the soil settles.

5. When the soil appears to have settled completely, record the finish time in your lab notebook.

6. Subtract the start time from the end time to determine the elapsed time needed for the specimen to settle completely. Record that elapsed time in your lab notebook.

7. Repeat steps 2 through 6 for specimens K1 through K5. Unless your specimens settle very quickly, you'll have time to start some or all of the remaining tubes before settling completes in the first tube.

> If a soil specimen contains clay, the finest particles may remain suspended for a very long time. We suggest that you observe settling time for no longer than one hour, by which time most or all of the soil may have settled out. If after that time the liquid remains cloudy, simply note that fact and move along.

8. Compare the settling times of the questioned specimen and the known specimens to determine if the questioned specimen is consistent with one or more of the known specimens.

PROCEDURE I-2-4: DETERMINE SOIL PARTICLE SIZE DISTRIBUTION

Particle size distribution is an important forensic characteristic of a soil specimen, providing one more piece of the "fingerprint" of that specimen. Some soils, particularly sandy soils, have relatively uniform particle sizes, with 90% or more of the particles contained within a size range of perhaps 5:1 or less. Other soils, particularly clay soils, may span a much broader range of particle sizes—10,000:1 or more—with the largest particles being pea-sized pieces of gravel and the smallest being barely large enough to discern under a microscope at high magnification.

> In a professional forensics lab, a soil specimen is passed through many mesh sizes—typically 20, 40, 60, 80, 100, 120, 150, and 200—and the mass percentage of each fraction is calculated and recorded for that specimen. But understanding the concept doesn't require a full set of sieves. What's important is not the exact mesh sizes used, but understanding the concept of separating soil specimens by particle size. We separated our specimens into only four fractions, using the fiberglass mesh from the kit, a discarded kitchen strainer with a mesh about half the size of the window screen, and a discarded permanent coffee filter made from a very fine metal mesh.

It's not necessary to use that many sieve sizes to understand the concept, or indeed to get useful information about our specimens. Using as little as one sieve (for example, the fiberglass mesh included in the kit) can provide useful data about the specimens, but if possible you should use additional sieve sizes to gather more data about the specimens. For example, in addition to the fiberglass mesh, we used an ordinary kitchen flour sifter (shown in Figure I-2-3) that had a sieve size smaller than the fiberglass mesh, and a metal coffee filter with a still finer mesh.

It's important to understand that the observed particle size distribution for a soil specimen depends on how that specimen is prepared and tested. For example, dry sieving is not normally used because particles in a dry specimen tend to adhere, creating deceptively large fractions of larger particle sizes. Those fractions can also vary noticeably, even within a single relatively small specimen. Wet sieving—adding water to convert the specimen to mud and using the sieve to separate the wet fractions—yields more reproducible results but requires drying the separated fractions thoroughly before weighing. Wet sieving, also reduces the overall mass of many soil specimens because any soluble salts present in the specimen are dissolved and so pass through even the finest mesh.

Figure I-2-3: *Wet sieving a soil specimen*

The presence of some fertilizers and other chemicals, natural or artificial, may also have an effect on observed particle size distribution. For example, any lime, limestone, or carbonates present in the specimen may function as a sort of glue that causes fine particles to aggregate into larger particles. Wet sieving eliminates some but not all of this aggregation. Treating the specimen with dilute hydrochloric acid (~3 M) eliminates most aggregation because the acid reacts with carbonates and other "glue" chemicals to produce gases and soluble compounds. Obviously, acid treatment may also reduce the mass of the specimen significantly.

In this part of the lab session, we'll use wet sieving to separate our soil specimens into several fractions. We'll then dry each fraction and determine its mass. By dividing the mass of each fraction by the total mass of all fractions, we'll determine the mass percentage for each fraction.

MAKING DO

This lab session can be quite time consuming and resource intensive because we'll be operating on six soil specimens—one questioned and five known—and separating each specimen into multiple fractions, each of which must be dried and weighed separately. For example, we used three mesh sizes—the fiberglass mesh provided with the kit, a kitchen flour sifter, and a permanent mesh coffee filter—so we ended up with a total of 24 fractions.

The exact procedure you use will depend on the physical form and other characteristics of your meshes. For example, our coarsest mesh was the fiberglass mesh supplied with the kit, which we placed flat on the top of a large funnel (which we made by cutting the top off a one liter soda bottle) and poured our specimens through, capturing the water and finer particles in a large container (we used a one-quart polypropylene container that our last take-out order of egg-drop soup arrived in). After pouring all of the wet initial specimen through the fiberglass mesh, we washed the mesh with additional water to rinse the captured particles and make sure that all of the finer particles had passed through the mesh. We then rinsed the captured particles off the mesh onto a saucer, and placed it in the oven to dry.

With that fraction isolated and drying, we then passed the remaining wet specimen through the kitchen flour sifter to capture the second fraction, which we again rinsed thoroughly to ensure that all particles that would pass through that mess did so. We transferred that wet fraction to another saucer and placed it in the oven to dry. We then passed the remaining wet specimen through our mesh coffee filter, isolated that fraction, and dried it. The final fraction—the particles small enough to pass through the coffee filter—were by then suspended in a large amount of water. We poured that last fraction into a large beaker for which we'd recorded its dry mass and heated the beaker to boil off the excess water and dry the remaining fraction. Subtracting the initial mass of the beaker from the mass of the beaker with the dried final fraction gave us the mass of the final fraction. Whew!

We then repeated that procedure for each of the five known specimens. All told, it took us three full days to complete the workup of all six specimens, although most of that time was spent waiting for specimens to dry. In terms of actual hands-on work, the entire lab session took less than two hours. Accordingly, we've modified the recommended procedure for this lab session to use natural drying by evaporation, which minimizes the amount of work required at the expense of extending the lab over a period of a week or so. Note that natural drying by evaporation does not completely dry a specimen, so using this method introduces some error. If you want your results to be as accurate as possible and are willing to put in the work to get those better results, you can reproduce the procedure described above.

1. If you have not already done so, put on your splash goggles, gloves, and protective clothing.
2. Assuming that you have six soil specimens, label six foam drink cups Q1, K1, K2, K3, K4, and K5.
3. Assuming that you will separate each soil specimen into four fractions, label four foam drink cups "Q1-F1" through "Q1-F4," four more cups "K1-F1" through K1-F4," and so on until you have 24 labeled fraction cups, four for each of the six soil specimens.
4. Weigh each fraction cup and record its mass to the maximum resolution of your balance. Write the mass of each cup on the cup itself.
5. Weigh about 200 g of the questioned specimen to the maximum resolution of your balance and record that mass in your lab notebook. (If the capacity of your balance is too small, simply weigh the 200 g of soil in multiple portions.)
6. Transfer the specimen to cup Q1 and add sufficient water to the cup to form a soupy mix.

7. Swirling the cup to keep the soil suspended, pour the suspension through your largest mesh, capturing the liquid and solids that pass through the mesh in another container. Add more water to the cup as necessary to make sure that all of the soil in the cup is rinsed into the mesh, but try to use as little water as possible while still transferring all of the soil.

8. Transfer the soil particles captured by the first mesh into cup Q1-F1. If necessary, rinse the particles off the mesh, but try to use as little water as possible.

9. The large particles captured by the first mesh should settle very quickly. Once those particles have settled, use a pipette to remove and discard as much water as possible from cup Q1-F1 to speed drying. Make sure not to remove any of the soil particles. Set cup Q1-F1 aside to dry.

10. Set up your second sieve, and pour the soil/water suspension that passed the first sieve through the second sieve, again capturing the liquid and solids that pass the mesh in another container. Make sure that all of the soil is transferred from the cup into the second sieve, using as little water as possible to rinse the soil into the sieve.

11. Transfer the soil particles captured by the second mesh into cup Q1-F2, again using as little water as possible to do a complete transfer. Once the particles have settled, again use a pipette to remove as much water as possible from the cup without removing any soil particles. Set cup Q1-F2 aside to dry.

12. Repeat the preceding steps with each of your sieves until you have isolated each fraction into its own fraction cup and set it aside to dry.

13. Repeat the preceding steps with soil specimens K1 through K5.

14. At this point, you have a large array of cups, all of which contain damp (or wet) soil specimen fractions. You can allow these specimens to dry naturally, which may require several days. Alternatively, you can dry them in an oven set to its lowest temperature (typically about 120°F or 50°C). Before you do that, perform a test with an empty cup to make sure it won't melt.

15. Once all of your specimens are dry, weigh each cup to the maximum resolution of your balance, as shown in Figure I-2-4. Subtract the empty mass of the cup from the mass of the cup with the specimen fraction to determine the mass of the specimen fraction, and record that value in your lab notebook.

> If your cups do hold up to drying in an oven at low temperature, you can improve accuracy by doing a final drying for an hour or so on all the cups immediately before weighing them.

Figure I-2-4: *Determining the mass of a soil specimen fraction*

> In our session, the mass of the foam cup was 4.03 g. With the dried Q1-F1 soil specimen fraction added, the mass was 131.93 g. Subtracting the mass of the empty cup gives us the mass of the Q1-F1 soil fraction, (131.93 – 4.03) = 127.90 g.

16. For each soil specimen, add the masses of fractions 1, 2, 3, and 4 and enter that total value in your lab notebook.

17. For each soil specimen fraction, divide the mass of that fraction by the total mass of all fractions, multiply that result by 100 to determine the fraction mass percentage, and enter that value in your lab notebook. Note that we are calculating the fraction mass percentages based on the total mass of all fractions isolated rather than on the initial mass of the specimen. That's because some of the material in the original specimen may have been soluble and so dissolved in the water we used to separate the fractions.

18. We have the data necessary to calculate one more possibly useful value, the insoluble mass percentage of each of our specimens. We know the original mass of each specimen, and we know the total mass of all of the fractions we isolated for each specimen. Divide the total mass of the isolated fractions of each specimen, divide that value by the original mass of that specimen, and multiply by 100 to determine the percentage of the original specimen that was insoluble in water. Record that value in your lab notebook.

REVIEW QUESTIONS

Q1: **Based on the physical characteristics you observed for your soil specimens, how well can you discriminate between the specimens?**

Q2: **Which of the physical tests were most and least useful in discriminating between specimens? Why?**

Q3: **Why is it useful to perform several different physical tests on each soil specimen?**

Examine the Microscopic Characteristics of Soil

Lab I-3

EQUIPMENT AND MATERIALS

You'll need the following items to complete this lab session. (The FK01 Forensic Science kit for this book, available from *http//:www.thehomescientist.com*, includes the items listed in the first group.)

MATERIALS FROM KIT

- Goggles
- Magnifier
- Slides, deep well
- Spatula

MATERIALS YOU PROVIDE

- Gloves
- Microscope (see text)
- Specimens: soil, known and questioned

WARNING

None of the activities in this lab session present any real hazard, but as a matter of good practice, you should always wear splash goggles, gloves, and protective clothing when working in the lab.

BACKGROUND

Forensic laboratories routinely use microscopic analysis to characterize soil specimens. Microscopic analysis is often used to identify and quantify the major mineral components of soil specimens, but the mix of non-mineral components present in a specimen is often at least as important for establishing a "fingerprint" of that specimen. For example, if a questioned soil specimen contains pollen or seeds, a forensic botanist may be able to identify unambiguously the plant or plants that produced the pollen or seeds. Known soil specimens taken from a vicinity where those plants are present can be compared against the questioned specimen to establish with high certainty that the questioned specimen originated in the same vicinity, particularly if the plant or mix of plants is relatively uncommon.

The initial microscopic examination is usually done with a stereo microscope at low magnification, typically 10X to 40X. This examination allows the forensic geologist to gain an overall impression of the soil specimen, note the presence and abundance of various natural and artificial non-mineral components such as seeds; bits of glass, plastic, or metal; and so on. The stereo microscope is used to view particles from as small as 10 micrometers (μm = 0.01 mm) to as large as the physical capacity of the microscope stage. The eyepieces used for such examination often include a reticle that allows measurement of individual particles in the specimen, or a grid that simplifies counting individual particles in the specimen to provide the data necessary to determine particle size distribution.

Soil specimens to be examined are usually placed in a metal tray, white for dark specimens and black for light specimens, and examined by incident (reflected) light. The tray may also contain a comparison grid, and is often treated with a sticky substance to hold the particles in fixed positions while they are being examined and counted.

Subsequent microscopic examination is usually done with a *petrographic microscope*, which is essentially a modified version of a standard compound biological microscope. A petrographic microscope makes provision for examining specimens by transmitted polarized light, and may also provide such features as an electrically heated stage, which is used in determining the density and index of refraction of individual soil particles.

Although it is not the ideal tool for the purpose, in a home lab, a standard compound microscope is a reasonable substitute for the stereo and petrographic microscopes used in professional forensics labs. To examine our soil specimens, we used our standard compound microscope with a gooseneck lamp to provide incident light, as shown in Figure I-3-1.

Figure I-3-1: *Standard compound microscope set up to examine soil specimens*

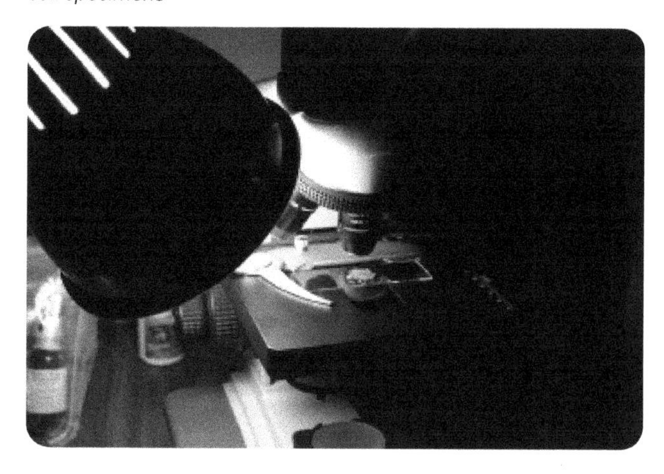

Under magnification, soil specimens that appear very similar or identical to the naked eye take on distinct individual characteristics. For example, Figure I-3-2 shows two soil specimen fractions taken within 50 cm of each other near a child's sandbox, and sieved with a kitchen flour sifter. Although it may not be obvious in the photograph, the specimen on the left appears *very* slightly redder and the specimen on the right *very* slightly browner, but otherwise, it is difficult to discriminate these specimens by naked eye.

Figure I-3-2: *Two soil specimen fractions that appear nearly identical to the naked eye*

The soil specimen fraction shown under low magnification in Figure I-3-3 is from near the edge of the sandbox. Individual sand grains are readily visible as glass-like, colorless quartz crystals. The specimen gathered half a meter away from the sandbox appears similar, but has many fewer sand grains visible.

Figure I-3-3: *Many sand grains are visible at 40X magnification*

PROCEDURE I-3-1: EXAMINE SOIL SPECIMENS UNDER MAGNIFICATION

1. If you have not already done so, put on your splash goggles, gloves, and protective clothing. (In this lab session, the purpose of these safety items is less to protect you from the specimens than to protect the specimens from you. Forensic technicians and scientists always wear protective gear to avoid contaminating specimens.)

2. Label six well slides Q1 and K1 through K5, and transfer small amounts of the corresponding soil specimens to each slide. You needn't fill the well completely. Ideally, you want just enough soil in each well to provide a single layer of particles.

3. Examine each specimen with the magnifier or at low magnification under a stereo microscope. Sometimes, similarities and differences between specimens are more clearly visible at lower magnification. At higher magnification, you may not be able to see the forest for the trees.

4. With the compound microscope set to 40X magnification, observe the questioned soil specimen. Record detailed observations for that specimen in your lab notebook. Here are some questions to keep in mind as you observe the specimen:

Particle size and distribution

Are the particles of relatively uniform size, or do they differ significantly? If the sizes differ, are particle sizes distributed relatively evenly from smallest to largest, or are particles of only a few distinct sizes visible? Do the particles appear to be discrete, or aggregates made up of smaller particles? If you have a reticle or grid eyepiece, use the known spacings of the reticle or grid and the known magnification to estimate the particle sizes. If possible, do a count of particles in different size ranges and use that count to estimate particle size distribution.

Particle structure

Are the particles smooth or rough? Spherical or elongated? Crystalline or amorphous? Is there any correlation between particle size and appearance, for example, that the smallest particles appear rough and jagged while larger particles appear smooth?

Particle color

What color are the particles? Are they transparent, translucent, or opaque? Is there any correlation between particle size or structure and particle color? Are individual particles of uniform color, or does the color vary within the particle?

Nonmineral matter

Does the specimen contain nonmineral matter, natural or artificial? If so, observe it carefully, using high magnification if necessary, and describe it as completely as possible. For example, one of our specimens contained several tiny, transparent particles that were almost perfect spheres. In a professional forensics lab, there would be comparison specimens available to help identify those objects. We have no such specimens, but we speculate that they are tiny glass spheres that were originally part of a reflective warning sign or movie projection screen. Another specimen contained objects that appeared to be seeds, probably from the grass that covers our yard. We're not botanists, so the most we can say is that those seeds appeared very similar to those in a bag of grass seed in the basement.

For bulk soil specimens in a well slide, the physical depth of the specimen means you won't be able to bring the entire specimen into focus at one time. Use the coarse and fine focus adjustments on your microscope to view all parts of the specimen.

5. When you finish examining the specimen at low magnification, switch to medium and then high magnification, which may reveal additional details.

If you have the necessary equipment, photograph each soil specimen at low, medium, and high magnification and append copies of those images to your report in your lab notebook.

6. Repeat steps 4 and 5 for each of the known specimens.

7. After you complete step 6, you should have a good idea of which of the known specimens, if any, is consistent with the questioned specimen. If you find a consistent known specimen, compare it directly with the questioned specimen, switching the slides in and out of the microscope stage. Record your observations during the direct comparison in your lab notebook.

REVIEW QUESTIONS

Q1: **Based on your microscopic observations, how well can you discriminate between the known soil specimens?**

Q2: **With what level of certainty can you determine if the questioned specimen is consistent with one of the known specimens?**

Q3: **Must a questioned specimen correspond exactly with a known specimen for you to declare that the two specimens are consistent?**

Assay Phosphate Concentrations in Soil Specimens

Lab I-4

EQUIPMENT AND MATERIALS

You'll need the following items to complete this lab session. (The standard kit for this book, available from *http://www.thehomescientist.com*, includes the items listed in the first group.)

MATERIALS FROM KIT

- Goggles
- Ascorbic acid tablet, 500 mg
- Centrifuge tubes, 15 mL
- Centrifuge tubes, 50 mL
- Graduated cylinder, 10 mL
- Graduated cylinder, 100 mL
- Molybdate reagent
- Phosphate extraction reagent, concentrate
- Phosphate standard, 1,000 ppm phosphate
- Pipettes
- Reaction plate, 24-well
- Stirring rod

MATERIALS YOU PROVIDE

- Gloves
- Balance (optional)
- Desk lamp or other light source
- Paper (white)
- Soft drink bottle, 500 mL (clean and empty)
- Water, distilled or deionized
- Specimens: soil, known and questioned

WARNING

Molybdate reagent and phosphate extraction reagent contain sulfuric acid, which is extremely corrosive. Wear goggles, gloves, and protective clothing.

BACKGROUND

Even if two soil specimens have very similar physical characteristics, their chemical characteristics may differ sufficiently to allow them to be discriminated from each other. Professional forensic labs use both instrumental and wet-chemistry quantitative analysis laboratory techniques to test soil specimens for the presence and concentration of various inorganic and organic compounds.

One major advantage of instrumental techniques is that forensic scientists don't have to decide ahead of time what to look for. Instrumental tests can provide a "fingerprint" of all of the chemical species present in a specimen—whether those species are common or rare—along with their concentrations. A wet-chemistry test, on the other hand, is used to identify the presence and concentration of a particular chemical species, which means the forensic analyst must decide ahead of time which species to look for. Despite that drawback, wet-chemistry tests (primarily color change tests) are still widely used because they are fast, cheap, sensitive, and accurate.

In this lab session, we'll use a standard wet-chemistry assay protocol called the molybdate test to determine the concentration of inorganic phosphate ions in our questioned and known soil specimens. In acidic solution, ammonium molybdate reacts with phosphate ions to produce a strongly colored phospho-molybdate complex. Ascorbic acid (vitamin C) is used to prevent that complex from oxidizing. That stabilizes the colors for an hour or more, allowing multiple comparisons to be made over the course of the lab session.

We'll use a phosphate extraction reagent made up of ammonium sulfate in sulfuric acid to extract the inorganic phosphate ions present in our soil specimens. We'll then set up an array of known concentrations of phosphate ions and react those and the extracts from the questioned and known soil specimens with the molybdate reagent. By comparing the intensity of the colors produced with the known phosphate concentrations against those produced by our specimens, we can determine the phosphate concentration in each of the specimens.

If the phosphate concentration in the questioned specimen differs greatly from the concentration in any of the known specimens, it's unlikely that the questioned specimen originated from the same source as any of the known specimens. If the phosphate concentration of the questioned specimen is similar to that of one of the knowns, that is one more piece of evidence that those two specimens are consistent.

FORMULARY

If you don't have the FK01 Forensic Science Kit, you can buy the necessary reagents or make them yourself. Wear gloves, goggles, and protective clothing when making up or handling these reagents.

Molybdate reagent Dissolve 0.5 g of ammonium molybdate in 10 mL of distilled or deionized water. Slowly and with swirling, add 8 mL of concentrated (98%) sulfuric acid. Allow the solution to cool, and make it up to 25 mL with distilled or deionized water.

Phosphate extraction reagent Dissolve 0.75 g of ammonium sulfate in 50 mL of distilled or deionized water. Slowly and with swirling, add 5 mL of concentrated (98%) sulfuric acid. Allow the solution to cool, and make it up to 250 mL with distilled or deionized water. This is a working strength solution.

Phosphate standard, 1,000 ppm phosphate Dissolve 1.433 g of potassium dihydrogen phosphate in distilled or deionized water and make up the solution to 1,000 mL to make a phosphate standard solution that is 1,000 ppm with respect to phosphate ion. Add a few drops of chloroform or a few crystals of thymol as a preservative.

PROCEDURE I-4-1: EXTRACT SOIL SPECIMENS

1. Label centrifuge tubes for each of your questioned and known soil specimens. If you have more soil specimens than tubes, you can substitute any small container. Ask a friend or lab partner to give you a small amount of one of your known specimens, which will be your questioned specimen.

2. Transfer 1 g of each soil specimen to the corresponding tube. If you don't have a balance, you can with some loss in accuracy use volumetric measures, such as 1/4 teaspoon of each soil. If you measure your specimens by volume, be careful to keep the volumes as consistent as possible between specimens.

Use dried, unprocessed soil specimens. A specimen that has been wet-sieved has lost some or most of the phosphates initially present. If you retained any of the washed and sieved known specimens from Procedure I-2-4, extract one or more of them as well. That will provide an interesting comparison of the phosphate concentrations in unwashed versus washed specimens of the same soil(s).

3. Add 10 mL of the working strength phosphate extraction reagent to each tube, cap the tube, and invert it several times over the course of 30 minutes or so to extract any inorganic phosphates present in the specimen.

The phosphate extraction reagent in the FK01 Forensic Science Kit is a 15X concentrate. Dilute the concentrate to working strength by adding one volume of the concentrate to 14 volumes of distilled or deionized water. For example, to prepare 75 mL of working solution (sufficient for a questioned specimen and six known specimens), add 5 mL of the concentrate to 70 mL of water.

4. Place the tubes aside to allow the soil to settle. The 50 mL tubes stand upright on their bases. The 15 mL tubes can be stood inverted on their caps.

These phosphate extract solutions are stable indefinitely. You can retain them until you are ready to perform the next procedure.

PROCEDURE I-4-2: ASSAY SOIL PHOSPHATE CONCENTRATIONS

The phosphate concentrations of soils can vary from greater than 1,000 parts per million (ppm) in recently fertilized specimens to less than 1 ppm in some phosphorus-poor soils. To cover this range, we'll populate 12 wells of the 24-well reaction plate with known phosphate concentrations ranging from 1,000 ppm to below 1 ppm. We'll then populate as many of the remaining wells as necessary with our questioned and known soil specimen extracts and react each of the populated wells with molybdate reagent. By observing the intensity of the blue color that results in each well, we can determine the phosphate concentrations in that well.

> Before you begin this procedure, crush a 500 mg vitamin C tablet and dissolve it in 20 mL of distilled or deionized water in a 50 mL centrifuge tube to produce a 25 mg/mL solution of ascorbic acid. (Don't be concerned if some solids refuse to dissolve. Tablets contain binders and other insoluble materials.) Prepare this solution immediately before you begin this procedure. Ascorbic acid solutions decompose in the presence of light and atmospheric oxygen.

1. Place the 24-well reaction plate on a sheet of white paper under a desk lamp or other strong light source.
2. Use a clean pipette to transfer 1 mL (the line on the pipette stem nearest the bulb) of distilled or deionized water to each of wells A2 through A6 and B1 through B6.
3. Use a clean pipette to transfer 1 mL of the 1,000 ppm phosphate standard solution to each of wells A1 and A2.
4. Mix the solution in well A2 by stirring it with the pipette tip and drawing up and expelling the solution with the pipette several times. Well A2 now contains 2 mL of 500 ppm phosphate solution.
5. Withdraw 1 mL of the 500 ppm phosphate solution from well A2, transfer it to well A3, and mix the solutions. Well A2 now contains 1 mL of 500 ppm phosphate solution, and well A3 contains 2 mL of 250 ppm phosphate solution.
6. Continue the serial dilution procedure until the wells are populated as follows:

   A1 – 1 mL of 1,000 ppm phosphate solution

   A2 – 1 mL of 500 ppm phosphate solution

   A3 – 1 mL of 250 ppm phosphate solution

   A4 – 1 mL of 125 ppm phosphate solution

   A5 – 1 mL of ~63 ppm phosphate solution

   A6 – 1 mL of ~31 ppm phosphate solution

   B1 – 1 mL of ~16 ppm phosphate solution

   B2 – 1 mL of ~8 ppm phosphate solution

   B3 – 1 mL of ~4 ppm phosphate solution

   B4 – 1 mL of ~2 ppm phosphate solution

   B5 – 1 mL of ~1 ppm phosphate solution

   B6 – 2 mL of ~0.5 ppm phosphate solution

7. Populate the wells in rows C and D with 1 mL each of your questioned and known original soil specimen extract solutions, one solution per well. If you have extracts from any washed and sieved specimens, populate wells with them as well.
8. Fill a clean pipette with the ascorbic acid (vitamin C) solution. Transfer five drops of this solution to each well.

9. Fill a clean pipette with molybdate reagent (CAUTION: CORROSIVE). Transfer five drops of this solution to each populated well.

10. Observe the populated wells for the appearance of the blue color that indicates the presence of the phospho-molybdate complex. Continue observing the wells until the color appears to have reached maximum intensity in each well. Depending on temperature and other factors, this may require 15 to 30 minutes or more, particularly in wells with more dilute phosphate solutions. Once developed, the color should be stable for at least an hour or two, so you can take your time doing the observations.

11. Compare the color intensity in the questioned specimen well against the wells in rows A and D. Determine the closest match and record the concentration of that well in your lab notebook.

12. Determine if the color intensity in the questioned well is closely similar to the color intensity in any of the wells that contain extracts of your known soil specimens. Record your observations in your lab notebook.

REVIEW QUESTIONS

Q1: **Based on the phosphate concentrations you observed for your soil specimens, how well can you discriminate between the specimens?**

Q2: **What other chemical tests might be useful in discriminating between specimens? Why?**

Q3: **Would it be more useful to test for chemical species that are widely distributed in different soils, or for chemical species that are not present in most soils? Why?**

Examine the Spectroscopic Characteristics of Soil

Lab I-5

EQUIPMENT AND MATERIALS

You'll need the following items to complete this lab session. (The standard kit for this book, available from *http://www.thehomescientist.com*, includes the items listed in the first group.)

MATERIALS FROM KIT

- Goggles
- Hydrochloric acid
- Inoculating loop
- Pipettes
- Spatula
- Spot plate
- Exemplar: spectroscopic soil specimen

MATERIALS YOU PROVIDE

- Gloves
- Burner, gas (see text)
- Digital voice recorder (optional)
- Spectrometer (Project Star or similar)
- Toothpicks, plastic
- Water, distilled or deionized
- Specimens: soil, known and questioned

WARNING

Hydrochloric acid is corrosive and emits strong fumes. The spectroscopic soil specimen supplied with the kit may contain toxic heavy metals. Wear splash goggles, gloves, and protective clothing when working in the lab.

BACKGROUND

Although wet-chemistry tests are still commonly used in professional forensics labs, various types of spectrometry are used extensively to supplement or replace these older wet-chemistry tests. Spectrometry has several advantages, notably that it is extremely sensitive, very accurate and precise, and can be used when only very small specimens are available for testing. The only real drawback to spectrometry is that professional-grade spectrometers are extremely expensive, costing from tens of thousands to literally millions of dollars.

Fortunately, the $40 Project Star spectrometer is sufficiently accurate and sensitive to provide useful data about our soil specimens. (Just out of curiosity, we tested the Project Star spectrometer by dissolving a few grains of table salt in a liter of distilled water and examining that specimen with the Project Star spectrometer; the yellow sodium lines in the spectrum were visible even at that tiny sodium concentration.)

In this lab session, we'll extract some of the ion species present in our soil specimens by treating them with 3 M hydrochloric acid, which reacts with many insoluble compounds to form soluble chlorides. Chloride salts are ideal for spectrometry because most of them vaporize at low temperatures relative to other inorganic compounds. We'll test those specimens with the Project Star spectrometer and determine which ion species we can detect in our soil specimens. The limitations of our inexpensive spectrometer mean that our results will be qualitative (present/not present) rather than quantitative, but even qualitative results may provide data useful for discriminating among our specimens.

This lab session has three procedures. In the first, we'll extract ion species from our known and questioned soil specimens by soaking them in dilute hydrochloric acid. In the second, we'll test those specimens with the spectrometer to record the emission lines from elements present in our soil specimens, and attempt to identify the species of interest present, if any. In the third, we'll prepare and test an exemplar that is known to contain significant levels of one or more species and attempt to identify those species.

FORMULARY

Professional forensics labs use spectrometers that are extraordinarily sensitive. Even tiny amounts of contaminants register on these devices, so professional labs prepare specimens for analysis using *spectroscopic-grade* hydrochloric acid, which is extremely expensive.

The hydrochloric acid supplied with the kit is pure, but nowhere near as pure as spectroscopic-grade acid. That's acceptable because our $45 spectrometer isn't nearly as sensitive as the $45,000 spectrometers that professional labs use.

If you don't have the kit, you can make up 3 M hydrochloric acid by adding one volume of concentrated (37%) reagent- or lab-grade hydrochloric acid to three volumes of distilled or deionized water. You can even use hardware store muriatic acid, which comes in various concentrations. It's not important that the concentration be exact. Anything reasonably close to 3 M (~9%) hydrochloric acid will work. For example, if you're using the common 31.45% muriatic acid, simply add one volume of that acid to about 2.4 volumes of water. If you have the 14% muriatic acid, add one volume of that acid to about half a volume of water.

With any acid less pure than spectroscopic-grade (including the acid supplied with the kit), you should run a blank to determine if contaminants are present in the acid at a high enough concentration to be visible with the spectroscope. For example, sodium is present in most acid specimens at a high enough level to be detected with the Project Star spectrometer, and iron is present in many samples at detectable levels.

PROCEDURE I-5-1: EXTRACT ION SPECIES FROM SOIL SPECIMENS

1. If you have not already done so, put on your splash goggles and gloves.
2. Use the spatula to transfer one level spoonful of your first known specimen to well #1 in the spot plate. Record the contents of that well in your lab notebook.
3. Wipe the spatula clean, and use it to transfer one level spoonful of your second known specimen to well #2. Record the contents of that well in your lab notebook.
4. Repeat step 2 to transfer each of your remaining knowns and your questioned specimen to empty wells of the spot plate.
5. Use a clean pipette to transfer 5 drops of distilled or deionized water to each of the populated wells and 10 drops of water to one of the vacant wells.
6. Use a clean pipette to transfer 5 drops of 6 M hydrochloric acid to each of the populated wells and 10 drops of 6 M acid to the well that contains only water.
7. Using a clean plastic toothpick for each wells, stir the contents of the well to mix them.

Allow the acid to operate on the soil specimens for at least a few minutes before you perform the next procedure.

PROCEDURE I-5-2: TEST SOIL SPECIMEN EXTRACTS WITH THE SPECTROMETER

In this procedure, we'll use a gas burner flame and the spectrometer to identify ions present in the various soil specimens. If possible, have an assistant for this and the following procedure. While one person observes with the spectroscope, the assistant should handle the specimens.

> You can use any type of gas burner. We used a gas microburner designed for lab use, but you can substitute a propane torch from the hardware store, a butane soldering torch from the hobby or hardware store, a propane camp stove, or a natural gas stove burner.

The gas flame vaporizes the ion species present in the specimen and excites the ions by raising electrons to higher energy levels. As those excited electrons return to their rest states, they emit photons at characteristic wavelengths. As you continue heating the specimen, it eventually "burns off" and stops emitting light at those wavelengths.

Depending on the species and its concentration, you may have only a few seconds to identify the wavelengths present. For that reason, we recommend using a digital voice recorder or similar device so that you can simply speak your observations without taking your eye from the spectrometer rather than having to take time to write them down.

1. If you have not already done so, put on your splash goggles and gloves.
2. Set up and ignite your gas burner.
3. Dip the tip of the inoculating loop in the well that contains the acid/water mixture. Position the tip of the loop at the tip of the flame to burn off any contaminants present on the loop.
4. Remove the loop from the flame and allow it to cool for a moment. Do not allow the tip of the loop to come into contact with the table or any other surface.

5. Dip the tip of the loop in the first specimen well. While viewing the flame through the spectroscope, as shown in Figure I-5-1, record the wavelength (in nanometers on the scale) and relative intensity of the emission lines visible. Concentrate first on the most intense lines. After you have recorded all of those, record the less intense lines. Record the relative intensity of each line as "very bright," "bright," "moderate," "dim," or "very dim." If there are an overwhelming number of emission lines present, record only those that you consider moderate or brighter.

Instead of manually recording the wavelengths and intensities of the spectral lines, you can shoot images of them. To do so, you'll need to make a clamp or bracket that holds the camera and spectrometer in fixed positions relative to the burner. If you use a digital camera, the images you capture will probably not match what you see visually. Digital cameras reproduce continuous analog color spectra in discrete digital chunks, so emission lines that are prominent visually may be much dimmer or even entirely absent photographically.

Figure I-5-1: *Barbara using the Project Star spectrometer to view emission spectra of soil specimens*

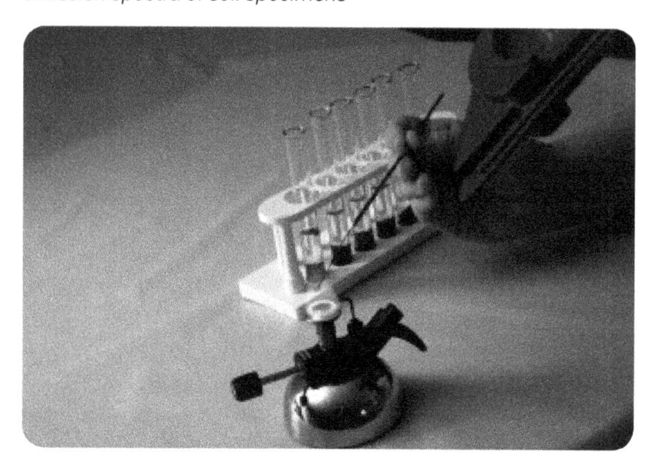

Depending on the specimen, your flame source, and other factors, you may need to re-dip the loop several times to gain enough observing time to record all of the lines visible. You needn't repeat the acid wash and burn-off if you're testing the same specimen repeatedly.

6. Repeat steps 3 through 5 for each of the remaining known and questioned specimens. (Retain the spot plate array for use in the next procedure.)

7. Using your observed data and the values in Table I-5-1, determine if each of the six species listed in the table are present in each of your specimens. Begin by looking for the prominent lines present for each species. For example, to determine if aluminum is present in a particular specimen, check your notes to see if a moderately intense line pair appeared at 394.4/396.2 nm. If that pair appears to be present, verify the presence of aluminum by checking to see if the remaining emission lines were also present. If the 394.4/396.2 nm was not present in the specimen, you can rule out the presence of aluminum (at least at levels detectable with the equipment you're using) and move on to the next species.

8. It's possible (even likely) that many of the emission lines present in your specimen, including some quite intense lines, represent species not listed in Table I-5-1. If you have time, attempt to identify some or all of these unknown species from the positions of those "extra" emission lines.

It's here that the difference between a $40 spectrometer and a $45,000 spectrometer becomes obvious. The professional instrument analyzes the elements present based on the observed spectrum, and displays or prints that information quantitatively. We have to do the comparison manually, which in practical terms limits us to comparing only a limited number of relatively bright spectral lines from a limited number of elements.

Table I-5-1 lists the prominent emission line wavelengths and relative intensities for the six elements we chose to test for. The relative intensities listed in parentheses beside each characteristic emission wavelength provide a semi-quantitative metric. For example, the sodium lines at 589.0 nm and 589.6 nm are extraordinarily bright, while the lines at 371.1 nm and 568.8 nm are much fainter, but bright enough to be easily visible. We established an arbitrary cutoff at a relative intensity of 450, which should be readily visible even at the extreme ends of the visible spectrum where your eyes are less sensitive.

Nearly every (we are tempted to say "every") soil contains sodium in amounts large enough to provide very prominent emission lines. Aluminum, although its emissions lines are fainter, is present in many soil specimens. Iron is present in some soils, particularly those with a reddish color, in amounts large enough to provide prominent emission lines. Barium and mercury are toxic heavy metals that (we hope) will not be present in your soil specimens, but nonetheless some soil specimens will contain sufficient quantities of either or both metals to provide spectra bright enough to be visible with our inexpensive spectrometer. (You can look at a mercury spectrum simply by pointing your spectrometer at a fluorescent lamp.) Strontium is not widely distributed in soils, but may be present in some specimens, particularly because it is commonly used in fireworks (including simple sparklers) and road flares, where it produces an intense red light.

With a professional spectrometer, the question is, "What elements are present in this specimen?" With our simple spectrometer, we have to do a bit more work by asking, "Is element X present in this specimen?" For example, we might decide to determine if sodium is present in one of our specimens. To do so, we look at Table I-5-1 to determine which very prominent lines are present in the sodium spectrum. Knowing that the 589.0 nm and 589.6 nm "sodium pair" are extremely prominent, we look for those wavelengths in our specimen. If we see bright lines at those wavelengths, we strongly suspect that sodium is present. We confirm that by looking for the fainter sodium lines at 371.1 nm and 568.8 nm. If those lines are also present in the spectrum, it's certain that sodium is present in the specimen. Repeating that procedure allows us to determine if aluminum, barium, iron, mercury, and/or strontium are present.

> Visible light ranges from about 380 nanometers (nm) (the border between deep violet and near ultraviolet) and 720 nm (the border between deep red and near infrared). The sensitivity of the human eye peaks in the green to yellow-green wavelengths, and falls off rapidly at both ends of the visible light range. Depending on the sensitivity of your own eyes, even relatively intense lines at the extremes of the visible range may appear dim or invisible.

Table I-5-1: *Prominent emission line wavelengths and relative intensities for selected elements ("finder" lines bolded for each species)*

Element	Prominent emission line wavelengths (relative intensities)
Aluminum	390.1 (450); **394.4 (4500)**; **396.2 (9000)**; 466.7 (550); 559.3 (450); 600.6 (450); 607.3 (450); 618.3 (450); 620.2 (450); 624.3 (450); 633.6 (450)
Barium	413.1 (910); 428.3 (530); **455.4 (9300)**; 458.0 (580); **493.4 (6900)**; 553.5 (1830); 577.8 (740); 582.6 (610); 597.2 (700); 599.7 (620); 601.9 (610); 606.3 (840); 611.1 (880); 614.2 (1510); 634.2 (900); 645.1 (580); 648.3 (1770); 649.7 (900); 649.9 (1060); 652.7 (890); 659.5 (740); 667.5 (462); 669.4 (454)
Iron	**404.6 (4000)**; 406.4 (1500); 407.2 (1200); 414.4 (800); 426.0 (800); 427.2 (1200); 428.2 (1200); 430.8 (1200); 432.6 (1500); 437.6 (800); **438.4 (3000)**; 440.5 (1200); 442.7 (600); 495.8 (1500); 516.7 (2500); 517.2 (500); 522.7 (1000); 527.0 (1200); 527.0+ (800); 532.8 (800); 534.1 (500)
Mercury	**404.7 (1800)**; **435.8 (2000)**; 546.1 (1100); 577.0 (1240); 579.0 (1100); 615.0 (1000)
Sodium	371.1 (850); 568.8 (560); **589.0 (80000)**; **589.6 (40000)**
Strontium	403.0 (1300); **407.8 (46000)**; **421.6 (32000)**; **460.7 (65000)**; 472.2 (3200); 474.2 (2200); 478.4 (1400); 481.2 (4800); 483.2 (3600); 485.5 (500); 486.9 (600); 487.2 (3000); 487.6 (600); 487.6 (2000); 489.2 (1000); 496.2 (8000); 496.8 (1300); 515.6 (800); 522.2 (1400); 522.5 (2000); 522.9 (2000); 523.9 (2800); 525.7 (4800); 545.1 (1500); 548.1 (7000); 548.6 (1100); 550.4 (3500); 552.2 (2600); 553.5 (2000); 554.0 (2000); 638.1 (1000); 638.7 (900); 638.8 (600); **640.8 (9000)**; 650.4 (5500); 654.7 (1000); 655.0 (1700); 661.7 (3000); 664.4 (800); 679.1 (1800); 687.8 (4800); 689.3 (1200)

PROCEDURE I-5-3: IDENTIFY IONS PRESENT IN EXEMPLAR

Professional forensic labs depend heavily on *exemplars*, which are known specimens used for comparison. Depending on the type of item, these exemplars may be actual physical samples, detailed macroscopic and/or microscopic images, spectrographs, and so on. For example, the impression analysis section probably doesn't have physical samples of every athletic shoe ever made, but what they do almost certainly have is references—printed, electronic, or both—that provide detailed images of the tread pattern of any shoe imaginable. Multiply that by the thousands of specimen types—from types of seeds and pollen to fabrics to tools, nuts, and bolts, and on and on—that must be kept on hand, and it's clear that massive amounts of reference material are necessary to the day-to-day functioning of any forensic lab.

The kit includes a specimen of one of our own exemplars, a soil specimen that is heavily contaminated with one or more of the ion species you tested for in the previous procedure. Because it's possible that your own known and questioned specimens do not contain any of the species tested for (except, almost certainly, sodium) at detectable levels, using a specimen known to have one or more of these species present gives you the opportunity to conduct an analysis that will yield positive results for at least one of the species. Ironically, in this case, what is for us a known exemplar will serve as a "known unknown" for you.

1. If you have not already done so, put on your splash goggles and gloves.
2. Transfer a level spatula spoonful of the exemplar to an empty well in the spot plate.
3. Add five drops of water and five drops of 6 M hydrochloric acid to the well, stir with a clean plastic toothpick, and allow several minutes for the acid to extract any ions present.
4. Set up and ignite your gas burner.
5. Dip the tip of the inoculating loop in the well that contains the acid/water mixture. Position the tip of the loop at the tip of the flame to burn off any contaminants present on the loop.
6. Remove the loop from the flame and allow it to cool for a moment. Do not allow the tip of the loop to come into contact with the table or any other surface.
7. Dip the tip of the loop in the exemplar well. While viewing the flame through the spectroscope, record the wavelength (in nanometers on the scale) and relative intensity of the emission lines visible.
8. The exemplar is known to contain one or more of the ion species listed in Table I-5-1 at levels detectable with the Project Star spectrometer. It may also contain detectable levels of other ions, including boron, calcium, cerium, or potassium, in any combination. Identify which of those 10 ions is present in the exemplar.

> Of course, we haven't provided emission line data for these four species, so you'll have to look them up in a printed reference or on the Internet. Such research is an important part of any forensic scientist's job.

REVIEW QUESTIONS

Q1: **Based on the spectra you observed for your soil specimens, how well can you discriminate between the specimens?**

Q2: **Which of the 10 species listed was present in the exemplar specimen?**

Q3: **Does it matter if you perform spectroscopic tests on an unwashed or washed soil specimen? Why?**

Hair and Fiber Analysis

Group II

Along with soil, hairs and fibers are the most common forms of trace evidence processed by forensic labs. In nearly all violent crimes, hairs and fibers are transferred from the criminal to the victim or the crime scene, and vice versa.

Hairs and fibers are easily transferred from people to people, people to objects, and vice versa, and they often cling tenaciously to the new environment. Hairs and fibers do not degrade quickly, which means that known specimens collected long after the fact can be compared successfully against questioned specimens collected weeks, months, or years before at a crime scene. Hairs and fibers are both ubiquitous and easily overlooked, so it's nearly impossible for a criminal to remove all of them from the crime scene, even if he is sufficiently well-informed to recognize their potential evidentiary value.

Because of their ubiquity and the diversity of their individual characteristics, hairs and fibers are an excellent form of class evidence. Hair may sometimes be individualized by DNA analysis, but even if a specimen cannot be individualized, it may be possible to categorize it quite specifically on the basis of its many class characteristics. Such evidence, although not conclusive, may be quite useful to establish guilt or innocence.

Although professional forensic labs often use instrumental analysis techniques to establish matches conclusively, wet chemistry tests and microscopy remain important techniques for hair and fiber analysis, both for initial screening and for establishing matches that will be used in court testimony. In this group, we'll perform many of the wet-chemistry and microscopic tests that real forensic labs do every day.

A DAY IN THE LIFE OF A T-SHIRT

If you want to do a real-world forensic hair and fiber analysis, begin with a new t-shirt, still in the plastic wrap. One day, unwrap the t-shirt and wear it all day long. Take notes throughout the day of the places you visit, the people and animals you encounter, where you sit, and so on.

At the end of the day, remove the t-shirt and place it on a clean surface. (A trash bag fresh from the box provides an uncontaminated work surface.) Under a strong light, use your magnifier or loupe to examine the entire surface of the t-shirt, front and back. Use forceps or tweezers to remove any hairs or fibers you find, and transfer them to a collection envelope.

Examine all of the questioned hair and fiber specimens using the procedures described in this group, and try to determine where they originated. When we tried this experiment, we found numerous hairs from ourselves and our dogs, fibers from our sofa, car seats, and office chairs, and many fibers for which we couldn't establish an origin, including several human and animal hairs and various fabric fibers.

Gathering Hair Specimens

EQUIPMENT AND MATERIALS

You'll need the following items to complete this lab session. (The standard kit for this book, available from *http://www.thehomescientist.com*, includes the items listed in the first group.)

MATERIALS FROM KIT

- Goggles
- Forceps
- Magnifier

MATERIALS YOU PROVIDE

- Gloves
- Light source (see text)
- Lifting tape (see text)
- Pen, permanent marking
- Transfer sheets (see text)
- Specimen storage containers (see text)
- Specimens: clothing (see text)

WARNING

Although none of the activities in this lab session present any significant risks, as a matter of good practice, you should always wear splash goggles, gloves, and protective clothing when working in the lab, if only to avoid contaminating specimens.

BACKGROUND

Hair is one of the most common and useful types of trace evidence, particularly in crimes such as assault, rape, and murder. Humans shed hair naturally, on average about 100 hairs per day. Even casual contact between two people ordinarily results in each of them transferring some hair to the other, and during a violent confrontation, the probability of hair being transferred from the victim to the attacker and vice versa is very high.

Several aspects of hair make it a very useful form of trace evidence:

- Hair degrades very slowly over time, so an old specimen is as useful as a recent one.
- Hair is easily shed and easily transferred from person to person or from person to clothing.
- Once transferred, hair often adheres persistently to the person or clothing to which it was transferred, so it is likely to survive until it can be collected by the forensic technician.
- Unlike bloodstains and other gross evidence, hair is small, light, and can easily go unnoticed by a criminal.
- Although, in the absence of DNA testing, hair is technically only class evidence, it possesses enough individual traits that it can often be characterized quite closely. For example, the color of a questioned hair may alone suffice to rule out a suspect.

DENNIS HILLIARD COMMENTS

Keep in mind that in some cases where a competent examiner may consider that two or more hairs have the same morphological characteristics and could have originated from the same source, DNA analysis has shown that they could not have originated from the same source. Project Innocence has used these results to discredit hair comparisons and this has reduced its value in court testimony. Hair analysis value in the lab is, as you state elsewhere, a screening tool to identify human versus animal origin and to exclude hairs from being DNA tested.

Questioned hair specimens may be collected at the crime scene by technicians, using vacuuming or other methods. But such specimens are also collected by forensic scientists in the lab, often from clothing or from a knife, club, or other weapon used in the commission of a crime.

DENNIS HILLIARD COMMENTS

Vacuuming for hairs and fibers is reserved for large areas such as vehicles. The vehicle is divided into sections, such as front right, front left, etc. The vacuum filter is changed for each section.

In this lab session, we'll use two of the methods commonly used in forensic labs for obtaining hair specimens: forceps and lifting tape.

You can use any suitable container for storing specimens, including vials, pill bottles, stoppered test tubes, small paper envelopes, or small resealable plastic bags. Whatever containers you use should be labeled as soon as you transfer the specimen to them, either by writing directly on the container or attaching a label to it.

A high-intensity desk lamp or a similar strong, directional light source is ideal for a close examination of the clothing specimens. You want the light to strike the surface of the specimen at an angle, which helps reveal hairs and other material that is not a part of the fabric itself. Placing the light source so that light strikes the specimen at a grazing angle is best for revealing hairs and other traces adhering to the clothing.

You can use any transparent, colorless sticky tape as lifting tape. We used packing tape.

For transfer sheets, you can use any transparent, colorless sheets of plastic. We used the stiff clear plastic sheets used as packing material to protect the screens of our computer displays. You can also use transparent notebook page protectors, theater gels, overhead transparency sheets, or any similar plastic sheets. If the lifting tape is small enough, microscope slides can be used as transfer sheets. (Make sure enough room remains uncovered by the tape to allow the slide to be labeled.)

Obtain unlaundered clothing specimens. Some of the best potential sources of hair specimens are the interior of hats, scarves, the collar and underarm areas of shirts, and the crotch area of panties or pantyhose and underpants.

PROCEDURE II-1-1: OBTAIN HAIR SPECIMENS WITH FORCEPS

Depending on the type of fabric and the nature of the clothing, it is often best to extract hair specimens by using a magnifier and forceps, as follows:

1. If you have not already done so, put on your splash goggles and gloves.
2. Place the clothing specimen on a flat surface and direct the light source at a grazing angle across the surface of the clothing.
3. Use the magnifier to examine the entire surface of the specimen thoroughly.
4. When you locate a hair or other piece of trace evidence, use the forceps carefully to remove the specimen and transfer it to an evidence container, as shown in Figure II-1-1. If there are many similar hairs, you can take only one or two of them as specimens, but be sure to continue examining the clothing specimen until you're certain that no other types of hairs are present.
5. As you transfer each specimen to its own container, initial and date that container and write a brief description of the contents. Include the clothing specimen number or description, the location on the clothing specimen from which the specimen was obtained, and any other pertinent information. Record full information about the specimen in your lab notebook.
6. Repeat steps 2 through 5 for additional clothing specimens, if applicable.

Figure II-1-1: *Using forceps to gather hair specimens from a clothing specimen*

PROCEDURE II-1-2: OBTAIN HAIR SPECIMENS WITH LIFT TAPE

Although it's generally preferable to extract hair specimens individually with forceps, there are times when using sticky tape to lift hair specimens may work better, particularly if the clothing has a rough texture. The other side of that coin, of course, is that using a tape lift often provides an embarrassing plethora of specimens, most of which are fibers from the clothing specimen itself.

One major advantage of the tape-lift method, shown in Figure II-1-2, is that it may capture very small trace-evidence specimens that may go unnoticed under low magnification. Even if you use forceps to extract specimens manually, you should follow that with a tape lift unless you're absolutely certain that no useful trace evidence remains on the clothing specimen.

Figure II-1-2: *Using lift tape to gather hair specimens from a clothing specimen*

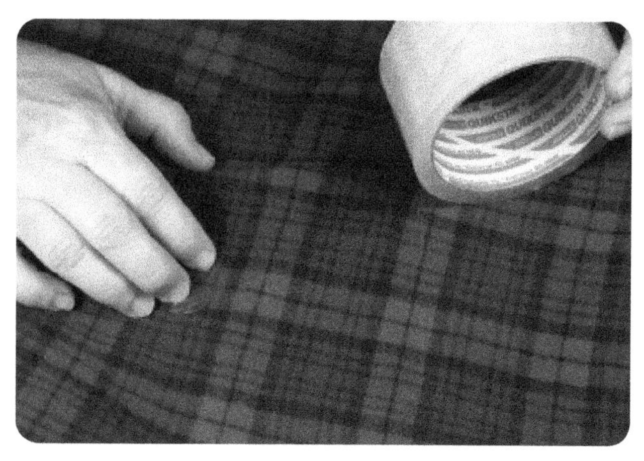

1. If you have not already done so, put on your splash goggles and gloves.
2. Place the clothing specimen on a flat surface. Spread it out and flatten it as much as possible.
3. Remove an appropriate amount of tape from the spool, being careful to avoid contaminating it with random fibers or other extraneous material.
4. Carefully place the tape sticky side down in contact with the clothing specimen, using the minimum amount of pressure needed to capture specimens from the clothing specimen. If the specimen cloth is smooth and hard-surfaced, very little pressure is needed to capture any specimens that are present. If the specimen cloth is rough and soft-surfaced, more pressure is needed (and correspondingly more of the cloth fibers will also adhere to the tape). Learning how much pressure to use is a matter of experience.
5. Peel the tape away from the clothing in one movement and immediately press the tape against the transfer sheet until it adheres. Label the lift as shown in Figure II-1-3. (In effect, the labeled lift takes the place of a labeled specimen storage container. You can subsequently remove individual hair specimens from the tape and examine them individually.) Also log all information about each lift to your lab notebook.
6. Repeat steps 2 through 5 for additional clothing specimens, if applicable.

Figure II-1-3: *Labeling tape-lifted hair specimens on a transfer sheet*

REVIEW QUESTIONS

Q1: **Can an individual hair found among the trimmings on the floor of a barbershop be individualized to the person from whom it originated? If so, how? If not, why not?**

Q2: **What are some possible disadvantages to using a vacuum to gather hair specimens?**

Study the Morphology of Human Scalp Hair

Lab II-2

EQUIPMENT AND MATERIALS

You'll need the following items to complete this lab session. (The standard kit for this book, available from *http://www.thehomescientist.com*, includes the items listed in the first group.)

MATERIALS FROM KIT

- Goggles
- Coverslips
- Forceps
- Glycerol
- Magnifier
- Pipettes
- Ruler
- Slides, flat

MATERIALS YOU PROVIDE

- Gloves
- Microscope (40X, 100X, and 400X)
- Ocular micrometer/reticle (optional)
- Specimens: human scalp hair (see text)

WARNING

Although none of the activities in this lab session present any significant risks, as a matter of good practice, you should always wear splash goggles, gloves, and protective clothing when working in the lab, if only to avoid contaminating specimens. Obviously, you may need to work without goggles when using a microscope or magnifier to examine specimens.

BACKGROUND

Historically, hair has been considered *class evidence* because a hair specimen could not be identified with certainty as having originated from a particular person. A forensic scientist who studied the *morphology* (shape, form, and structure) of a hair specimen could testify as to the gross physical characteristics of the hair (color, degree of curl, etc.), its internal and external structural characteristics, the likely somatic region from which the hair originated (scalp, beard, pubic, axillary, etc.), and—at least for scalp hair and often for pubic hair—the probable race of the person from whom the hair originated. But all of those are class characteristics rather than individual characteristics, so the most the forensic scientist can state based on morphological examination is that a hair specimen is "consistent with" or "similar in all respects to" another specimen.

That changed with the advent of forensic DNA testing. Initially, DNA testing of hair focused on *nuclear DNA*, which can identify an individual unambiguously but is present only in the *follicle* that surrounds the *root* of the hair (see Figure II-2-2). For this reason, many people still believe, incorrectly, that DNA testing of hair requires a hair specimen that has been pulled out by the root and includes the follicle.

Figure II-2-1: *Major structural components of hair shaft*

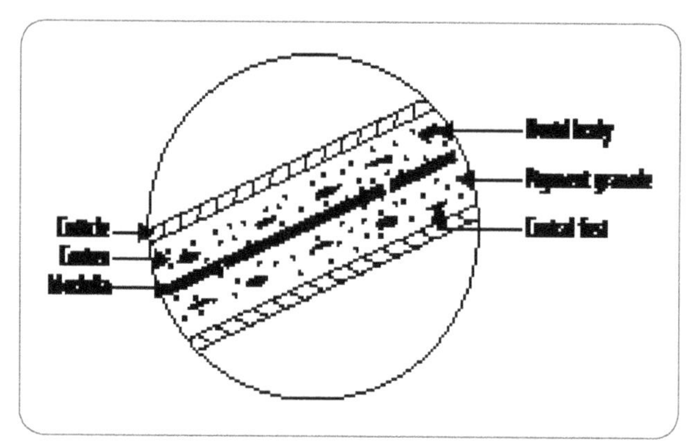

In fact, DNA tests can be done on a hair specimen that includes only the shaft, including specimens that have been cut rather than pulled from the scalp. Although the shaft does not contain nuclear DNA, it does contain *mitochondrial DNA* (*mtDNA*). Unlike nuclear DNA, mtDNA is not unique to an individual because it is inherited directly from the individual's mother rather than from both parents. Technically, that means that mtDNA can provide only class evidence rather than individual characteristic evidence, but an mtDNA match greatly reduces the size of the class in question. Despite this limitation, mtDNA tests are frequently done on hair specimens because mtDNA testing is faster and cheaper than nuclear DNA testing.

Despite the availability of DNA testing, morphological classification of hair specimens remains an important part of the work of any forensics lab. Relative to morphological examination, DNA testing is very expensive and time consuming. Accordingly, morphological examination is used to do preliminary screening of hair specimens. Morphological examination can definitively rule out matches between some specimens, so only specimens for which a match cannot be ruled out morphologically need be submitted for DNA testing.

This preliminary morphological screening is done in two phases:

Macroscopic examination

Macroscopic examination consists of observing specimens with the naked eye and at low magnification with a stereo microscope or magnifier to note such characteristics as color, length, and degree of curl. Ideally, this examination should be done by both reflected and transmitted light, although often only reflected light is used. The purpose of this examination is preliminary screening of a questioned specimen against known specimens. For example, if the questioned hair specimen is red, all knowns that are not red can immediately be eliminated from consideration.

Macroscopic examination may eliminate none, some, or all of the known specimens from consideration. The remaining specimens—those that are consistent macroscopically with the questioned specimen—can then be subjected to additional testing, including microscopic examination and possibly DNA testing.

Characteristics such as degree of curl are best observed with unmounted specimens at low magnification. After examining the unmounted specimens, you may if you wish dry-mount the specimens, either simply by covering them with a coverslip or by securing them to the slide with two drops of nail polish or other adhesive, one on either side of the cover slip.

Unmounted or dry-mounted specimens are best for observing the external characteristics of the specimens. Later in the lab session, we'll wet-mount specimens to view their internal structures.

Microscopic examination

Microscopic examination consists of observing specimens with a compound or comparison microscope at medium to high magnification (typically, 40X or 100X to 400X or more) to note fine structural details. For this phase, the specimen is wet-mounted by using a mounting fluid with a refractive index close to 1.52, that of keratin, the protein that is the chief structural component of hair. Wet-mounting reveals internal structure that is invisible with a dry mount. Additional knowns can often be eliminated based upon observed microscopic characteristics, leaving fewer specimens that require DNA analysis.

For temporary wet mounts, you can use distilled water (RI ~1.33), although glycerol (RI ~1.47), castor oil (RI ~1.48), or clove oil (RI ~1.54) is a closer match for the refractive index of the keratin that makes up the hair. Canada balsam or a similar mounting fluid matches the RI of keratin very closely, and can be used to make permanent mounts.

In this lab session, we'll examine the gross morphology of human scalp hair under low magnification. We'll then wet-mount the specimens and observe the fine morphology under medium to high magnification.

The most convenient source of human scalp hair specimens is, of course, yourself. Obtain at least half a dozen specimens from different areas of your scalp and label each with the source (crown, side, front, back, etc.). At least one or two of those specimens should be plucked, so that you can observe the root structure. Other specimens can be cut. You can also obtain undifferentiated specimens from your hair brush.

But don't limit your specimens to only your own hair. Obtain as many specimens as possible from family, friends, and others, and compare and contrast those specimens against each other. It's particularly useful to obtain specimens from as diverse a population as possible, including those of different ages (from children to elderly), both sexes, different races, and so on.

PROCEDURE II-2-1: MACROSCOPIC EXAMINATION OF HUMAN SCALP HAIR

1. Examine each specimen, by eye and with the magnifier, a stereo microscope, or other low-magnification optical aid, unmounted and, optionally, dry-mounted, by both reflected and transmitted light. (You can use your compound microscope at its lowest magnification, usually 20X or 40X.)
2. Record the following information about each specimen in your lab notebook: source, somatic region, color, uniformity of color from root to tip, length, degree of curl (straight, wavy, slightly curled, tightly curled, kinky, etc.), description of proximal (root) end (presence or absence of root, etc.), description of distal (tip) end (tapered, square cut, angle cut, split end, etc.).

PROCEDURE II-2-2: WET-MOUNT HAIR SPECIMENS

Wet-mounting is used to prepare specimens for microscopic examination. Using a mounting fluid with a refractive index close to that of the hair reveals the medulla, cortical fusi, pigment granules, and other internal structure of the hair. You can use distilled water, glycerol (glycerin), castor oil, or clove oil to make a temporary wet mount. For a permanent or semipermanent wet mount, use Canada balsam, Melt Mount, or a similar mounting fluid.

The traditional method of wet-mounting a hair specimen in a loop is sometimes called the figure-8 mount or infinity mount, because the mounted specimen often resembles the numeral eight or an infinity symbol. To wet-mount your specimens, proceed as follows:

1. Label a microscope slide with a description of the specimen.
2. Place a drop or two of mounting fluid in the center of the microscope slide.
3. Using forceps, place the center of the specimen in the mounting fluid, and loop back the ends until they adhere to the mounting fluid, as shown in Figure II-2-2.
4. Carefully place a coverslip over the specimen and press it into place, making sure no air bubbles are trapped beneath it. If there are air bubbles that can't be removed with gentle pressure on the coverslip, use a disposable pipette to add another small drop of mounting fluid at the edge of the coverslip, which should be drawn under the coverslip. Slight additional pressure should dislodge the air bubbles.
5. Use the disposable pipette or the corner of a paper towel to remove any excess mounting fluid from around the coverslip. Be careful not to draw off so much mounting fluid that it is pulled from under the coverslip.
6. Repeat steps 1 through 5 for each remaining specimen.

Figure II-2-2: *Barbara wet-mounting a hair specimen*

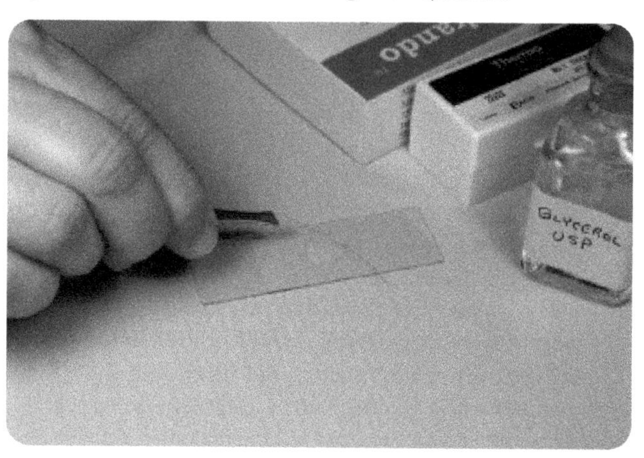

STORING WET MOUNTS

Always store wet-mounted specimens flat to prevent the coverslip from shifting. Specimens wet-mounted with distilled water remain usable for at least a few hours and perhaps overnight or longer. Specimens wet-mounted with glycerol, oils, or other less volatile mounting fluids may remain usable for several weeks or longer. Specimens wet-mounted with Canada balsam, Melt Mount, and similar permanent or semi-permanent mountants may remain usable indefinitely if they are stored properly.

PROCEDURE II-2-3: MICROSCOPIC EXAMINATION OF HUMAN SCALP HAIR

As useful as macroscopic examination is for preliminary screening, examining a specimen at low magnification by reflected light reveals little about the internal structural features of the specimen. In the microscopic examination, we'll examine each specimen by transmitted light at medium (100X) and high (400X) magnification. We'll observe, measure, and record the following microscopic characteristics about each specimen:

Shaft

Note the shape of the hair shaft as round, oval, oblate, triangular, or other. Use the ocular micrometer to measure the minimum and maximum diameters of the main body of the hair shaft. (Human hair ranges from about 20 to 175 µm in diameter; the diameter may vary substantially between somatic regions and even within the same somatic region from hair to hair.) Note any special appearance of the shaft, such as whether it is split, undulated, invaginated, buckled, shouldered, or convoluted.

Color

Although we've already recorded the color based on macroscopic examination, microscopic examination often reveals more detail. Note the color of the hair as colorless, blond, red, brown, black, or other, by both reflected and transmitted light. A sharp change in color near the proximal (root) end of the hair indicates that the hair is dyed, as does coloration that is relatively evenly distributed throughout the cortex. Note if the hair appears to have been bleached (overall yellowish tinge to the cortex), dyed, or otherwise treated.

Pigment bodies

Note the presence of pigment bodies (absent, few, abundant) and the size of the individual pigment bodies as small, medium, or large (this obviously requires comparison with other specimens or with a reference book). Also note the degree of aggregation (uniformly distributed, patches, streaks, clumps) and the size of these aggregates (small, medium, or large). Describe the density of the pigment bodies as opaque, heavy, medium, light, or sparse, and their distribution as uniform, peripheral (clustering near the cuticle), one-sided, or otherwise. Pigment bodies are also called pigment granules.

Medulla

Note the absence or presence of the medulla. If present, note whether the medulla is continuous, broken, or fragmentary. Note its appearance as opaque, translucent, amorphous, or cellular. Note any other characteristics, such as a split, division, doubling, or twist. Use the ocular micrometer to measure the width of the medulla and use that to calculate the medullar index (MI).

> The medullar index (MI) is simply the ratio of the diameters of the medulla and the shaft. For example, if medulla diameter is 15 µm and shaft diameter 100 µm, record the MI as 0.15. MI is one easy way to discriminate human hair from animal hair. Human hair has an MI of 0.35 or lower (usually much lower) while animal hair has a high MI value.

Cortex

Note the texture of the cortex as coarse, medium, or fine. If the cortex has an unusual appearance (e.g., cellular or striated), note that as well.

Cortical fusi

Cortical fusi are tiny air bubbles in the cortex, which appear black under the microscope. Cortical fusi, if present, are most common near the root of human hair, although they may be found anywhere within the cortex, and are typically larger than pigment granules. Note the size, shape, abundance, and distribution of cortical fusi.

Ovoid bodies

Ovoid bodies are large (much larger than pigment bodies and larger than most cortical fusi) round or oval bodies with sharp, regular edges. Ovoid bodies are more commonly found in cattle, dog, and other nonhuman hair, but are sometimes found in human hair. Note the size, abundance, and distribution of any ovoid bodies visible in the specimen.

Cuticle

Note the absence or presence of the cuticle. If the cuticle is present, note the appearance of the outer cuticle margin (exterior surface of the hair) as flat, smooth, cracked, or serrated. If scale detail is visible at high magnification, note the appearance of the scales. Note the appearance of the inner cuticle margin (where the cuticle touches the cortex) as distinct, diffuse, or otherwise.

Ends

Note the presence or absence of a root on the proximal end. If a root is present, describe its appearance. Note the appearance of the distal end (tip) as tapered (natural or razor-cut), scissors- or clipper-cut (square or angled), split, frayed, abraded, crushed, broken, or otherwise.

Complete your microscopic examination of the hair specimen as follows:

1. Place the wet-mounted specimen on the microscope stage and examine it at 100X, as shown in Figure II-2-3, working from one end to the opposite end. (It doesn't matter if you start from the proximal or distal end, but be consistent.) Record your observations in your lab notebook.

> Here's an example of how to record your observations, quoted directly from Robert's lab notebook:
>
> "round, smooth shaft w/ uniform diameter ~45 µm ffl ~2 µm; uniformly red by reflected light w/ evenly distributed small pigment bodies; amorphous medulla with frequent small breaks; MI ~0.2; fine homogeneous cortex; few, small, random cortical fusi near proximal end; no ovoid bodies visible; cuticle present, w/ smooth outer margin w/ no scale pattern visible and diffuse inner margin"

2. Increase magnification to 400X and examine the fine internal structure of the hair, noting any details that were not visible at 100X.

3. If you have the necessary equipment, shoot at least three images of the hair specimen: proximal tip, distal tip, and main body. Record the details, including the specimen number, by image filename for each image.

4. Repeat steps 1 through 3 for each hair specimen.

Figure II-2-3: *A human scalp hair at 100X*

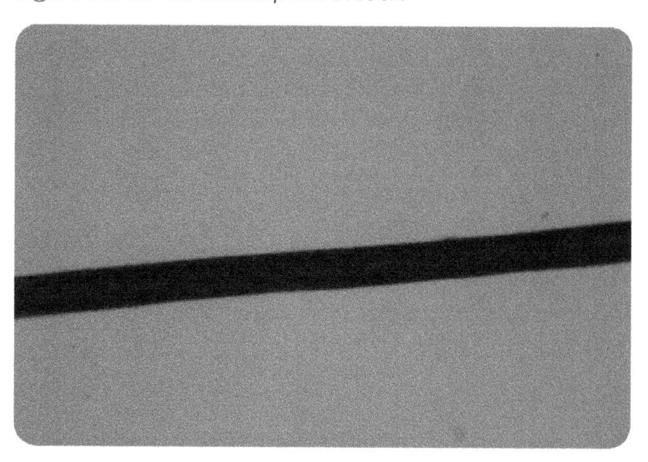

REVIEW QUESTIONS

Q1: **Which two somatic regions yield hairs that are most significant forensically? Why?**

Q2: **What are the advantages and limitations of nuclear DNA testing versus mitochondrial DNA testing?**

Q3: **What is the primary consideration in choosing a mounting fluid for a hair specimen, and why?**

Q4: **Why did we recommend preparing wet mounts of all of your specimens before observing any of them, rather than just preparing the wet mount for each specimen as you're ready to observe it? What, if any, disadvantage is there to preparing all the specimens before observing any of them, and how would you avoid that disadvantage?**

Q5: **What two characteristics suggest that a hair specimen has been dyed?**

Make Scale Casts of Hair Specimens

Lab II-3

EQUIPMENT AND MATERIALS

You'll need the following items to complete this lab session. (The standard kit for this book, available from *http://www.thehomescientist.com*, includes the items listed in the first group.)

MATERIALS FROM KIT

- Goggles
- Forceps
- Slides, flat

MATERIALS YOU PROVIDE

- Gloves
- Microscope (40X, 100X, and 400X)
- Nail polish (colorless)
- Nail polish remover (or acetone)
- Specimen(s): human hair

WARNING

Although none of the activities in this lab session present any significant risks, as a matter of good practice, you should always wear splash goggles, gloves, and protective clothing when working in the lab, if only to avoid contaminating specimens.

BACKGROUND

The *cuticle* is made up of overlapping plates or scales of keratin arrayed in characteristic patterns. Although these scale patterns may be visible on a wet-mounted specimen at high magnification, it is often difficult or impossible to discern the scale pattern if the refractive index of the scales is very close to that of the mounting fluid. One way around this problem is to make a cast of the exterior surface of the hair and examine that cast under high magnification.

Figure II-3-1 shows the three major types of scale patterns. The *imbricate scale pattern* is a flattened wavy pattern that is commonly found on human hair and many types of animal hair. The *coronal scale pattern* is a crown-like pattern that resembles a stack of paper cups, and is normally found only on very fine hair. Coronal scales are found on many types of animal hair and are very rarely present on human hair. The *spinous scale pattern* is a petal-like pattern made up of triangular scales that protrude from the cuticle. Spinous scales are found in the proximal (root) region of the fur hair of some animals, including bobcat, chinchilla, fox, lynx, mink, mouse, otter, raccoon, rat, sable, seal, and sea lion. Spinous scales are never found in human hair.

Figure II-3-1: *Imbricate, coronal, and spinous scale patterns*

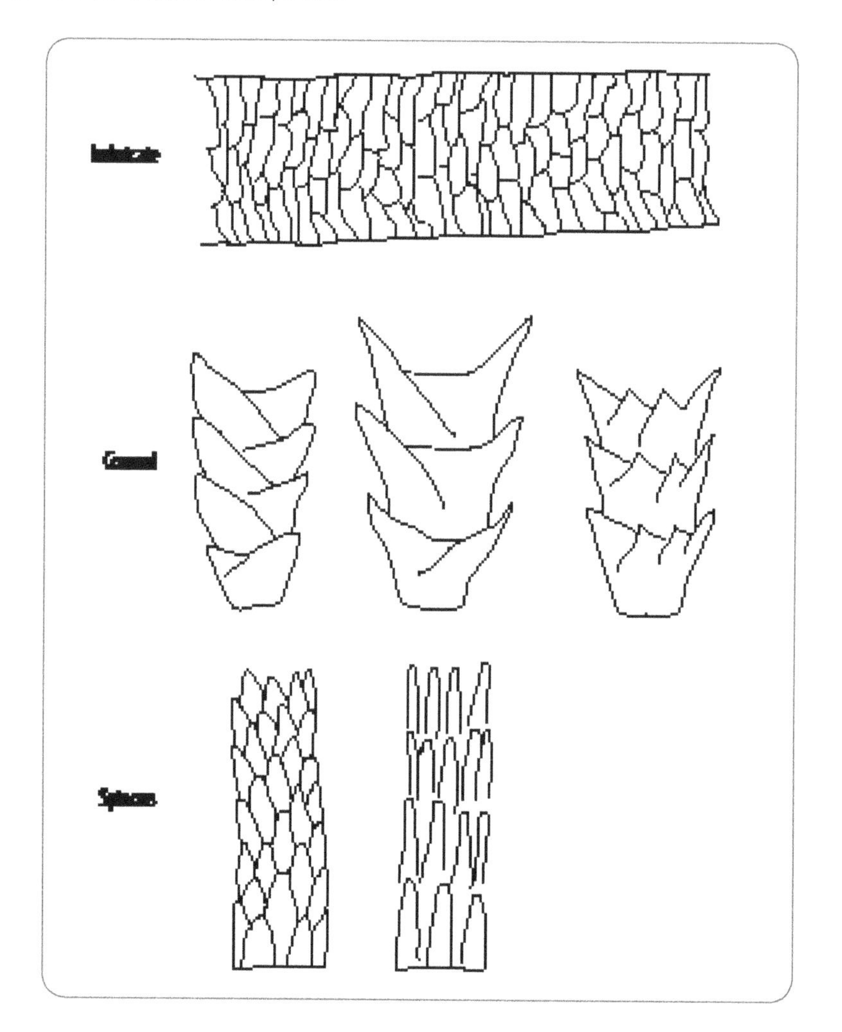

Although scale patterns are seldom useful for characterizing human hair specimens, they are important for discriminating human hair from animal hair and for determining the type of animal from which a specimen originated. For much more information about scale patterns, read *Microscopy of Hair Part II: A Practical Guide and Manual for Animal Hairs*, Forensic Science Communications, July 2004, Volume 6, Number 3 (*http://www.fbi.gov/hq/lab/fsc/backissu/july2004/research/2004_03_research02.htm*).

In this lab session, we'll make scale casts using ordinary colorless nail polish as a casting medium, the same method used by professional forensic labs. (Yes, they actually use the same colorless nail polish you can buy at the drugstore.)

PROCEDURE II-3-1: MAKE AND OBSERVE SCALE CASTS OF HUMAN HAIR

1. Brush a thin layer of clear nail polish onto the middle third of a microscope slide.
2. Carefully press the hair specimen into the tacky nail polish until it adheres.
3. Allow the nail polish to dry.
4. Using the forceps, carefully pull the hair specimen away from the slide in one smooth motion, as shown in Figure II-3-2.

Figure II-3-2: *Using forceps to remove the hair specimen from the dried nail polish*

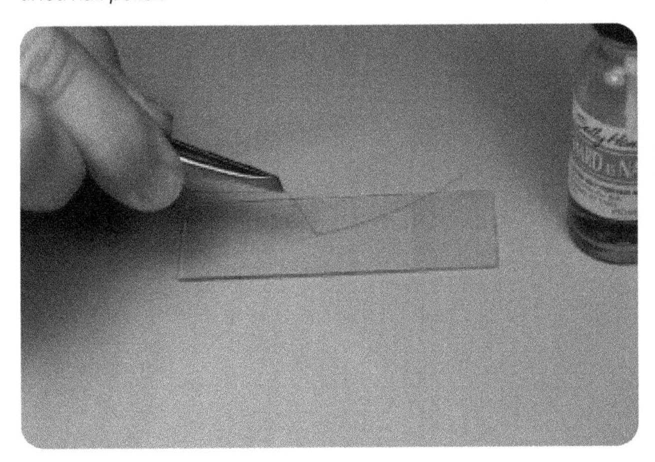

5. Examine the scale cast under high magnification by both transmitted and incident light, adjusting the brightness, angle, and contrast of the lighting to optimize visibility of the scale pattern. Use your ocular micrometer to measure the size of the scales, and note any variations in the pattern or size of the scales over the length of the shaft. Record your observations in your lab notebook.
6. Repeat steps 1 through 5 for each specimen.
7. If you have the necessary equipment, shoot an image of a representative scale pattern, as shown in Figure II-3-3, for your lab notebook. Otherwise, make a sketch and paste it into your notebook.

Figure II-3-3: *A scale cast of a canine hair at 400X*

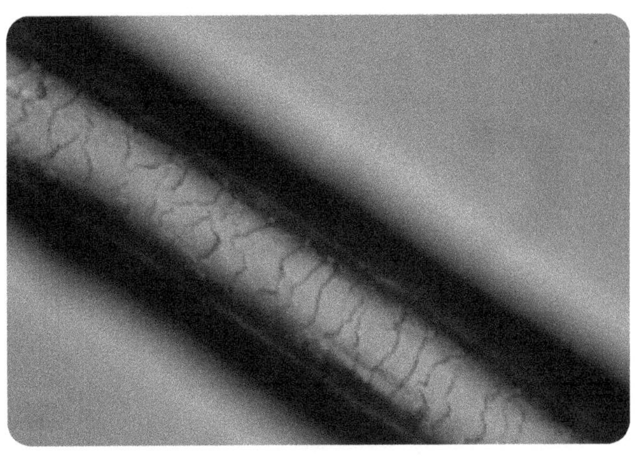

Scale casts may last for several days to several months, depending on how they are stored. High temperature and high humidity reduce the useful lifetime of such casts. When you are finished using a cast, you can use nail polish remover or acetone to clean the nail polish from the slide.

REVIEW QUESTIONS

Q1: **The presence of which of the three scale patterns rules out a hair specimen as human?**

Q2: **What is the primary forensic value of determining the scale pattern on a hair specimen?**

Study the Morphology of Animal Hair

EQUIPMENT AND MATERIALS

You'll need the following items to complete this lab session. (The standard kit for this book, available from *http://www.thehomescientist.com*, includes the items listed in the first group.)

MATERIALS FROM KIT

- Goggles
- Coverslips
- Forceps
- Glycerol
- Magnifier
- Pipettes
- Ruler
- Slides, flat

MATERIALS YOU PROVIDE

- Gloves
- Clove oil (optional)
- Microscope (40X, 100X, and 400X)
- Nail polish (colorless)
- Nail polish remover (or acetone)
- Specimens: animal hair

WARNING

Although none of the activities in this lab session present any significant risks, as a matter of good practice, you should always wear splash goggles, gloves, and protective clothing when working in the lab, if only to avoid contaminating specimens. Obviously, you may need to work without goggles when using a microscope or magnifier to examine specimens.

BACKGROUND

As you might expect, hair from non-human mammals can be difficult or impossible to discriminate from human hair under macroscopic examination. Fortunately, under microscopic examination, it's relatively easy to discriminate most non-human animal hair from human hair. (Hair from some non-human primates is the major exception.)

Animal hair, particularly pet hair, is frequently examined in forensics labs. Although in the past animal hair was of limited evidentiary value, the advent of DNA testing—which can as easily be applied to animal hair as to human hair—has made animal hair testing an important part of the work of modern forensics labs. It is, for example, now possible to convict criminals based on DNA matches between animal hairs from the criminal's pet that were found on the victim, or vice versa.

As is true of human hair, DNA testing of animal hair is relatively expensive and time-consuming, so macroscopic and microscopic examination is used for preliminary screening. In fact, DNA testing of animal hairs is still relatively rare. As late as 2004, no state, federal, or private forensic labs were performing DNA testing of animal hair, and as of early 2012, only a few private forensics labs routinely do DNA testing of animal hair.

Animal hairs are more differentiated by somatic region and purpose than human hair. Animal hairs are classified as members of four broad types:

- *Guard hairs* form the outer coat of the animal, shed water, and protect the inner hair and skin.
- *Fur* or *wool hairs* form the inner coat and provide insulation.
- *Tactile hairs*, also called *whiskers*, are found on the head (the snout or ears), where they provide sensory functions.
- *Special-purpose hairs*, such as tail hairs and mane hairs, have a morphology that may differ substantially from the main body hairs of the animal.

Human hair differs noticeably from animal hair in the following respects:

- Most human hairs are consistent in color for the entire length of the hair shaft, while many animal hairs vary significantly in color over relatively short sections of the shaft, a phenomenon called *banding*.
- The pigment granules in human hair are generally relatively evenly distributed throughout the cortex, with perhaps some concentration toward the cuticle, while the pigment granules in most animal hair are generally strongly concentrated centrally in close proximity to the medulla.
- The medulla in human hair, if present, is generally amorphous, often broken or fragmented, and occupies a third or less the width of the hair shaft, while the medulla in animal hair is generally present, continuous, sharply defined and structured, and occupies one third or more (sometimes nearly the entire) width of the shaft.
- Cuticle scaling in human hair, if present, is often subtle and usually of the imbricate pattern (see the preceding lab session) or, much more rarely, the coronal pattern, while cuticle scaling in animal hair is often more readily visible and may be imbricate, coronal, or spinous pattern, which is never found in human hair.

In this lab session, we'll examine various animal hair specimens macroscopically and microscopically and produce scale casts to learn how to differentiate animal hair from human hair.

For temporary wet mounts, the glycerol (RI ~1.47) supplied with the kit gives good differentiation of the internal structures of hair, although clove oil (RI ~1.54) is a closer match for the 1.56 refractive index of the keratin that makes up hair. Colorless nail polish or a similar mounting fluid can be used to make permanent mounts. Unless you are short of slides and coverslips, you may want to label and mount your animal hair specimens permanently for later use as reference standards.

Obtain at least one shed or (ideally) plucked hair specimen from as many animals as possible. Dog and cat hair are readily available, from your own pets or from those of family members, friends, and neighbors. You may also be able to obtain hair specimens from someone who keeps a rabbit, rat, ferret, gerbil, hamster, or other small mammal as a pet. Specimens from wild animals such as squirrels, raccoons, opossums, and others can sometimes be obtained from road kill. Store each specimen you obtain in a labeled container.

PROCEDURE II-4-1: OBSERVE ANIMAL HAIR

The procedures for this lab session are identical those of the preceding two lab sessions.

1. Examine an animal hair specimen under low magnification and record your observations in your lab notebook.
2. Wet-mount the animal hair specimen using glycerol (or clove oil, if you have it).
3. Examine the animal hair specimen under medium and high magnification (100X and 400X), as shown in Figure II-4-1, and record your observations in your lab notebook. Compare and contrast these microscopic observations with those you made of your scalp hair.

> Professional forensics labs use comparison microscopes to do side-by-side comparisons of hair specimens. We can't afford the $5,000 or more cost of a decent comparison microscope, but there is a workaround that we've found to be just as useful for learning purposes. Simply wet-mount two or more hair specimens side-by-side and as close together as possible on one slide. That way, you can do a comparison of both or all specimens in one field of view. You can use this method to compare human hairs against each other, human hairs against animal hairs, or different types of animal hairs.

4. Make a scale cast of the animal hair specimen, and examine the scale cast under high magnification, recording your observations in your lab notebook. If you have the necessary equipment, shoot an image of the scale cast for your lab notebook
5. Repeat steps 1 through 4 for each of your animal hair specimens, comparing and contrasting your various animal hair specimens with each other.

Figure II-4-1: *A canine hair at 100X*

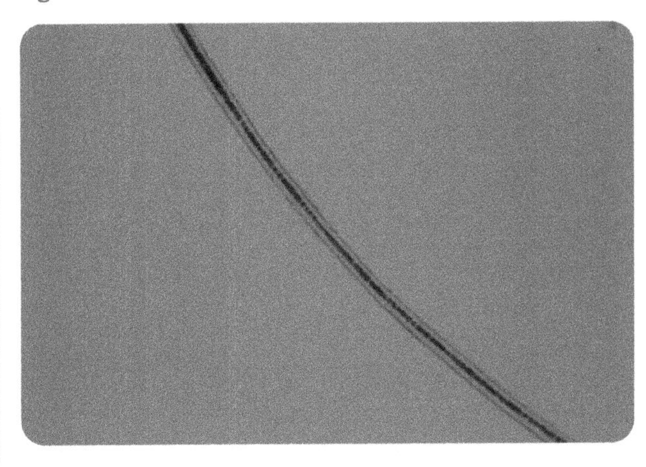

If you want to retain your animal hair specimens as comparison standards for later use, mount a sample of each specimen using colorless nail polish, Permount, or a similar mounting fluid.

REVIEW QUESTION

Q1: **What microscopic characteristics can be used to readily differentiate an animal hair specimen from a human hair specimen?**

Individualize Human Hair Specimens

Lab II-5

EQUIPMENT AND MATERIALS

You'll need the following items to complete this lab session. (The standard kit for this book, available from *http://www.thehomescientist.com*, includes the items listed in the first group.)

MATERIALS FROM KIT

- Goggles
- Coverslips
- Forceps
- Glycerol
- Magnifier
- Pipettes
- Ruler
- Slides, flat

MATERIALS YOU PROVIDE

- Gloves
- Clove oil (optional)
- Microscope (40X, 100X, and 400X)
- Nail polish (colorless)
- Nail polish remover (or acetone)
- Specimens: human hair (see text)

WARNING

Although none of the activities in this lab session present any significant risks, as a matter of good practice, you should always wear splash goggles, gloves, and protective clothing when working in the lab, if only to avoid contaminating specimens. Obviously, you may need to work without goggles when using a microscope or magnifier to examine specimens.

BACKGROUND

The title of this laboratory session is actually a misnomer. In the absence of DNA testing, hair can never be more than class evidence. Based purely on visual examination, the most that a forensic scientist can state definitively is that two hair specimens are consistent with each other in every respect, but not that those specimens are certain to have originated from the same individual. Individualizing hair specimens requires not just a DNA match, but a nuclear DNA match.

Despite the fact that visual examination of hair technically provides only class evidence rather than individual evidence, hair evidence testimony has been widely accepted by courts since the early 20th century and has been used to convict thousands of criminals of serious crimes, including murder and kidnapping. Many of those criminals were subsequently executed or sentenced to long prison terms.

Was accepting this evidence a miscarriage of justice? Not necessarily, because although hair cannot be individualized without DNA testing, it can be closely classified by comparing the specific characteristics of known and questioned specimens. If the questioned and known specimens are entirely consistent, a forensic scientist may testify to that fact, which becomes just one more supporting piece of evidence in the prosecution's case. (Although no one should ever be convicted based solely or primarily on such class evidence as hair, there are unfortunately many such convictions in the literature; these were truly miscarriages of justice.)

Because non-DNA hair evidence testimony inevitably depends strongly on the subjective opinion of the expert witness, it's important that the witness have extensive experience in visual comparison of hair specimens—not just hundreds of comparisons but thousands or tens of thousands. Only with such experience is it possible to make valid judgments, because the morphology of hair varies, not just between individuals or somatic regions, but between specimens from the same somatic region of the same individual. Understanding these variations is key to learning to interpret visual hair evidence properly.

Individual characteristics are most pronounced in hair from the scalp. As we learned in Laboratory II-2, the macroscopic and microscopic characteristics of scalp hair can differ noticeably from region to region. A hair from your crown, for example, is likely to differ visually from a hair from your forehead, and even two closely adjacent hairs may exhibit noticeable variations.

Despite this variation, scalp hair is prized by forensic scientists because it often has sufficient individual characteristics to make it easily classifiable. For example, by examining scalp hair, forensic scientists can often make a strong inference about the race of the individual from whom the specimen was obtained. Such characterizations are often more difficult or impossible for pubic hairs, and in particular for hairs from other somatic regions, such as axillary hairs. Figure II-5-1 shows a typical human scalp hair at 400X magnification.

Figure II-5-1: *A human scalp hair at 400X*

In this lab session, we'll examine the characteristics of hair specimens from different individuals and different somatic regions. This lab session is open-ended. You can learn a great deal in only one or two lab periods, but you can easily spend literally years without exhausting the learning opportunities. Even if you devote only one lab period to this work, you'll gain an appreciation for the expertise that is developed over the course of years by professional forensic hair examiners.

You'll need a lot of microscope slides and coverslips for this lab session, whether or not you decide to mount some or all of your specimens permanently for later use as reference standards. Even if you create temporary wet mounts, you'll want to retain them until you've finished the lab session so that you can compare all specimens, so plan on having 25 to 50 or more microscope slides and coverslips available.

For temporary wet mounts, use the glycerol supplied with the kit. If you have clove oil available, you can use it to reveal even more of the internal structural detail of your specimens. You can make permanent mounts with colorless nail polish, Permount, or a similar mounting fluid.

PROCEDURE II-5-1: OBTAIN HAIR SPECIMENS

Request hair specimens from several people. Use the following guidelines in deciding which specimens to request:

- Maximize the range of specimens by requesting them from persons of different ages, sexes, and races. (If possible, obtain infant hair specimens for comparison with adult specimens.)
- Try to obtain specimens from at least one person with brown or black hair, one with red or blond hair, one with gray hair, and one with dyed hair. Make sure to ask each person who provides scalp hair specimens if his or her hair color is natural, and whether or not any treatment such as a permanent has been applied to the hair specimen.
- Clipped specimens are fine for most of your specimens. Ask the person providing the specimen to trim the hair as closely as possible to the skin.
- From each person, request specimens from as many somatic regions as the person is willing to provide—scalp, mustache, beard, axillary, arm, chest, abdomen, pubic, and leg, as applicable.
- Some people who willingly provide a specimen of scalp hair may balk at providing specimens from other areas of their bodies; use your best judgment in deciding who and how to ask, emphasize that you are asking in the spirit of scientific inquiry, and accept gracefully what is offered.
- Although one hair is acceptable as an exemplar, ideally you want at least two or three hairs of each type, and more is better.

Provide each volunteer with as many specimen containers as necessary. Coin envelopes, small plastic resealable bags, or similar containers work well. Pre-label each group of containers with a number to identify the person who is providing the specimens. (You can also use the person's initials, but many people will feel more comfortable with a nice anonymous number.) Within the numbered group, pre-label each container with the somatic region of the hair it will contain. Record the details of each specimen donor, including sex, age, race, national origin, and any other factors that may be pertinent.

PROCEDURE II-5-2: OBSERVE AND CHARACTERIZE HAIR SPECIMENS

1. Begin by performing full macroscopic comparisons of specimens from each of your own somatic regions, followed by full wet-mount microscopic comparisons and, if you have time, scale-cast comparisons. Log detailed descriptions of each specimen in your lab notebook and, if possible, shoot images for subsequent comparison.
2. Repeat step 1 for each other person from whom you were able to obtain specimens from two or more somatic regions.
3. Compare and contrast the differences and similarities of hair specimens across somatic regions from the same individual.

4. Compare scalp hair specimens from the different individuals, noting any characteristics shared by the scalp hair specimens.

5. Repeat step 4 for specimens from at least one other somatic region of different individuals, ideally pubic hair, which is often forensically significant.

6. Compare and contrast the differences and similarities of hair specimens across individuals from the same somatic region.

REVIEW QUESTIONS

Q1: **What macroscopic characteristics did you find most useful for differentiating human hair specimens from different subjects?**

Q2: **What microscopic characteristics did you find most useful for differentiating human hair specimens from different subjects?**

Q3: **Based on examining specimens from various subjects, do you think it is possible to estimate the approximate age of the person from whom the specimen was obtained? If so, how?**

Physical and Chemical Tests of Fibers

Lab II-6

EQUIPMENT AND MATERIALS

You'll need the following items to complete this lab session. (The standard kit for this book, available from *http://www.thehomescientist.com*, includes the items listed in the first group.)

MATERIALS FROM KIT

- Goggles
- Beaker, 250 mL
- Cylinder, graduated, 100 mL
- Coverslips
- Forceps
- Hydrochloric acid, 6 M
- Magnifier
- Pipettes
- Slides, deep-cavity
- Slides, flat
- Sodium dithionite
- Spatula
- Stirring rod
- Test tubes
- Test tube clamp
- TIS #1 dye
- TIS #3A dye
- Specimens: Fabrics A, B, C, D, and E
- Specimen: TFI Multifiber Fabric #43

MATERIALS YOU PROVIDE

- Gloves
- Ammonia, household (clear, non-sudsy)
- Bleach, chlorine laundry
- Burner, butane lighter, or other flame source
- Dishwashing detergent (Dawn or similar)
- Microscope
- Microwave oven or hotplate
- Nail polish remover (or acetone)
- Paper towels
- Scissors
- Soft drink bottles, small (temporary storage)
- Vinegar, distilled white
- Water, distilled or deionized
- Specimens: additional fibers (optional)

BACKGROUND

Although instrumental analysis is an important part of forensic fiber analysis, wet-chemistry techniques are still widely used by forensic labs to identify questioned fibers. The most commonly used physical and chemical fiber tests include the following:

Burning tests

Burning tests are just what they sound like: a small specimen of the questioned fiber is ignited, and the forensic scientist observes the results. How does the fiber burn, if at all? What odor is emitted by the burning fiber? Does the fiber continue burning when removed from the flame source? What does the residue look like? And so on. Results from burning tests can identify the specific fiber, or at least narrow down the possibilities.

Solubility tests

Solubility tests determine the effect, if any, of various solvents on a questioned fiber. A questioned fiber may be unaffected by one solvent, but readily soluble in a second solvent. By determining the effect of several solvents on known and questioned specimens, the forensic scientist may at best be able to determine the specific fiber type, and at worst to eliminate many fiber types from consideration.

Dye-stripping tests

Many questioned fabric specimens are dyed. By dye-stripping (bleaching) known and questioned specimens with different stripping solutions that have different effects on different dyes, a forensic scientist may be able to identify the specific fiber type and dye type of the questioned specimen. Even if this is not possible, it is always possible to determine whether a questioned specimen behaves consistently with a known specimen, which by itself may be valuable information.

Differential dyeing tests

A questioned fabric specimen that is white (or has been dye-stripped) can be treated with textile identification stains (TIS) that provide differential results with different fiber types. By using multiple stains on a questioned specimen and comparing the results with the different stains, it's often possible to identify the fiber type unambiguously.

In this lab session, we'll use all four of these methods to attempt to characterize five questioned fabric specimens with as much specificity as possible.

PROCEDURE II-6-1: TEST FIBER SPECIMENS BY BURNING

WARNING

Although none of the activities in this procedure present any significant risks, as a matter of good practice, you should always wear splash goggles, gloves, and protective clothing when working in the lab, if only to avoid contaminating specimens.

In the preceding lab sessions, we've been examining hair, which of course is a type of fiber. But hair is by no means the only type of fiber that may have forensic significance. Many natural and artificial fibers are used in clothing, rugs, carpets, draperies, and other home furnishings, packing materials, building materials, rope and cord, and so on. Such fibers may be transferred between the victim and the criminal, or between a person and the environment at a crime scene or other significant location.

There are five main classes of fibers that are of interest to forensic scientists:

Animal fibers
: *Animal fibers*, such as wool or mohair, are processed versions of raw animal hair and retain some or all of the characteristics of the raw animal hair. Silk, which is the dried exudate produced by silkworms, is not technically a hair but is classed as an animal fiber and shares many of the characteristics of fibers produced from animal hair. (Although spider silk is very similar to the silk produced by silkworms, we were unable to find any cases in the literature in which the forensic examination of spider silk played a prominent role.)

Plant fibers
: *Plant fibers*, such as cotton, linen, sisal, and hemp are very widely used, and the forensic examination of these fibers frequently yields important evidence in criminal cases. Plant fibers may undergo little processing, such as the hemp used in rope and the sisal used for baggage, or they may be highly processed, such as the mercerized cotton used in clothing. Regardless of the production method used, plant fibers retain characteristic chemical and physical properties that can be used by forensic examiners to identify and match specimens.

Artificial fibers
: *Artificial fibers*, also called *synthetic fibers*, include such familiar fibers as nylon and polyester. These fibers are essentially plastics in fiber form. They are widely used in clothing and other fabrics as substitutes or supplements for natural fibers, either because they are less expensive than the natural fiber they replace or because their physical characteristics are superior to those of natural fibers. Artificial fibers may mimic the properties of natural fibers quite closely, and may be difficult to distinguish from similar natural fibers without close examination. Under microscopic examination, artificial fibers are easily discriminated from natural fibers, because the artificial fiber, being machine-made, is absolutely consistent and does not show the variations present in all natural fibers.

Reconstituted fibers
: *Reconstituted fibers*, also called *semi-synthetic fibers*, such as rayon, are manufactured from cellulose and other natural raw materials that have been processed into a raw liquid that is then reconstituted as fibers using the same types of equipment used to produce artificial fibers. Accordingly, reconstituted fibers resemble natural fibers chemically, but resemble artificial fibers microscopically.

Mineral fibers
: *Mineral fibers*, notably asbestos, are sometimes important forensically. These fibers are easily discriminated from all other types of fibers based on their physical properties alone. In essence, they are rock in fiber form. Although fiberglass is a manufactured product, it is often considered a mineral fiber.

The first task a forensic scientist faces when presented with a questioned fiber is to identify the class to which that fiber belongs. The oldest method for determining fiber class, and one that is still sometimes used today, is to burn a small specimen of that questioned fiber to determine what the smoke, if any, smells like and to observe the charred end of the fiber. Using this test (as shown in Figure II-6-1), you can readily discriminate among most of these classes of fiber, as follows:

- Animal fibers burn readily and produce the characteristic ammonia-like odor of burning hair
- Plant fibers burn readily and produce an odor similar to that of burning paper
- Artificial fibers melt as they burn and produce an acrid odor of burning plastic
- Reconstituted fibers burn with the same odor as their natural precursors (typically cellulose) and so cannot be discriminated from natural plant fibers
- Mineral fibers do not burn or produce any odor.

Using the burning test eliminates all but one or two fiber classes from further consideration. If your nose is sensitive, you may even be able to discriminate different fibers within the same class. For example, after some practice, most people can discriminate the odor of burning wool from that of burning silk or that of burning nylon from that of burning polyester. If your nose is very sensitive, you may even be able to identify blends, such as a 60/40 cotton/polyester shirt fabric.

Figure II-6-1: *Barbara test burning a fiber specimen*

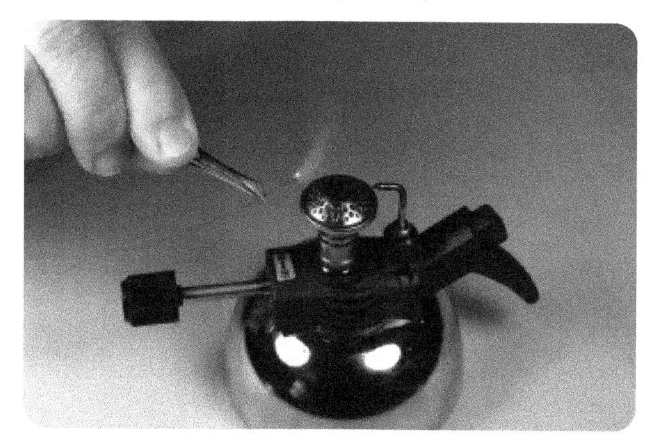

> In case you were wondering, the color of a fiber has little or no discernible effect on the burning test. Even heavily dyed fabrics contain relatively little dye by mass percentage, so the dye seldom contributes noticeably to the odor when the specimen is burned.

In this lab session, we'll burn fiber specimens and record our observations. To do that, we'll need exemplars of various known fibers. The FK01 kit includes a swatch of Multifiber Fabric #43, the fiber standard used by real forensic laboratories. It's manufactured by Testfibers, Inc. and includes 8 mm stripes of the following 13 fibers:

- Spun diacetate
- SEF (modacrylic)
- Filament triacetate
- Bleached cotton
- Creslan 61 (acrylic)
- Dacron 54 (polyester)
- Dacron 64 (polyester)
- Nylon 66 (polyamide)
- Orlon 75 (acrylic)
- Spun silk
- Polypropylene (polyolefin)
- Viscose (rayon)
- Worsted wool

If you don't have the kit, obtain as many as possible of the following known fiber specimens:

- Acetate (Acele, Aviscon, Celanese, Chromspun, Estron)
- Acrylics (Acrilan, Courtelle, Creslan, Dralon, Orlon, Zefran)

- Cotton
- Modacrylic (Dynel, Kanecaron, Monsanto SEF, Verel)
- Nylon
- Nytrils (Darvan)
- Polyesters (Avlin, Beaunit, Blue C, Dacron, Diolen, Encron, Fortrel, Kodel, Quintess, Spectran, Trevira, Vectran, Vyoron)
- Rayons (cellulose reconstituted by the cuprammonium or viscose process, including Avril, Avron, Cordenka, Dynacor, Enka, Fiber 700, Fibro, Nupron, Rayflex, Suprenka, Tyrex, Tyron, Zantrel)
- Silk
- Triacetate (Arnel)
- Vinyons (polyvinyl chlorides: Avisco, Clevyl, Rhovyl, Thermovyl, Volpex)
- Wool

Ideally, each specimen should be of a single type of fiber, although you can also test blends. Obtain specimens large enough for the other tests in the following lab sessions—at least a few square centimeters—and label each specimen or store it in a labeled container. One good source of known fiber specimens is discarded clothing, which is nearly always labeled with the fiber content. You can also obtain specimens from hems and other hidden areas of your wardrobe. Craft and fabric stores are another good source of known fiber specimens.

Fibers can usually be burn-tested successfully without any preliminary treatment. In some cases, however, the presence of oil, starch, wax, or some other surface coating may interfere with the test. Preliminary cleaning of the fibers, called *boiling off*, is usually a simple matter of boiling the fibers for a few minutes in distilled or deionized water. If that treatment fails, try using a warm dilute (~0.1 M) solution of hydrochloric acid or sodium hydroxide, followed by a thorough rinse.

1. If you have not done so already, put on your goggles, gloves, and protective clothing.
2. Grasp the first Multifiber Fabric #43 specimen with your forceps. Burning even one fiber may provide a strong enough odor to be characterized by an experienced examiner, but until you get some experience we recommend using at least several fibers.
3. Ignite the burner or lighter, and bring the fiber close to, but not into contact with, the flame. Does the fiber ignite, curl, or melt? Note your observations in your lab notebook in a table that resembles Table II-6-1.
4. Touch the end of the fiber to the flame. Does the fiber ignite immediately, slowly, or not at all? Does it simply melt, or is there no apparent change? If it burns, does it burn quickly or slowly, smoothly or with a sputtering flame? Note your observations in your lab notebook in a table that resembles Table II-6-1.
5. Remove the fiber from the flame. Does it continue burning or extinguish? If it continues burning, is there an open flame or does it smolder with a glowing tip? Note your observations in your lab notebook in a table that resembles Table II-6-1.
6. Repeat steps 2 through 5 for each of your Multifiber Fabric #43 known fiber specimens.
7. Remove a few threads from questioned fiber specimen A and repeat the burning test. Record your observations in your lab notebook. By comparing your results to those you obtained from the known specimens, attempt to identify the fabric type or types present in specimen A.
8. Repeat step 7 for questioned fiber specimens B, C, D, and E.

Olfactory memory is notoriously unreliable, so don't hesitate to repeat burning tests on your known specimens as necessary when you are attempting to identify the questioned specimens.

Table II-6-1: *Burning characteristics of fibers*

Fiber	Melts?	Curls?	Burns?	Extinguishes?	Residuum
Acetate					
Acrylic					
Aramid					
Azlon					
Cotton					
Glass/mineral					
Modacrylic					
Nylon					
Nytril					
Olefin					
Polyester					
Rayon					
Saran					
Silk					
Triacetate					
Vinal					
Vinyon					
Wool					
Specimen A					
Specimen B					
Specimen C					
Specimen D					
Specimen E					

PROCEDURE II-6-2: TEST FIBER SPECIMENS BY SOLUBILITY

WARNING

Read the MSDS for each of the chemicals you use and follow the handling precautions noted. Acetone is extremely flammable. Avoid open flames and other ignition sources. Hydrochloric acid is corrosive and produces strong fumes. Use adequate ventilation or work outdoors. Sodium hypochlorite is corrosive, and reacts with acids to form toxic chlorine gas. Wear splash goggles, gloves, and protective clothing at all times.

Although it sounds odd, one of the best ways to discriminate fiber specimens is by attempting to dissolve them in various solvents. For example, you may have identified a questioned fiber as artificial by the burning test, but you do not yet know what type of artificial fiber it is. By attempting to dissolve a small specimen of that fiber in each of several solvents, you can determine if it is, for example, nylon, acetate, or polyester.

Solubility tests are particularly important for discriminating artificial and reconstituted fibers because they are produced by machinery and so lack the microscopic variations characteristic of natural fibers. For example, although it is easily possible to discriminate wool or cotton from an artificial fiber microscopically, it may be impossible to determine whether an artificial fiber is acetate or polyester without performing solubility or other chemical tests.

One major advantage of solubility testing is that it can be performed on tiny specimens. For example, the question may arise if a single short thread found snagged at a crime scene is consistent with clothing worn by the suspect. After microscopic examination, that thread may be subjected to solubility tests to determine if it is consistent with one of the fiber types found in the suspect's clothing. With only such a tiny specimen available for testing, the technicians would first identify the fiber types found in the clothing to narrow the range of possibilities, and then might use solubility tests to determine if the thread is consistent with any of those fiber types. (If the thread is not consistent, that provides exculpatory evidence favoring the suspect; if the thread is consistent, the suspect *may* have been present at the crime scene.)

Incidentally, forensic scientists don't accept evidence that they have not confirmed themselves. For example, although a shirt submitted for examination may be labeled as "60% cotton, 40% polyester," labels do not always reflect reality. The lab would test that shirt to confirm that it is in fact a cotton/polyester blend.

Many of the solvents used for fiber solubility testing are extremely hazardous to handle, even at room temperature, and more so at high temperatures. Here are some of the solvents routinely used by forensic technicians for fiber solubility testing:

Acetic acid
: At 20°C, glacial (100%) acetic acid dissolves acetates. Boiling glacial acetic acid dissolves acetates and triacetates. Nylon dissolves, but very slowly.

Acetone
: At 20°C, acetone dissolves acetates, triacetates, modacrylics, and vinyons.

Chlorobenzene
: Boiling chlorobenzene dissolves Saran and vinyon.

Chloroform
: At 20°C, chloroform dissolves acetate, Arnel, and vinyons.

Cresol
: At 80 to 100°C, cresols (methylphenols) dissolve acetates, Arnel, Fortisan, nylon, Orlon, silk, vinyon, and viscose. Dynel forms characteristic clumps, and Saran decomposes (slowly at 80°C and more rapidly at 100°C).

Dimethylformamide
: At 20°C, 60% dimethylformamide dissolves acetate, Acrylan, Arnel, Dacron, Darvan, Dynel, nylon, Orlon, Saran, Verel, and vinyon.

Formic acid
: At 20°C, concentrated (88%) formic acid dissolves acetate, Arnel, and Fortisan, and softens nylon. At 40°C, this acid dissolves nylon.

Hydrochloric acid

At 20°C, concentrated (37%) hydrochloric acid dissolves acetates, triacetates, saponified cellulose acetate (Fortisan), nylon, and silk. Rayons dissolve slowly. Less concentrated (6 M) hydrochloric acid dissolves nylon, but not the other fibers listed.

Nitric acid

At 20°C, concentrated (68%) nitric acid dissolves acetate, Arcylan, Arnel, Creslan, Darvan, nylon, Orlon, vinyon, and Zefran.

Phenol

At 20°C, a 90% aqueous solution of phenol dissolves acetate, Arnel, Dynel, and nylon. Dacron softens, and Fortisan softens and dissolves slowly. Dacron dissolves at 40 °C. At boiling, this solvent also dissolves vinyon and viscose.

Sodium hydroxide

At 100°C, a 5% solution of sodium hydroxide dissolves nytrils, wool and other animal hair, and cultivated silk. Acetates, triacetates, wild (Tussah) silk, and reconstituted protein fibers dissolve only partially and very slowly. At 100°C, a 50% solution of sodium hydroxide dissolves all of the above fibers readily, and slowly (30 to 60 minutes) dissolves polyesters. Acrylics dissolve partially and very slowly. Modacrylics and sarans (Enjay, Saran) melt but do not dissolve.

Sodium hypochlorite

At 20°C, 5.25% sodium hypochlorite (chlorine laundry bleach) dissolves only animal fibers, including hair, silk, wool, and other protein fibers.

Sulfuric acid

At 20°C, 60% sulfuric acid dissolves acetates, triacetates, rayons (cuprammonium and viscose), silk, and nylon, but not cotton, linen, or other cellulosic plant fibers. At 20°C, 70% sulfuric acid dissolves all of the above fibers. The 70% acid can be used to discriminate reconstituted cellulose fibers from cotton and other natural cellulose fibers.

Zinc chloride

At 45°C, 75% zinc chloride solution dissolves acetates, triacetates, acrylics, silk, cotton that has not been mercerized, and rayons (cuprammonium and viscose).

In the normal course of a solubility test, particularly if only a small specimen is available for testing, the forensic scientist might first use microscopic examination and other means to narrow the range of possibilities as far as possible and then follow a flowchart, using various solvents in a particular order to eliminate possibilities. For example, if a particular solvent dissolves only artificial fiber type 1, while a second solvent dissolves fiber types 1, 2, 3, and 4, the first solvent would be used first. If the specimen dissolves in that solvent, the forensic scientist can state that the fiber was probably of type 1. If the scientist had used the second solvent first, she could state only that the specimen was probably of type 1, 2, 3, or 4. Because the test consumed the specimen, no further tests could be done on it.

Even if a larger specimen is available, solubility tests are ordinarily done on the smallest possible specimen while observing it with a microscope. Fibers are tiny things, so working microscopically allows more information to be obtained more easily than working macroscopically. Accordingly, in this lab session we'll perform solubility testing under the microscope. We'll use deep-cavity slides as our reaction vessels, with coverslips in place to protect us and our microscope from the corrosive solvents we'll be using.

For our purposes, it's not necessary to do exhaustive solubility testing using a full range of solvents. Instead, we'll learn the essentials of solubility testing using only three representative solvents: acetone, chlorine laundry bleach (5.25% sodium hypochlorite), and 6 M hydrochloric acid. Despite the tiny amounts of solvent we'll use, you should follow full safety precautions, wearing gloves and goggles. Acetone is extremely flammable, but the few drops we'll use present little real hazard. Conversely, the 6 M hydrochloric acid solution is extremely corrosive, and the bleach solution can blind you in literally seconds if it gets in your eyes

1. If you have not done so already, put on your goggles, gloves, and protective clothing.

2. Transfer a few drops of 5.25% chlorine bleach to a deep-cavity slide and place the slide on the stage of your microscope. (Be careful. Chlorine bleach is corrosive. Do not allow it to contact your eyes or skin. It may damage your microscope if it comes into contact with it.)

3. Use the forceps and scissors to remove a short (~5 mm) single thread of the spun diacetate from the Multifiber Fabric #43 specimen. Transfer that thread to the well of the cavity slide, using the forceps to make sure the thread is immersed in the bleach solution.

> The Multifiber Fabric #43 contains the following fabrics in the following order, starting on the side with the black loop-stitch thread: spun diacetate, SEF (modacrylic), filament triacetate, bleached cotton, Creslan 61 (acrylic), Dacron 54 (polyester), Dacron 64 (polyester), Nylon 66 (polyamide), Orlon 75 (acrylic), spun silk, polypropylene (polyolefin), viscose (rayon), and worsted wool.

4. With the low-magnification objective in position, adjust the focus until at least a part of the thread is sharply focused. Adjust the brightness and diaphragm for optimum contrast.

5. Observe the thread and record any visible changes in your lab notebook. If no visible change has occurred after two minutes, record that fact and use the forceps to remove that thread from the well. (Because no reaction has occurred, you can use the same solution for the next fabric test.) If the thread dissolves, swells, clumps, or undergoes any other visible change, record the details of that change and continue observing the slide for several minutes to see if additional changes occur. For example, after two minutes the fiber may have swollen, but if you allow the test to continue for several minutes the fiber may dissolve completely. When no more visible changes are occurring, empty the contents of the well, rinse the slide, and transfer fresh solution to the well.

> If you have multiple deep-cavity slides available, you can save a lot of time by running multiple solubility tests simultaneously. For example, if you have a dozen deep-cavity slides, you might start by testing one fabric type against all three solvents. After you rinse that slide, place threads from each of the other 12 fabrics in your 12 deep-cavity slides, one fabric type per slide, and transfer several drops of the first solvent to each well. Observe those 12 slides, noting the effects of that solvent on each of the fabric types. Then rinse all 12 slides, place threads from each of the 12 fabrics in the wells, and repeat the test with the second solvent. Repeat for the third solvent.

6. Repeat steps 3 through 5 for each of the other 12 fabric types.

7. Repeat steps 2 through 6 using 6 M hydrochloric acid.

8. Repeat steps 2 through 6 using acetone or nail polish remover.

Now that you've determined the effect of each of your solvents on each of the known fiber types, you can examine your questioned fiber specimens to determine the effects of each solvent on each of the questioned specimens. The questioned specimens may be made up of only one fiber or of several different fibers, so rather than test just a single thread of each questioned specimen, we'll test a tiny swatch to make sure that we're testing all fiber types present in the specimen.

1. Cut three tiny swatches (each about 5 mm square) of questioned Fiber Specimen A and transfer one of them to the well of a deep-cavity slide.

2. While observing the swatch at low magnification, transfer a sufficient amount of the first solvent to fill the well about half full. Make sure the fiber specimen is immersed in the solvent. Observe the specimen for one minute or until no further visible changes occur. Record your observations in your lab notebook.

3. Repeat step 10 with the second swatch in a clean deep-cavity side, using your second solvent.

4. Repeat step 10 with the third swatch in a clean deep-cavity side, using your third solvent.

5. Repeat steps 10 through 12 with each of your other questioned fiber specimens, B, C, D, and E.

6. By comparing your results from your known specimens to those of the five questioned specimens, attempt to identify the fabric type or types present in specimens A, B, C, D, and E.

PROCEDURE II-6-3: TEST FIBER SPECIMENS BY DYE STRIPPING

WARNING

Read the MSDS for each of the chemicals you use and follow the handling precautions noted. Do not mix these strippers. Chlorine bleach produces toxic gases when mixed with acids or bases. Wear splash goggles, gloves, and protective clothing at all times.

Stripping (bleaching) dyes from fibers is a common procedure in forensics labs. It is done for three reasons:

- The presence of dyes may interfere with some chemical fabric tests.
- Comparing the action of various stripping agents on a questioned specimen against the action of those same stripping agents on known specimens may allow the questioned specimen to be characterized. For example, navy-blue known and questioned specimens may react similarly to one stripping agent, but very differently to a second stripping agent, which establishes that the fabrics and/or dyes in the two specimens are not identical. Conversely, if the two specimens react similarly to all stripping agents, the forensic scientist can state that the two specimens are entirely consistent within the limitations of the stripping test.
- Any existing dye or dyes must be stripped before the fiber specimen can be subjected to the dyeing tests described in the following procedure.

Professional forensics labs use many different stripping agents to discriminate fibers and dyes, some of which contain very hazardous chemicals. (Modern dyes are surprisingly hard to remove from fabrics.) In this procedure, we'll use the following dye-stripping agents to test known and questioned fibers.

Dilute acetic acid

Ordinary distilled white vinegar is another name for 5% acetic acid. This mildly acid stripper can be used straight from the bottle to remove basic dyes from silk, wool, and other animal fibers.

Dilute aqueous ammonia

Clear household ammonia is generally 5% to 10% concentration. To make up 100 mL of the dilute ammonia stripping solution, dilute 20 mL of clear, non-sudsy household ammonia solution with 80 mL of tap water to produce a 1% to 2% ammonia solution. This mildly basic stripper can be used to remove acid dyes from silk, wool, and other animal fibers.

Sodium dithionite

Sodium dithionite (also called *sodium hydrosulfite*) is widely used in forensics labs for dye stripping. Dithionite solutions are good general strippers for most plant fibers and many synthetic fibers, although they work poorly or not at all on acetate and triacetate fibers and most animal fibers. To make up this stripper, stir about 5 g (a rounded half teaspoon) of sodium dithionite into 100 mL of distilled or deionized water.

Chlorine laundry bleach

Ordinary chlorine laundry bleach used undiluted is an extremely powerful basic stripper, too powerful for most applications. Forensics labs generally use a solution of 0.75% to 2% sodium hypochlorite, either unmodified or adjusted to pH 10 to 11 with sodium hydroxide. To make up 100 mL of 2% sodium hypochlorite, add 38 mL of standard 5.25% chlorine laundry bleach to 62 mL of tap water.

Acidified 2% chlorine bleach

Acidified chlorine bleach may be effective in stripping black or other very dark dyes from a specimen if none of the other listed stripping agents works. To make up 100 mL of this stripping agent, carefully add 44 mL of distilled white vinegar to 38 mL of 5.25% chlorine laundry bleach and make up the solution to 100 mL with tap water. This solution should be made up as needed immediately before use, and any remaining fresh or spent solution should be discarded by flushing it down the drain with plenty of water.

DON'T GAS YOURSELF

Mixing acids with chlorine bleach is ordinarily a major no-no, because it produces toxic chlorine or chlorine dioxide gas. With a dilute solution of chlorine bleach and acetic acid, which is a weak acid, the gases produced remain (mostly) in solution, although you should still perform this activity only in a well-ventilated area.

During this lab session, you'll heat this solution, which may cause chlorine dioxide gas to be evolved. If that happens, add 3% hydrogen peroxide solution dropwise until the evolution of gas ceases.

1. If you have not done so already, put on your goggles, gloves, and protective clothing.

2. All dye-stripping solutions work faster when hot rather than at room temperature. Prepare a hot water bath by filling the 250 mL beaker about half full of boiling tap water.

3. Label five test tubes, one for each of the dye-stripping solutions.

4. Cut 10 small swatches, about 5 mm square, of fabric specimen A, and add two of the swatches to each of the test tubes.

5. Transfer about 5 mL of the corresponding dye-stripping solution to each tube and place all five tubes in the hot water bath. Make sure each swatch is fully immersed in the solution.

6. Observe the specimens carefully, agitating the tubes occasionally. Some specimens may begin losing color very quickly, while others may appear unaffected even after several minutes in the solution. After five minutes or when no further changes are evident, remove the test tubes from the hot water bath and allow them to cool.

7. Flush the solutions from the test tubes down the drain with plenty of water, being careful to retain the bleached specimens in the tubes.

8. Fill each tube with tap water and use the stirring rod to agitate the fiber specimens to remove excess stripping solution. Repeat this wash several times with fresh water.

9. When the specimens are thoroughly washed, remove them from the test tubes and set them aside on a paper towel until they have dried completely. (Make sure not to lose track of which specimen is which.)

10. Repeat steps 4 through 9 for each of the questioned fabric specimens, B, C, D, and E.

11. When all specimens have dried thoroughly, observe each of them closely and note their appearance in your lab notebook. Compare the stripped specimens side-by-side with the original material, and use a magnifier, loupe, or stereo microscope to examine the fabric closely to observe, for example, if the fibers have been bleached evenly or unevenly or whether fibers running in one direction reacted differently from fibers running in the cross direction.

Retain the dye-stripped specimens for use in the following procedure. Figure II-6-2 shows one fabric specimen before dye-stripping. Figure II-6-3 shows the same specimen after stripping.

Figure II-6-2: *Fabric specimen before dye stripping*

Figure II-6-3: *The same fabric specimen after dye stripping*

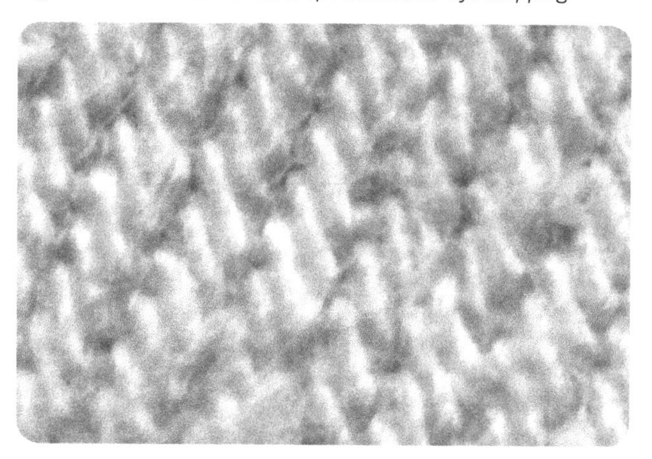

PROCEDURE II-6-4: TEST FIBER SPECIMENS BY DYEING

This procedure uses hot liquids, which must be handled carefully to avoid burns. The TIS dyes also stain skin and clothing. Wear splash goggles, gloves, and protective clothing, and use forceps to handle the specimens.

Different dyes vary in their effect on different fibers. A dye that strongly tints animal fibers may impart little or no color to vegetable fibers or artificial fibers, and vice versa. This differential effect of dyes on fibers is used by forensic scientists to classify questioned fibers.

Testfabrics, Inc. (*http://www.testfabrics.com*) produces a series of standard *Testfabrics Identification Stains* (TIS), which are mixtures of dyes of different colors chosen for their differing effects on various types of fibers. In use, one or more TIS solutions are used to dye a white or previously dye-stripped questioned fiber specimen as well as a series of known fiber specimens. When the dyeing is complete, the specimens are rinsed and dried, after which the questioned specimen can be compared against the knowns to identify the questioned fiber. Using multiple TIS solutions makes it easier to discriminate questioned fibers because it offers multiple points of comparison.

In this lab session, we'll use TIS #1, which is designed to discriminate various natural fibers, and TIS #3A, which is designed to discriminate artificial fibers. Despite these somewhat arbitrary distinctions, TIS #1 stains some artificial fibers and TIS #3A stains some natural fibers, so using both dyes on both types of fibers provides additional information that is useful for discriminating either type of fiber.

For our series of known fibers, we'll use strips of Testfabrics, Inc. Multifiber Fabric #43, which includes specimens of spun diacetate, SEF (modacrylic), filament triacetate, bleached cotton, Creslan 61 (acrylic), Dacron 54 (polyester), Dacron 64 (polyester), Nylon 66 (polyamide), Orlon 75 (acrylic), spun silk, polypropylene (polyolefin), viscose (rayon), and worsted wool. We'll stain one multi-fiber strip with TIS #1 and another with TIS #3A and then compare the results against questioned fiber specimens that have been stained in one or the other TIS solution to determine the type of fiber of the questioned specimen.

1. If you have not done so already, put on your goggles, gloves, and protective clothing.

2. Cut two narrow (~1 cm) strips across the width of the Multifiber Fabric #43 specimen, making sure that all 13 fabric types are present in each strip.

3. Label five test tubes, one for each of the five dye-stripped fabric specimens, A, B, C, D, and E.

4. Transfer about 150 mL of tap water to the 250 mL beaker. Using the hotplate or microwave oven, bring the water to a boil. Stir in one heaping spatula spoonful of TIS #1 dye powder. Keep the beaker at a gentle boil for several minutes, stirring occasionally, to dissolve the powder.

5. Transfer one of the Multifiber Fabric #43 strips to the beaker. Keep the beaker at a gentle boil, stirring occasionally, for five minutes.

6. Remove the beaker from the heat. Use the forceps to remove the Multifiber Fabric #43 strip from the beaker and rinse the strip under running water to remove excess dye. Place the strip aside on a clean paper towel and allow it to dry.

7. When the contents of the beaker have cooled sufficiently for safe handling, transfer about 10 mL of the dye solution to each of the five test tubes. Discard the remaining dye solution or save it in a capped container for later use.

8. Rinse the beaker, transfer all five of the test tubes to the beaker, and fill the beaker with tap water about 2/3 full. Place the beaker on the hotplate or in the microwave oven and bring the water to a gentle boil.

9. Place one of the dye-stripped swatches you retained from the preceding procedure in each tube, making sure that swatch A goes in tube A, swatch B in tube B, and so on.

10. Keep the beaker at a gentle boil for five minutes, making sure that each fabric swatche remains immersed in the dye in its test tube. After five minutes, remove the beaker from the heat.

11. Use the test tube clamp to remove each test tube from the beaker. Carefully pour the dye from each tube down the sink, making sure to retain the dyed fabric swatch. Rinse each fabric swatch under running water to remove excess dye. Place the swatch on a clean paper towel to dry. (Don't lose track of which swatch is which; you can label its identity directly on the paper towel.)

To give us differential results, we'll repeat this procedure using the second dye, TIS #3A, using a slightly different procedure.

1. Transfer about 125 mL of tap water to the 250 mL beaker. Using the hotplate or microwave oven, bring the water to a boil. Transfer a full pipette (~2.5 mL) of TIS #3A dye concentrate solution to the beaker. Keep the beaker at a gentle boil for five minutes.

2. Transfer about 25 mL of distilled white vinegar to the beaker, again bring the beaker to a boil, and continue boiling for five minutes.

3. Transfer the second Multifiber Fabric #43 strip to the beaker. Keep the beaker at a gentle boil, stirring occasionally, for five minutes.

4. Remove the beaker from the heat. Use the forceps to remove the Multifiber Fabric #43 strip from the beaker and rinse the strip under running hot water to remove excess dye. Transfer the strip to a container of warm tap water with a few drops of dishwashing detergent added and allow the strip to soak for five minutes with occasional stirring. Remove the strip from the detergent solution, rinse in hot running tap water, and place the strip aside on a clean paper towel to dry.

5. When the contents of the beaker have cooled sufficiently for safe handling, transfer about 10 mL of the TIS #3A dye solution to each of the five test tubes. Discard the remaining dye solution or save it in a capped container for later use.

6. Rinse the beaker, transfer all five of the test tubes to the beaker, and fill the beaker with tap water about 2/3 full. Place the beaker on the hotplate or in the microwave oven and bring the water to a gentle boil.

7. Place one of the dye-stripped swatches you retained from the preceding procedure in each tube, making sure that the swatch A goes in tube A, swatch B in tube B, and so on.

8. Keep the beaker at a gentle boil for five minutes, making sure that each fabric swatche remains immersed in the dye in its test tube. After five minutes, remove the beaker from the heat.

9. Use the test tube clamp to remove each test tube from the beaker. Carefully pour the dye from each tube down the sink, making sure to retain the dyed fabric swatch. Rinse each fabric swatch under hot running water to remove excess dye, soak each swatch in warm detergent solution for five minutes, rinse the swatch again in hot running tap water, and place the swatch on a clean paper towel to dry.

10. By comparing your results from your known specimens to those of the five questioned specimens, attempt to identify the fabric type or types present in specimens A, B, C, D, and E.

> Your known specimens are each made up of a single fiber type. Your questioned specimens may contain one or multiple fiber types, so examine the questioned specimens carefully with the magnifier or microscope to determine their makeup.

Figure II-6-4 shows specimens of Multifiber Fabric #43 before dyeing, dyed with TIS #1 stain, and dyed with TIS #3A stain.

Figure II-6-4: *Multifiber Fabric #43 specimens before dyeing (left), dyed with TIS #1 stain (center), and dyed with TIS #3A stain*

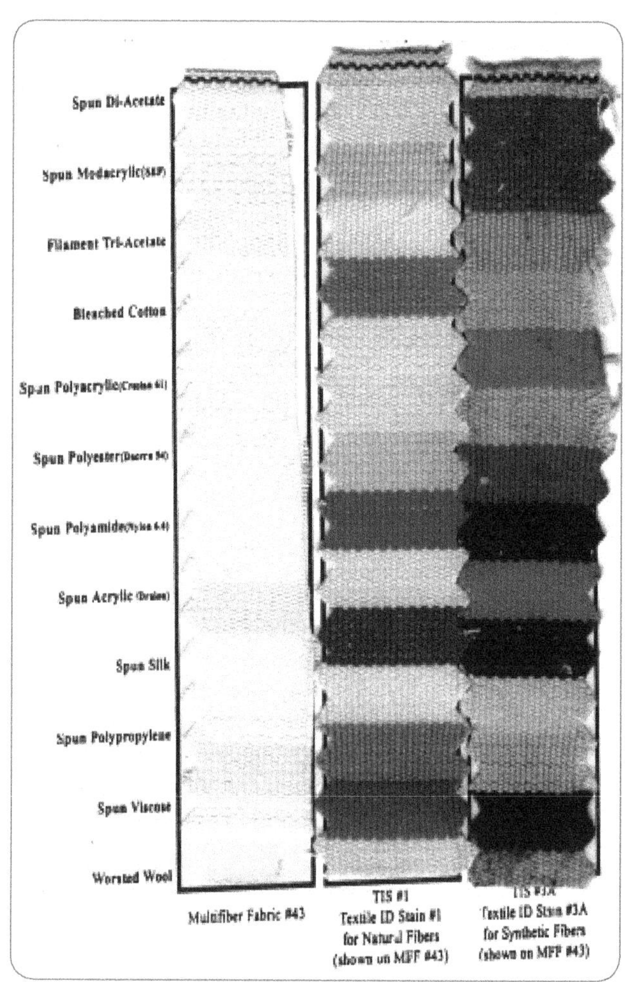

REVIEW QUESTIONS

Q1: From your completed table in Procedure II-6-1, what are the burning characteristics of each fiber you tested?

Q2: **Were you able to identify the questioned fiber using burning tests?**

Q3: **You burn a fiber and detect an odor similar to household ammonia. What type of fiber do you suspect?**

Q4: **You burn a fiber and detect an odor similar to burning paper. What type or types of fiber do you suspect?**

Q5: **You burn a fiber, which does not melt as it approaches the flame. It curls as it burns, and it self-extinguishes when you remove it from the flame. A hard black bead forms at the burned end of the fiber. What specific fiber do you suspect?**

Q6: **At room temperature, a questioned fiber specimen dissolves in 6 M hydrochloric acid. What type of fiber do you suspect?**

Q7: **At room temperature, a questioned fiber specimen dissolves in acetone, but not in glacial acetic acid. What type of fiber do you suspect?**

Q8: **A questioned fiber specimen proves to be insoluble in all of your room-temperature reagents. What fiber type do you suspect, and what reagent would you use to confirm your suspicion?**

Q9: **Burning and solubility tests have established that your questioned fiber is synthetic. Which stripping agents would you rule out immediately, and why?**

Q10: **Burning and solubility tests have established that your questioned fiber is triacetate. Which of the listed stripping agents would you use to remove dyes from this specimen and in what order? Why?**

Q11: **Given the availability of burning, solubility, and dye-stripping tests, what value (if any) do fabric dyeing tests have and why?**

Q12: **Which fiber or fibers make up questioned fabric specimens A, B, C, D, and E? How do you know?**

Study the Morphology of Fibers and Fabrics

EQUIPMENT AND MATERIALS

You'll need the following items to complete this lab session. (The standard kit for this book, available from *http://www.thehomescientist.com*, includes the items listed in the first group.)

MATERIALS FROM KIT

- Goggles
- Centrifuge tubes, 1.5 mL (14)
- Coverslips
- Filter, plane-Polarizing
- Forceps
- Glycerol
- Magnifier
- Oil, cassia
- Oil, olive
- Pipettes
- Ruler
- Scalpel
- Slide, cross-sectioning
- Slides, deep-cavity
- Slides, flat
- Specimens: fabrics A, B, C, D, and E
- Specimen: TFI Multifiber Fabric #43

MATERIALS YOU PROVIDE

- Gloves
- Marking pen, ultra-fine, permanent
- Microscope
- Scanner and scanning software (optional)
- Scissors
- Water, distilled or deionized
- Specimens: additional fabrics/fibers (optional)

BACKGROUND

WARNING

Although none of the activities in this lab session present any significant risks, as a matter of good practice, you should always wear goggles and gloves when working in the lab, if only to avoid contaminating specimens.

Yes, forensic scientists sometimes work without goggles when using a microscope or magnifier to examine non-hazardous specimens. They may also work without gloves if the specimens and reagents are non-hazardous and there is no danger of contaminating specimens. That doesn't mean you should go without goggles or gloves, at least until you have sufficient experience to judge when it is appropriate and safe to do so.

Fibers are class evidence because a questioned fiber cannot be matched conclusively against a known specimen. In the absence of a gross physical match—a torn questioned specimen that exactly matches the missing portion of a known specimen—the most that a forensic scientist can say with certainty is that the questioned specimen is consistent in every way with a known specimen.

CASHMERE DNA?

Although we have found no such cases in the literature, there is no obvious reason why wool or other animal hair fibers could not be tested for an mtDNA match. We speculate that mtDNA testing is not used on animal-hair fibers from fabrics for two reasons: First, there is much less genetic variation among heavily bred and selected livestock than among humans. Within relatively few generations, hundreds or thousands of individual animals may trace their lineage to a single champion female ancestor. Second, most animal-hair fabrics are made up of hair obtained in bulk from hundreds or thousands of animals in one batch. (Think sheep shearing.) It's entirely possible, even probable, that one animal-hair fiber from a piece of fabric comes from a different animal than the fibers immediately adjacent to it in the fabric.

Although the chemical fiber tests described in the preceding lab session are important tools for initial screening among types of fibers, morphological examination can tell the forensic scientist a great deal about the individual characteristics of a questioned fiber. Morphological matching of questioned fibers to known fibers may provide compelling (although not definitive) evidence, particularly if one or more of the questioned fibers is relatively uncommon.

Morphology is a collective term that encompasses the form and structural characteristics of a specimen, such as size, shape, and color. Different types of specimens have different morphological characteristics. For example, height and weight are two morphological characteristics of people, while thread count is a morphological characteristic of a fabric specimen.

Assume that a murder victim has been found, wrapped in an old blanket, and dumped in a ditch along a back road. Upon examination, fibers are found adhering to the victim and to the blanket. It is determined that those fibers come from a particular type of carpet in a particular color that was used only in a certain make and model of car over a two-year period, and only in those cars that used a particular color of exterior paint.

The police have many potential suspects, but one of those suspects owns a car of the correct make, model, year, and color to match the questioned fibers found on the victim. After obtaining a search warrant, the police impound the car and submit it to forensic testing. Carpet fibers from the car are consistent in all respects with fibers found on the victim.

By itself, this datum is suggestive, but not conclusive. After all, thousands of cars may have this particular carpet. But those thousands of cars are probably relatively evenly distributed across the country, so only a small percentage are likely to be found locally, which narrows things down considerably.

And we aren't yet finished with the fiber evidence. Remember the blanket in which the body was wrapped. If fibers from that blanket can also be matched to fibers present in the car trunk, it becomes extremely likely that that car was used in commission of that crime.

Probabilities can be multiplied. For example, if it is determined that, say, only 0.01% (0.0001) of cars have that particular carpet and that only 0.1% of blankets use that particular fiber in that particular color, the probability that both fibers would be found randomly becomes (0.0001 * 0.001) = 0.0000001, or one in ten million. (Actually, less than that, because not all cars have blankets in their trunks.) Furthermore, if fibers from the victim's clothing can be matched to fibers in the car trunk—particularly if there are several different types of clothing fibers—it is extremely likely, nearly certain, that that victim was at some time present in that car trunk.

Although real forensics labs use instrumental analysis for definitive characterization of fibers, morphology remains an important aspect of fiber analysis. In this lab session, we'll use many of the same techniques we used earlier in the group for human and animal hair to examine the morphology of natural and artificial fibers.

We'll also use one additional technique, refractive index testing with polarized light, to reveal additional information about our artificial fiber specimens. In plane-polarized light, some artificial fibers are *isotropic*, which means they have only one refractive index, regardless of the polarization plane of the light used to view them. Other artificial fibers are *anisotropic* (also called *birefringent*), which means their refractive indices differ as the polarization plane is rotated relative to the specimen.

Of course, determining the refractive index of a single fiber is not a trivial matter, particularly using only the equipment practical for a home lab. Fortunately, there's a work-around that, although it doesn't provide a numeric value for refractive index, does allow us to determine the *relative refractive index* (*RRI*) of a fiber as we change the polarization plane.

The first step in this procedure is to determine whether the fiber is isotropic or anisotropic, which can be done simply by viewing the wet-mounted fiber using light polarized in one plane and then rotating the polarizing filter 90°. If the fiber is isotropic, its appearance does not change as you rotate the polarization plane. If the fiber is anisotropic, its appearance changes as you rotate the polarization plane.

If the fiber is determined to be anisotropic, the next step is to determine the RRI. "Relative" in this case means relative to the refractive index of the wet-mounting medium. By choosing a wet-mounting medium with a refractive index close to that of the fiber (typically, about 1.54), you can easily determine if the RI of the fiber is higher or lower than that of the mounting medium. If you have time, you can mount the fiber with several mounting media of differing RI and determine an actual numeric value for the RI of the fiber.

Forensic scientists use expensive *polarizing microscopes* (also called *petrographic microscopes*, because such microscopes are commonly used by petroleum geologists) to determine RRI for fibers. We'll use our standard compound microscope with an inexpensive piece of polarizing film to make the same determination.

This lab session has five procedures. In the first, we'll perform a macroscopic examination of fibers and fabrics. In the second, we'll do a microscopic examination of whole-mount specimens. In the third, we'll make cross sections of individual fibers and examine them microscopically. In the fifth and final procedure, we'll determine whether various fiber specimens are isotropic or anisotropic and determine the RRI for each anisotropic specimen.

PROCEDURE II-7-1: MACROSCOPIC EXAMINATION OF FABRICS

1. Examine each of your fabric specimens, by eye and with the magnifier, a stereo microscope, or other low-magnification optical aid. If you are equipped to do so, shoot an image or make a scan of each fabric specimen.
2. As you examine it, record the following information about each fabric specimen in your lab notebook: source/identity of the specimen; color and general appearance of the fabric; weave type; thread count; differences in color, size, or other characteristics of the warp and weft threads; and any other visible characteristics that apply to the fabric (as opposed to the individual fibers).

One increasingly popular technique that bridges macroscopic and microscopic examination is the use of an ordinary scanner to image fabrics. Even inexpensive scanners may provide 4,800 dpi or higher resolution, which translates to a 23 megapixel image of each square inch. Scanned image files typically include embedded metadata that records the resolution of the scan and other capture parameters, which means the image files are inherently calibrated.

Figures II-7-1 and II-7-2 show a 4,800 dpi scan of a few square centimeters of a fabric specimen. The first image shows the entire 44 megapixel scanned image file, and the second shows a small part of that image file as it appears displayed at full size on a computer monitor.

Figure II-7-1: *A 44 megapixel scan of a fabric sample (full image)*

Figure II-7-2: *A 44 megapixel scan of a fabric sample (detail as it appears onscreen)*

One major advantage of using scanned image files to record the appearance of fabric specimens is that those image files can be modified with Photoshop, The GIMP, or similar image processing software to enhance contrast, apply edge filters, and use other techniques that reveal details that are invisible to the naked eye.

PROCEDURE II-7-2: MICROSCOPIC EXAMINATION OF FIBERS AND FABRICS

Microscopic examination focuses primarily on individual fibers rather than fabrics, but can provide useful information about fabrics—such as thread counts—that are difficult or impossible to discriminate under macroscopic examination.

The four major classes of fibers—animal, plant, mineral, and artificial/reconstituted—each share common class characteristics that determine the type and amount of information that can be gained by microscopic examination.

- Animal fibers are simply processed animal hair, and can be characterized on the same basis.

- Plant fibers are often ribbon shaped, may have twists at regular or irregular intervals, and often have little or no visible internal structure.

- Mineral fibers appear rock-like or glass-like and often have jagged or fractured tips.

- Artificial and reconstituted fibers are notable for their regular appearance and absence of variability in structure, and often have a multi-lobular structure and/or regular longitudinal striations that are never found in nature.

1. Wet-mount each fiber as described in Laboratory II-2, using glycerol (or another mounting fluid with an RI near 1.54) as the mounting medium.

2. Examine each specimen by transmitted and reflected light at medium (~100X) and high (~400X) magnification. For animal fibers, observe and note the characteristics described in Lab Sessions II-2 and II-4 in your lab notebook, as well as any of the other characteristics described below. For other fibers, observe, measure, and record the following microscopic characteristics about each specimen:

   - Light transmission (transparent, translucent, or opaque).

   - The color of the fiber by transmitted and reflected light, and whether the color is a pigment (discrete color bodies) or a dye (continuous color).

   - The presence or absence of any delustrant (non-reflective particles, usually white) or other surface treatment.

   - The longitudinal morphology of the fiber (e.g., smooth, striated, serrated, twisted, kinked, or other), and a description of its surface texture (if any) and any embedded features such as air bubbles or voids.

   - The appearance of the ends, such as square-cut, angle-cut, tapered, frayed, crimped, melted, etc.

   - If you can determine the cross-sectional shape of the fiber from the longitudinal view, note it (round, oval, square, pentagonal, lobed, etc.) and measure the diameter of any visible lobes.

Figures II-7-3, II-7-4, and II-7-5 show typical plant (cotton), animal (wool), and synthetic (polyester) fibers, respectively, at 100X magnification. Cotton fibers, as shown in Figure II-7-3, are flat ribbons with a regular twist.

Figure II-7-3: *A cotton fiber at 100X magnification*

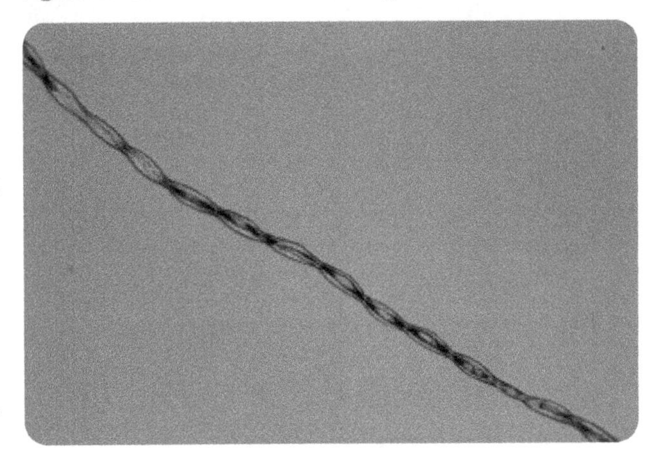

As you might expect, wool fibers show features similar to those of other mammalian hairs, but are often dyed. The wool fiber shown in Figure II-7-4 (from a Pendleton woolen scarf) has clearly been dyed and shows the prominent, broken medulla and diameter variations typical of wool.

Figure II-7-4: *A wool fiber at 100X magnification*

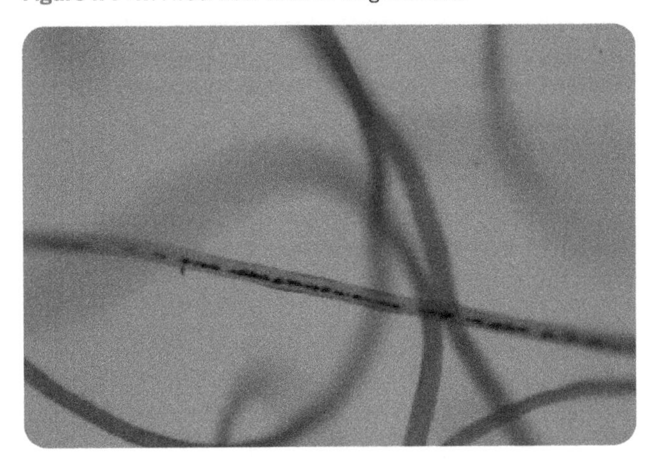

The polyester fiber shown in Figure II-7-5 is obviously too regular in every respect to be anything other than artificial.

Figure II-7-5: *A polyester fiber at 100X magnification*

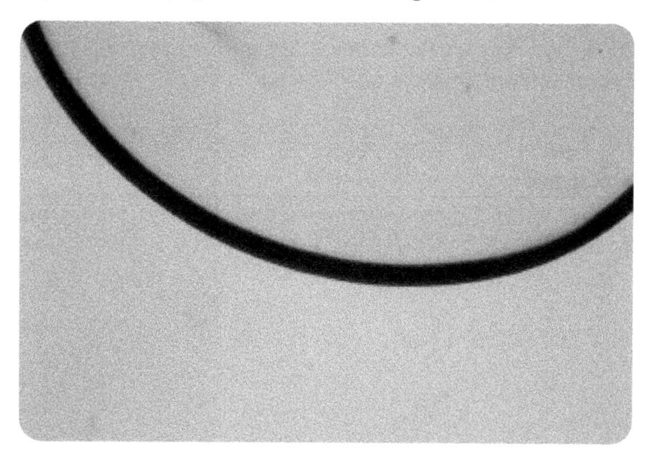

PROCEDURE II-7-3: CROSS-SECTIONAL EXAMINATION OF FIBER SPECIMENS

Until now, we've been observing whole-mounted fibers longitudinally. As useful as a longitudinal view is, it doesn't reveal all of the visible information about a specimen. To complete our microscopic examination, we need to prepare and view *cross sections* (also called *transverse sections*) of fiber specimens.

Because the surface of a specimen is seldom perfectly transparent, a transverse section may reveal details about the internal structure of a specimen that are obscured in a longitudinal view. For example, it may be obvious from a longitudinal view that a fiber has been dyed, but the "end-on" view of a transverse section allows us to determine whether the dye is spread uniformly throughout the fiber or concentrated near the surface. Similarly, it can be difficult to determine the cross-sectional shape of a fiber from a longitudinal view. With a transverse section, it's immediately obvious if the cross section is circular, oval, oblate, kidney-shaped, triangular, lobed, irregular, or otherwise.

Producing usable fiber cross sections in a home lab requires some effort. It's easy enough, of course, to cut a hair or fiber. The hard part is getting a thin transverse section mounted and correctly oriented on a slide. Several means can be used to obtain transverse sections of hairs and fibers, including the following:

Cross-section test slide method

Professional forensics labs generally use special metal microscope slides designed for sectioning hair and fiber specimens. These slides are made of 0.25 mm thick stainless steel, punctured with numerous 0.9 mm holes. In use, a fiber bundle is drawn through a hole using a short length of thin, soft copper wire (about 34 AWG) to pull the looped fiber bundle through the hole. If necessary, the fiber bundle can be bulked out to fill the hole by using known fibers. Using a scalpel, craft knife, or single-edge razor blade, the fiber bundle is sliced off flush with the top and bottom surfaces of the plate, leaving 0.25 mm thick fiber sections that can be viewed by reflected or transmitted light. These slides are quite fragile and are often discarded after one use. Purchased in quantity, they cost a dollar or so each, so they're a bit expensive for a home lab.

Plastic slide method

You can accomplish much the same thing with inexpensive materials by drilling small holes in plastic slides using a drill and small bit, as shown in Figure II-7-6. Bits from about

#61 (0.039" or 0.991 mm) through about #75 (0.021" or 0.533 mm) make holes of useful sizes for fibers of differing thicknesses. If you don't have a drill, you can use a heated needle to melt holes in the slide, but plan to waste one slide for practice before you master melting holes that don't have excessively raised edges.

Once you have a slide with usable holes, simply draw fiber bundles through each hole using a needle or thin copper wire and then use a scalpel to slice off the excess fiber flush with both surfaces of the slide. Depending on the thickness of the plastic slide, this method yields thicker sections, typically 0.5 mm to 1.0 mm, which may be too thick to view by transmitted light.

Figure II-7-6: *Drilling a plastic microscope slide for use as a sectioning slide*

Using cross-section slides is faster if you have several specimens to section, but you can also section specimens individually using one of the following methods.

Commercial microtome

Lab supply vendors sell commercial microtomes and casting mediums suitable for making thin sections. Student-grade microtomes are available for $35 or so, and are precise enough to produce sections as thin or thinner than those produced with cross-section slides.

Home-made microtome

You can make a microtome using a fine-thread bolt and a matching flat nut. To use this method, thread the bolt one or two turns into the nut, leaving the remaining part of the nut to form a well. Using forceps, hold the fiber vertically in that well as you drip molten candle wax into the well, filling it slightly overfull. Allow the wax to solidify and then use a scalpel, craft knife, or single-edge razor blade to trim off the excess wax flush with the face of the nut. Discard this first section, and then turn the nut slightly to drive a thin section above the face of the nut. Use the scalpel to slice off that thin section and mount it on a slide. Used with care, this method can produce sections of 0.25 mm or thinner.

Drinking straw method

The drinking straw method is simple, requires only items found around the house, and can give good results. It is also fussy, requires an assistant for best results, and may require multiple attempts to get a good section. Cut a short (~5 cm) length of plastic drinking straw. Lubricate the inside surface with a very small amount of oil, WD-40, or similar lubricant. Place the straw section vertically against a safe work surface, using the bottom edge of the straw to pin one end of the fiber against the work surface. Using forceps, hold the other end of the fiber as nearly vertical as possible centered in the straw and have your assistant drip candle wax into the straw until it is nearly full, being careful not to burn the fiber or yourself. Allow the wax to harden for a minute or so, and then use a dowel or similar object to slide the wax plug until the end is just past flush with the end of the straw. Use the scalpel to trim off the first section and discard it. Press the plug a bit further out of the straw and carefully use the scalpel to cut a section as thin as possible.

Dripping wax method

One of the easiest ways we found to make a fiber cross section is to drip melted candle wax onto a microscope slide until you have a mound built up about 3 mm thick. Carefully embed the fiber into the wax before it hardens completely. Then, drip more wax onto the fiber to build up the mound another 3 mm or so, as shown in Figure II-7-7. Once the mound hardens completely, use a scalpel or razor knife to cut thin vertical sections through the wax.

Figure II-7-7: *Making a fiber cross section with melted candle wax*

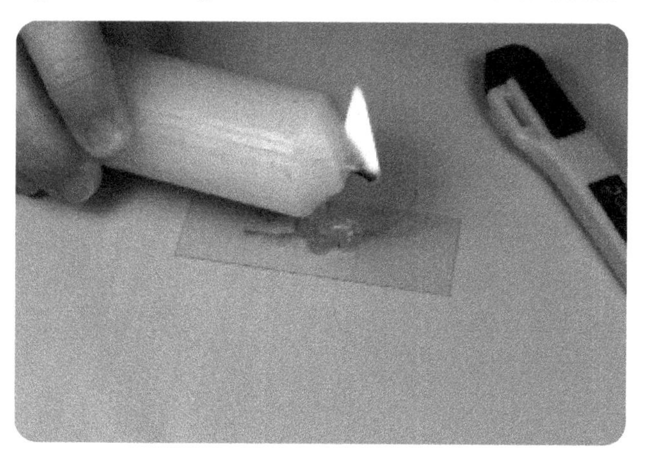

Cork method

We include this method for completeness, although we haven't had much luck with it. Begin with a high-quality, fine-grain cork section about 2.5 cm thick. Press a thin needle through the cork, using pliers if necessary, until the eye of the needle is just visible above the surface of the cork. Thread your fiber specimen through the eye of the needle, and draw the needle all the way through the cork, stopping before you pull the fiber specimen completely free of the cork. Use a scalpel to cut the fiber free from the needle, flush with the surface of the cork. Then use the scalpel to cut a thin section of the cork with the fiber specimen embedded. The problem we had was that the cork crumbled when we tried to cut thin sections. Perhaps our corks just weren't good enough.

In this procedure, we'll make cross sections of our known and questioned fiber specimens and examine those sections microscopically to learn what we can see from the cross-sectional perspective that is invisible with longitudinal whole-mount specimens.

Make and observe cross sections for as many known and questioned fibers as you have time to do. Make and label sketches of each cross section in your lab notebook, or shoot images of each cross section and attach labeled copies of each image in your lab notebook. If time allows, make and observe cross sections of your various human and animal hair specimens as well.

PROCEDURE II-7-4: DETERMINE THE REFRACTIVE INDEX OF FIBERS WITH RI MATCHING LIQUIDS

The *refractive index* (also called *index of refraction*) of a material specifies the reduction of the speed of light in that material relative to the speed of light in a vacuum, which is assigned the value 1. For example, water has a refractive index of about 1.333, which means that in water, light travels at 1.000/1.333 = 0.75, or 75% the speed of light in a vacuum. When a light beam traveling through a transparent medium of low refractive index (such as vacuum or air) encounters a transparent material with a higher refractive index (such as a fiber), that light beam is bent (or refracted) to a different angle.

The refractive index (abbreviated *n* or, more casually, *RI*) for a specific material varies significantly with the temperature and the wavelength of the light used to measure RI. RI values for many materials can be found in tables that list those values at various temperatures and wavelengths of light. If only one value is listed for RI, it is usually for 20°C using monochromatic yellow light at 589.3 nm, the sodium D-Line.

In ordinary (as opposed to polarized) light, most common fibers have RI values in the 1.47 to 1.61 range. The easiest way (and often the only practical way) to determine the refractive index of a fiber specimen is by immersing it in a liquid of known refractive index. If the refractive indices of a colorless fiber specimen and the liquid are the same, the fiber becomes invisible in the liquid. (If the fiber is colored, the fiber appears as a line of color in the liquid, with structural details ill-defined or absent.)

You can demonstrate this phenomenon by placing Pyrex stirring rods in beakers of water and olive oil, as shown in Figure II-7-8. In water, the rod remains easily visible because the RI of the glass (~1.47) is quite different from the RI of the water (~1.33). Conversely, when you immerse the rod in olive oil (RI ~1.47), the RIs of the glass and liquid are so similar that the rod becomes invisible, or nearly so.

Figure II-7-8: *Glass rods in water (left) and olive oil*

Forensic scientists use this phenomenon to determine the RI of fiber (and glass and plastic) specimens very accurately. Professional forensics labs use refractive index matching liquids sold by Cargille Laboratories, which are made and calibrated to very tight standards and are, as you might expect, very expensive. For example, the RF-1 set of Cargill Refractive Index Matching Liquids covers the range from 1.4 to 1.7 with intervals of 0.002 with accuracies of 0.0001. A set of 7 mL bottles of these 151 liquids costs more than $2,300.

Cargille liquids are supplied with graphs that list the actual RI of each liquid at various temperatures. Professional forensics labs use temperature-controlled microscope stages that allow the temperature of the liquid to be adjusted over a wide range to within 0.2°C or better, until the RI of the liquid exactly matches the RI of the specimen. By choosing a Cargille liquid with an RI just a bit higher than that of the specimen and then heating the liquid until the RI of the specimen and liquid are identical, a forensic scientist can determine the RI of the specimen very precisely by looking up the RI of the liquid at the temperature where the specimen disappears from view.

Obviously, Cargille liquids, temperature-controlled microscope stages, and sodium D-line monochromatic light sources are too costly for our purposes. Fortunately, we can get reasonably good RI values for our fiber specimens without any of those things. In this procedure, we'll make up our own refractive index matching liquids. The FK01 Forensic Science Kit includes a set of 1.5 mL snap-top micro-centrifuge tubes that you can use to store your refractive index liquids after you've made them up. We'll use these liquids drop-wise under the microscope, so even 1 mL is enough to do 20 to 30 comparisons.

You can make up refractive index matching liquids using combinations of oils that are readily available from supermarkets, drugstores, health food stores, herbalists, and aromatherapy vendors. Among the most useful of those are olive oil (RI = 1.47) and and cassia oil (RI = 1.61), both of which are included in the FK01 Forensic Science Kit.

Mixtures of these oils have refractive indices that fall between the refractive indices of the pure oils. For our purposes, we'll assume that the refractive index of mixed oils is proportionate to the volumes of each oil used. For example, if we mix 10 drops of olive oil (RI = 1.47) with 20 drops of cassia oil (RI = 1.61) to yield 30 drops of mixed oils, we can calculate the approximate refractive index of the mixed oils as follows:

$$(10 * 1.47) + (20 * 1.61) = (30 * x)$$

$$14.7 + 32.2 = 46.9 = (30 * x)$$

$$x = (46.9/30) = 1.56$$

The actual refractive indices of the mixed oils may not be exactly as calculated, both because there is some variation in the RI of any particular oil from different sources and because volumes are not necessarily additive. (For example, adding 5 mL of one oil to 5 mL of another oil may not yield exactly 10 mL of the mixed oils; the final volume may be only, say, 9.96 mL, or it may be 10.04 mL.) Still, the calculated RI values should be accurate at least to the second decimal place.

Because nearly all common fibers have refractive indices between 1.47 and 1.60, we decided to use olive oil and cassia oil to make up our refractive index matching liquids. Olive oil is conveniently close to the low end of that range, and cassia oil to the high end. Making up 14 mixed oils allows us to cover the entire range in increments of about 0.01, which is sufficient for our purposes. We used the ratios of drops listed in Table II-7-1 to make up our oil mixtures. You'll need only a drop of a mixed oil to test a specimen, so make up at most 1 or 2 mL of each testing liquid. If you use a different pair of oils, modify your calculations accordingly.

Retain your refractive index matching liquids after you finish this procedure. We'll use them in the next lab session for testing the refractive indices of glasses and plastics.

Table II-7-1: *Nominal refractive indices for mixtures of olive and cassia oils*

Olive oil	Cassia oil	Nominal RI
All	None	~1.467
28 drops	2 drops	~1.477
26 drops	4 drops	~1.486
24 drops	6 drops	~1.496
22 drops	8 drops	~1.506
20 drops	10 drops	~1.515
18 drops	12 drops	~1.525
16 drops	14 drops	~1.535
14 drops	16 drops	~1.544
12 drops	18 drops	~1.554
10 drops	20 drops	~1.564
8 drops	22 drops	~1.573
6 drops	24 drops	~1.583
4 drops	26 drops	~1.593
2 drops	28 drops	~1.602
None	All	~1.612

To make up your own RI matching liquids, proceed as follows:

1. Use the ultra-fine point permanent marker to label fourteen 1.5 mL micro-centrifuge tubes. (We recommend labeling them 1 through 14 or A through N and recording the actual contents of the oil mixtures they'll contain in your lab notebook and on the container you'll use to store the tubes.)
2. Using a clean pipette, transfer the required number of drops of the olive oil to each tube.
3. Using a second clean pipette (to avoid contaminating the oils), transfer the required number of drops of the cassia oil to each tube. Swirl each tube to mix the oils thoroughly.

These uncalibrated RI liquids are accurate enough for learning purposes, but if you want better accuracy, you can calibrate your liquids using the Becke Line method described in the following procedure by immersing crystals of some of the common lab chemicals listed in Table II-7-2 in each of your fluids.

1. Identify the compound listed in Table II-7-2 whose RI most closely matches the calculated RI of one of your RI matching fluids.
2. Place one drop of the fluid and one crystal of the compound in a well slide.

If the RI of a colorless crystal matches the RI of the fluid, the crystal becomes invisible in the fluid because both refract light identically. (If the colors of the crystal and fluid differ significantly, the crystal remains visible because of the color contrast, but its structure is no longer visible.)

3. At 100X magnification using (ideally) monochromatic sodium light, focus critically on the top edges of the crystal. Note the position of the Becke line.
4. With the fine focus knob, open focus very slightly. If the RI of the crystal is higher than the RI of the fluid, the Becke line moves inward, past the edge of the crystal and toward the center. If the RI of the crystal is lower than the RI of the fluid, the Becke line moves outward.
5. If you determine that the RI of the fluid is lower than the RI of the crystal, repeat the test using a crystal with a slightly lower RI. Conversely, if the RI of the fluid is higher than that of the crystal, repeat the test using a crystal with a slightly higher RI. By this method, you can determine the actual RI of the fluid (at ambient temperature) to at least three decimal places.

Figure II-7-9 shows sodium chloride crystals (RI = 1.544) in olive oil (RI = 1.467). Because the RI of these two materials is significantly different, the crystals are readily visible. Figure II-7-10 shows sodium chloride crystals in clove oil (RI = 1.543). Because the RI of these two materials is very similar, the sodium chloride crystals show very little relief against the surrounding fluid.

Figure II-7-9: *Sodium chloride crystals (RI = 1.544) in olive oil (RI = 1.467), showing high relief*

Figure II-7-10: *Sodium chloride crystals (RI = 1.544) in clove oil (RI = 1.543), showing low relief*

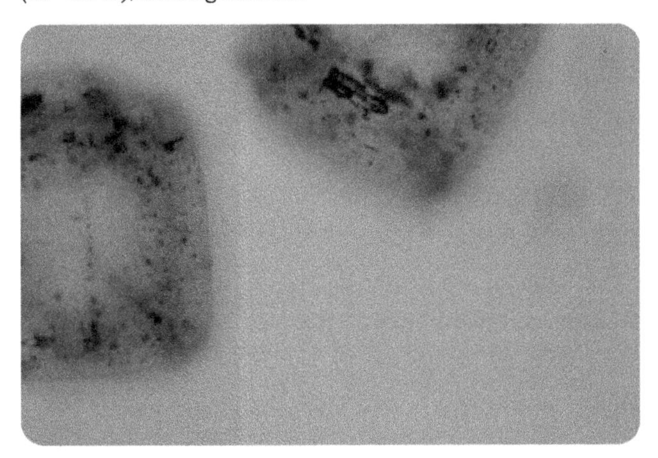

Table II-7-2: *Solids for calibrating refractive index liquids*

Compound	n^{20}_D
Sodium tetraborate decahydrate (borax)	1.469
Potassium perchlorate	1.474
Sodium sulfate	1.477
Ammonium hydrogen sulfate	1.480
Zinc sulfate heptahydrate	1.480
Potassium bicarbonate	1.482
Ammonium perchlorate	1.483
Sodium monophosphate monohydrate	1.485
Potassium chloride	1.490
Magnesium acetate tetrahydrate	1.491
Ferrous ammonium sulfate	1.492
Zinc acetate dihydrate	1.494
Potassium sulfate	1.495
Calcium nitrate	1.498
Sodium bicarbonate	1.500
Magnesium carbonate trihydrate	1.501
Ammonium persulfate	1.502
Barium hydroxide	1.502
Potassium nitrate	1.504
Magnesium chloride hexahydrate	1.507
Sodium thiosulfate pentahydrate	1.508
Potassium monophosphate	1.510
Nickel sulfate hexahydrate	1.511
Sodium chlorate	1.515
Barium acetate	1.517
Potassium chlorate	1.517

Continued

Table II-7-2: *Solids for calibrating refractive index liquids (continued)*

Compound	n_D^{20}
Sodium metasilicate	1.520
Ammonium sulfate	1.523
Calcium sulfate (gypsum)	1.523
Ammonium monophosphate	1.525
Potassium tartrate	1.526
Sodium bisulfite	1.526
Potassium carbonate anhydrous	1.531
Sodium carbonate anhydrous	1.535
Copper(II) sulfate pentahydrate	1.537
Sodium chloride	1.544
Potassium bromide	1.559
Aluminum chloride hexahydrate	1.560
Magnesium hydroxide	1.562
Sodium sulfite	1.565
Ferrous chloride	1.567
Lithium carbonate	1.567
Potassium ferricyanide	1.569
Calcium hydroxide	1.574
Potassium ferrocyanide trihydrate	1.577
Sodium nitrate	1.587
Strontium nitrate anhydrous	1.588
Manganous sulfate monohydrate	1.595
Ammonium nitrate	1.611

With your refractive index matching liquids prepared, test your known and questioned fiber specimens as follows:

1. Label several well slides or deep-cavity slides, one for each of your RI matching liquids. (If you don't have enough slides to allocate one to each liquid, simply do your testing in stages, cleaning the slides each time before you use a different liquid.)

2. Transfer several drops of the appropriate RI matching liquid to each corresponding labeled slide.

3. Remove several short fibers (~10 mm) from your first known specimen.

4. Place the slide containing one of the RI matching liquids on the stage. Using the forceps, and while looking through the microscope eyepiece at 40X or 100X, dip one end of the fiber into the liquid in the well. If the RI of the fiber is an exact match for the RI of the liquid, the structure of the immersed portion of the fiber will be invisible (if the fiber is colored, the fiber will be visible as only a colored streak, without any structural detail visible). If the RI of the fiber is close to that of the liquid, the edges of the fiber will be poorly defined. If the RI of the fiber and the liquid differ significantly, the edges of the fiber will stand out against the liquid.

> By working smart, you can avoid having to test each fiber against all of the RI matching liquids. For example, you might start with a liquid in the middle of the range. If the fiber and liquid are a reasonably close match, you don't have to test against the liquids on either end of the range. Try to narrow down the possibilities this way, and you may have to test a particular fiber against only three or four liquids to determine the closest match.

5. Repeat your tests on the first fiber until you have established the closest match, and record that value in your lab notebook.

> In a professional forensics lab, there may be only a single fiber to test, so the technician will take great care to make the best use of that single fiber. That's why professional labs use heated microscope stages and professional RI matching liquids. In our case, we have enough of our fibers available that we can use a fresh fiber for each liquid test. If we had only a single fiber, after testing in one RI matching liquid, we would remove any remnants of that liquid by dipping the fiber in a solvent or washing it before using it with the next RI matching liquid.

6. Once you have established the RI of the first fiber as closely as possible, repeat the tests for your other known and questioned fibers, recording the values you obtain for each in your lab notebook.

7. Referring to your notes, attempt to identify each of your questioned fibers as consistent with one of the known fibers.

PROCEDURE II-7-5: EXAMINING FIBERS BY POLARIZED LIGHT

In this procedure, we'll determine the RRI of each fiber specimen, as well as its birefringence.

1. Insert one of your wet-mounted fiber specimens into the slide holder.

2. Place a polarizing filter under the microscope condenser, with its polarization plane oriented parallel (N‖) to the longitudinal direction of the mounted fiber specimen.

3. Close the substage condenser diaphragm to provide axial illumination.

4. If the fiber is flat, focus critically on the edges of the fiber. If the fiber is roughly cylindrical, focus critically on the top surface of the fiber.

5. While observing the fiber at 400X, use the fine-focus adjustment to increase the separation between the objective lens and the fiber very slightly (if your microscope is tube-focusing, raise the tube; if it is stage-focusing, lower the stage.)

If the refractive index of the fiber is higher than the RI of the mounting fluid, you will see a bright line (the Becke line) move toward the center of a cylindrical fiber as you increase separation between the objective lens and the fiber. If the RI of the fiber is lower than the RI of the mounting fluid, the bright line will spread as you increase focus distance, and the center of the fiber will become darker. For flat fibers, the Becke line movement occurs at both edges of the fiber, but the movement is the same in either case: a shift toward the medium with the higher RI as the focus distance is increased.

> You use only one polarizing filter in this lab session because the fiber itself effectively acts as the second polarizing filter.

Figures II-7-11 and II-7-12 show the movement of the Becke line toward the center of a cotton fiber (RI = 1.52+) wet-mounted in glycerol (RI = 1.47) as we open focus.

Figure II-7-11: *At focus, the Becke line appears near the edge of the fiber*

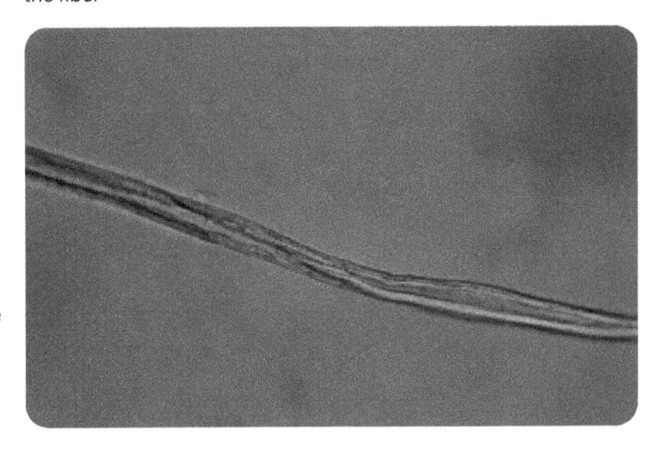

Figure II-7-12: *As focus is opened, the Becke line moves toward the center of the fiber*

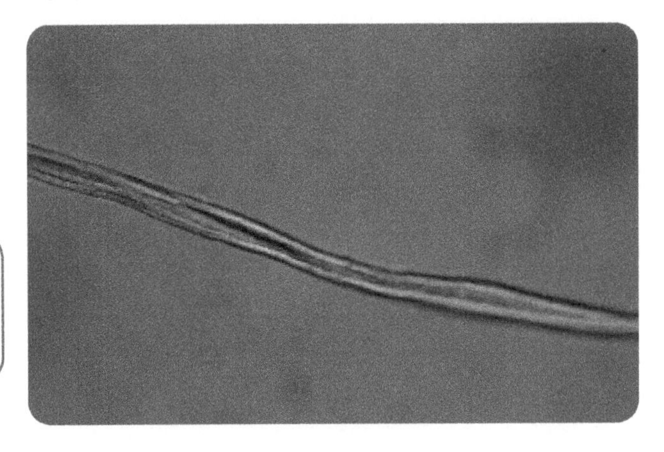

1. As you observe the fiber, slowly rotate the polarizing filter by 90° (until the plane of polarization is perpendicular (N┼) to the length of the fiber) and note the effect on the contrast between the fiber and the mounting fluid. If there is no apparent change, the fiber is isotropic, which means it has only one RI in polarized light. If the fiber appears to darken or lighten relative to the mounting fluid, that fiber is anisotropic (or birefringent), which means it has two or more refractive indices in polarized light.

2. Note the degree of relief visible at each position. High relief means the fiber is distinctly visible with high contrast against the mounting fluid, and indicates that the refractive indices of the fiber and mounting fluid differ significantly. Low relief means that the fiber tends to blend into the mounting fluid, showing only a low-contrast image, and indicates that the refractive indices of the fiber and mounting fluid are similar. (If the refractive indices are identical, a colorless fiber disappears entirely; a colored fiber is visible only as a streak of color.)

3. If the fiber shows birefringence, repeat steps 4 and 5 with the polarizing filter perpendicular (N┼) to the length of the fiber to determine relative RIs for the fiber.

4. Note your observations in your lab notebook. Use the information in Table II-7-3 to determine which fiber material or materials is consistent with the characteristics of the fiber you are testing.

5. Repeat these tests for each your known and questioned fiber specimens.

Figures II-7-13 and II-7-14 show birefringence exhibited by a cotton fiber with the polarizing filter oriented to provide minimum contrast (Figure II-7-13) and maximum contrast (Figure II-7-14).

Figure II-7-13: *A cotton fiber viewed by plane-polarized light showing minimum contrast with the mounting fluid*

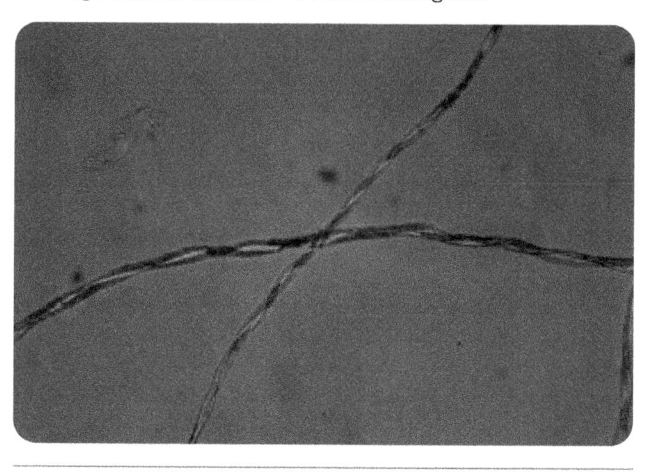

Figure II-7-14: *A cotton fiber viewed by plane-polarized light showing maximum contrast with the mounting fluid*

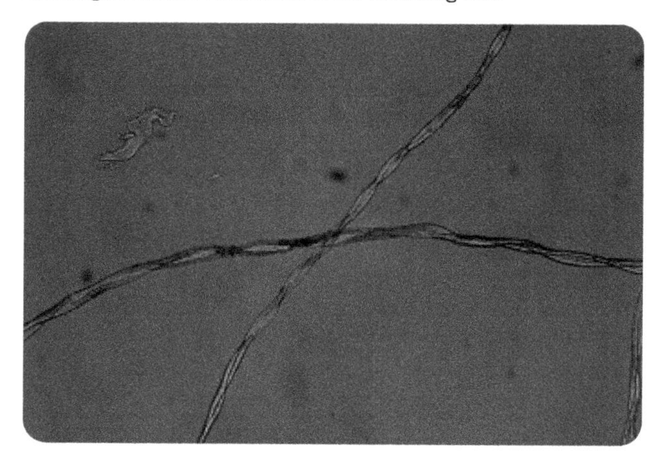

Table II-7-3: *Optical characteristics of fibers*

Fiber	N‖	N⊥	Birefringence
Acetate	1.47–1.48	1.47–1.48	Weak
Acrylic	1.50–1.52	1.50–1.52	Weak
Cotton	1.58–1.60	1.52–1.53	Strong
Glass/mineral	1.55	1.55	None
Modacrylic	1.54	1.53	Weak
Nylon 6	1.57	1.51	Strong
Nylon 66	1.58	1.52	Strong
Nytril	1.48	1.48	None
Polyester	1.63 or 1.71–1.73	1.53–1.54	Very strong
Rayon	1.54–1.56	1.51–1.53	Strong
Saran	1.61	1.61	Weak
Silk	1.59	1.54	Weak
Triacetate	1.47–1.48	1.47–1.48	Weak
Vinal	1.55	1.52	Strong
Vinyon	1.53–1.54	1.53	Weak
Wool	1.55–1.56	1.55	Weak

REVIEW QUESTIONS

Q1: **What one morphological characteristic most easily discriminates a natural fiber from a manufactured fiber?**

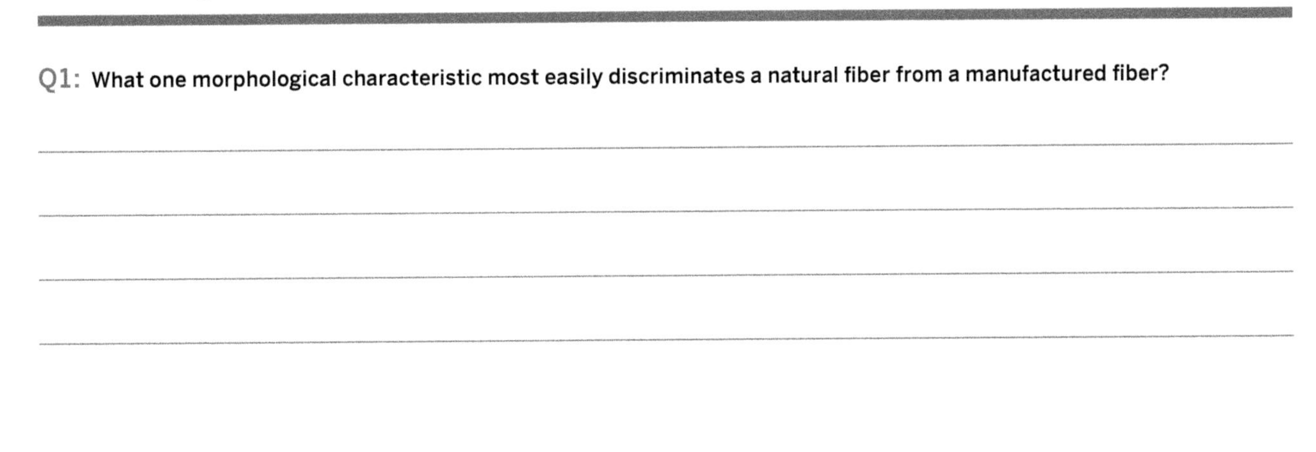

Q2: Would you expect an acetate fiber to exhibit low or high relief when wet-mounted in glycerol (RI = 1.47)? What if it were wet-mounted in clove oil (RI = 1.54)? What degree of relief would you expect for an acrylic fiber wet-mounted in these two fluids? Why?

Q3: What advantages do fiber cross sections have relative to longitudinal sections?

Q4: You are examining a birefringent fiber by polarized light. With the polarizer oriented to provide maximum relief, you slowly rotate the polarizer by 90°. The relief decreases until it reaches a minimum at about 45°. The relief then begins to increase again as you continue rotating the polarizing filter, reaching another maximum at about 90°. What do you conclude about the refractive indices of the mounting fluid relative to the perpendicular and parallel refractive indices of the birefringent fiber?

Glass and Plastic Analysis

Group III

Analysis of glass and plastic specimens is an important part of the workload of any forensics lab. For example, a forensic scientist may be asked to examine plastic taillight fragments found at the scene of a hit-and-run accident to determine the make and model of vehicle from which those fragments originated, or to match glass fragments found embedded in the clothes or shoes of a suspect against broken glass found at a crime scene.

By its nature, glass and plastic evidence is usually class evidence because these materials are ubiquitous. Millions of tons of many common types of glass and plastic are manufactured every year. It is usually impossible, for example, to individualize a fragment of window glass, simply because so many other windows are made from exactly the same glass. In such cases, establishing that a questioned specimen is consistent with a known specimen in every respect may be of little evidentiary value because the class is so large.

Despite this problem, forensic labs routinely analyze the physical, optical, and chemical characteristics of questioned glass and plastic specimens, because by doing so they may be able to rule out a match with known specimens. In some circumstances, establishing that no match exists may be as useful as establishing a positive match.

All of that said, forensic labs are sometimes able to individualize glass or plastic evidence based on gross and fine morphology of the specimens. A piece of glass or plastic can shatter in a nearly infinite number of different ways, and those shatter patterns can allow a questioned specimen to be individualized against known specimens.

If a case is important enough to justify devoting the necessary time and effort, forensic scientists may be able to reassemble the shards like a jigsaw puzzle (albeit, often with pieces missing). If the edges of two shards fit together perfectly, courts accept that as establishing that those two pieces were once part of a whole. Furthermore, using a comparison microscope to match the microscopic shatter patterns on the edges of the two fragments can establish beyond reasonable doubt that the two fragments are parts of a whole. This is why for serious crimes forensic technicians take pains to collect every bit of glass and plastic evidence rather than simply collecting representative specimens.

If such a physical match between shard edges is not possible, other methods are used. Professional forensic labs use various instrumental analysis techniques to identify glass and plastic specimens, but older analysis methods are still widely used. Among those, the two most important are measuring the *density* (also called *specific gravity*) and the *refractive index* (RI) of glass and plastic specimens.

In this group of lab sessions, we'll apply all three of these methods to characterize various known and questioned specimens of glass and plastic.

Determine Densities of Glass and Plastic Specimens

EQUIPMENT AND MATERIALS

You'll need the following items to complete this lab session. (The standard kit for this book, available from *http://www.thehomescientist.com*, includes the items listed in the first group.)

MATERIALS FROM KIT

- Goggles
- Graduated cylinder, 10 mL
- Forceps
- Pipettes
- Test tube(s)
- Test tube rack

MATERIALS YOU PROVIDE

- Gloves
- Balance
- Isopropanol, 99% or acetone
- Specimens: glass and plastic (see text)

WARNING

Glass and plastic fragments have sharp edges and may cut you. Handle them with forceps. Isopropanol is flammable. As a matter of good practice, you should always wear splash goggles, gloves, and protective clothing when working in the lab.

BACKGROUND

The density of a material is its mass per unit volume. The official (SI) units for density are kilograms per cubic meter (kg/m^3) and grams per cubic centimeter (g/cm^3). Because milliliters and cubic centimeters are the same for all practical purposes, density is also sometimes specified in grams per milliliter (g/mL). Density values for most common materials have been established very accurately. For example, at 20 °C, the density of lead is about 11.342 g/cm^3, that of ethanol about 0.78945 g/cm^3, and that of water about 0.99821 g/cm^3.

> Specific gravity is a dimensionless value that specifies the density of one material relative to the density of some other material. For example, the specific gravity of lead relative to ethanol is about 14.367 (11.342 g/cm^3 divided by 0.78945 g/cm^3); the specific gravity of lead relative to water is about 11.362 (11.342 g/cm^3 divided by 0.99821 g/cm^3); and the specific gravity of ethanol relative to lead is about 0.06960 (0.78945 g/cm^3 divided by 11.342 g/cm^3. Unless otherwise stated, specific gravity values are always specified relative to water at 20 °C and standard pressure. For practical purposes, the density of water is often taken to be 1.000 g/CC, so the specific gravity and density of a material are equal.

The mass of a specimen is easily determined by putting it on a balance, but to determine the density of that specimen, we also need to know its volume. If the specimen has a regular shape, such as spherical or cubical, you can determine its volume by measuring its dimensions and doing some simple calculations. But most glass and plastic specimens do not have regular shapes, so we need some other method to determine their volumes.

One way to determine the volume of an irregular specimen is to submerge it in a liquid and determine the volume of liquid that the specimen displaces. You can do that by partially filling a graduated cylinder with water or some other liquid, noting the initial volume, submerging the specimen in the liquid, and noting the final volume. Subtracting the final volume from the initial volume gives the volume of the specimen.

For example, assume you have a small chunk of some unknown metal. You weigh it and find that its mass is 4.05 grams. You fill a 10 mL graduated cylinder with water to the 5 mL mark, and then drop the specimen into the graduated cylinder. With the specimen fully submerged, the graduated cylinder reads 6.5 mL, so the volume of the specimen is (6.5 mL – 5.0 mL) = 1.5 mL. Dividing the 4.05 g mass by the 1.50 mL volume gives you a value of 2.70 g/mL for the density of that specimen. Looking up that density in a table tells you that that specimen is aluminum metal. We can use the same method to determine the densities of known and questioned glass and plastic specimens, assuming those specimens are large enough to cause an accurately measurable change in volume and small enough to fit into the cylinder.

Although water is the most common choice, you can use nearly any liquid to measure displacement. The density of the liquid is immaterial, except that it must be lower than the density of the specimen. Otherwise, the specimen floats in the liquid, making it impossible to determine its volume. This issue may arise when you test plastic specimens, because some plastics are less dense than water. To determine density for such specimens, substitute another liquid, such as 99% isopropanol or pure acetone, that is less dense than water.

For very small specimens, determining density by displacement is difficult or impossible because the mass and (particularly) the volume of the specimen may be too small to measure accurately. Fortunately, there's an alternative, called the *flotation method*, that can provide accurate density values for such specimens.

If a solid object is denser than a surrounding liquid, the object sinks to the bottom. If the object is less dense, it rises to the top. If the object is the same density as the liquid, it remains suspended, neither sinking nor rising. By placing a specimen in liquids of different densities—or by adjusting the density of the liquid in which the specimen is immersed—we can obtain an accurate value for the density of even a tiny specimen.

Historically, forensics labs used solutions with different proportions of a heavy liquid, usually bromoform (2.89 g/mL) and a lighter liquid, usually bromobenzene (1.52 g/mL). By adjusting the proportions of those two liquids, a forensic scientist can produce solutions of known density in the range of 1.52 g/mL to 2.89 g/mL, which encompasses the densities of most common glasses and the denser types of plastic.

In recent years, the use of brominated organic compounds has declined because they are expensive, toxic, require distillation to be separated for reuse, and present a hazardous waste disposal problem. Most professional forensics labs now use special heavy liquids, most of which are solutions of tungsten salts, such as sodium polytungstate and lithium metatungstate. These tungstate salts are not hazardous, allow solution densities up to 3 g/mL or slightly higher, can be reused indefinitely, and can be reconcentrated simply by evaporating excess water. They are, however, too expensive for a home lab. (The least expensive source we found for sodium polytungstate was about $600 per kilo.)

The least dense glasses have densities of about 2.1 g/mL, and typical glasses have densities ranging from about 2.2 g/mL to about 2.8 g/mL. Unfortunately, there is no inexpensive, readily available, nontoxic chemical that can be used to produce solutions in this density range. Fortunately, we can learn just as much by using the flotation method to determine the density of plastic specimens. Polyethylene and polypropylene are readily available, and have typical densities of about 0.89 g/mL to 0.96 g/mL. That means we can use water (1 g/mL) as our "heavy" liquid and concentrated isopropanol (~0.79 g/mL, depending on exact concentration) or acetone (0.791 g/mL) as our miscible light liquid to make solutions that encompass the range of densities of polyethylenes and polypropylenes.

In this lab session, we'll use the displacement method to determine the densities of various known and questioned glass and plastic specimens. We'll determine densities for various polyethylene and polypropylene specimens using the flotation method, by adjusting the liquid density until it matches the density of each specimen and then weighing a known volume of that liquid to determine its density and therefore the density of the specimen.

Before you begin this lab session, obtain specimens of as many types of glass as possible, which will also be used in the next lab session. Some possible sources are: bottles of various types; drinking glasses; lead crystal; windows; Pyrex labware (a broken test tube or stirring rod); car windows and headlights (junkyard); broken or discarded eyeglasses; and various optical glasses from old or broken telescopes, binoculars, or other optical instruments. You can obtain suitable fragment sizes by wrapping the glass object in several layers of paper grocery bags and using a hammer to break up the material into small fragments. Wear heavy gloves and eye protection while doing this.

Plastics are ubiquitous, and many plastic objects are conveniently labeled with the Plastic Identification Code number within a triangle:

01 = polyethylene terephthalate (PET, PETE)

02 = high-density polyethylene (HDPE)

03 = polyvinyl chloride (PVC)

04 = low-density polyethylene (LDPE)

05 = polypropylene (PP)

06 = polystyrene (PS)

07 = other (usually polycarbonate or ABS)

PET is used for soft drink bottles. HDPE is used for kitchen containers, laundry detergent bottles, milk jugs, and so on. PVC is used for plastic pipes, storage bins, furniture, toys, and in many structural applications. LDPE is used for food-storage containers, six-pack rings, reusable bottles, and so on. PP is widely used for storage containers, lids, and so on. PS is used for disposable cutlery, CD and DVD jewel cases, and many other applications where a hard, transparent plastic is desirable. Depending on the form of plastic, you may be able to shatter it with a hammer, or you may need to use a knife, scissors, snips, or a saw to cut pieces of appropriate size.

The isopropanol sold in drugstores is usually 70%, which is too dilute for our purposes. Many drugstores also carry 99% isopropanol, but you may have to ask for it. You may also substitute pure acetone, which is available in any hardware store, paint store, or DIY home center.

PROCEDURE III-1-1: DETERMINE DENSITY BY DISPLACEMENT

1. Place a weighing paper on your balance pan. Weigh about 10 g of your first glass specimen and record its mass to the resolution of your balance in your lab notebook. The specimen may be one piece—as long as it will fit into the graduated cylinder—or several smaller pieces.

2. Use a pipette to transfer about 4 mL of water to the 10 mL graduated cylinder. Record the initial volume as accurately as possible in your lab notebook.

3. Carefully add the specimen to the graduated cylinder, making sure that you transfer all of the specimen to the cylinder and that the specimen is fully submerged in the liquid. Record the final volume as accurately as possible in your lab notebook. (If the final volume exceeds 10 mL, repeat the procedure using a smaller specimen.)

For visibility, we used a glass graduated cylinder for Figure III-1-1. The polypropylene graduated cylinder supplied with the kit is actually better for density determinations because it has no meniscus (the curved shape of the top of the liquid, caused by the glass cylinder attracting water and other liquids, drawing them slightly up the inside surface of the cylinder). The lack of a meniscus with the polypropylene cylinder means that the top of the liquid forms a flat line, making it easier to read volumes accurately.

Figure III-1-1: *Determining the amount of liquid displaced by a small specimen*

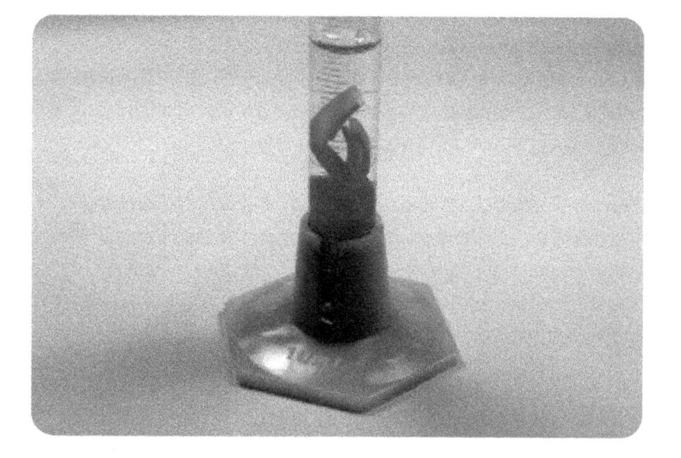

4. Subtract the initial volume from the final volume to give the displaced volume, and enter that value in your lab notebook.

5. Divide the mass of the specimen by the displaced volume to determine the density of the specimen, and enter that value in your lab notebook.

6. Repeat steps 1 through 5 for each of your known and questioned glass specimens.

7. Repeat steps 1 through 5 for each of your known and questioned plastic specimens, but use alcohol instead of water and use only about 4 g of each plastic specimen rather than 10 g. (Typical glass has a density of about 2.5 g/mL, while typical plastics have densities in the 0.8 g/mL to 1.6 g/mL range.)

8. Compare the density of your questioned specimen or specimens against the density values you obtained for each of your known specimens to determine if you can eliminate any or all of the known specimens and if you can make a tentative match for your questioned specimen(s) with one or more of the knowns based on their densities.

9. Make a tentative identification of the type of plastic or glass for each of your specimens using the data in Tables III-1-1 and III-1-2.

Table III-1-1: *Densities of some common plastics (g/cm^3)*

Plastic type	Density
Acrylic	1.17–1.20
Acrylonitrile/butadiene/sytrene (ABS)	1.01–1.08
Alkyds	1.30–1.40
Epoxies	1.11–1.48
Melamines	1.47–2.00
Phenolic resins	1.24–1.32
Polyacetal	1.41–1.42
Polyamide (nylon)	1.07–1.08
Polybutylene terephthalate (PBT)	1.30–1.38
Polycarbonate	1.20–1.21
Polyethylene, high-density (HDPE)	0.94–0.96
Polyethylene, low-density (LDPE)	0.89–0.94
Polyethylene terephthalate (PET, PETE)	1.29–1.40
Polyesters, unsaturated	1.01–1.46
Polypropylene (PP)	0.89–0.91
Polystyrene (PS)	1.04–1.08
Polyurethane	1.17–1.28
Polyvinyl chloride (PVC), flexible	1.10–1.35
Polyvinyl chloride (PVC), rigid	1.30–1.58
Styrene/acrylonitrile (SAN)	1.02–1.08

Table III-1-2: *Densities of some common glasses (g/cm^3)*

Glass type	Density
Automobile headlight glass	2.20–2.29
Automobile window glass	2.53–2.75
Fused quartz/silica (Corning 7980, Dynasil 1100, GE 124, Vycor 7913)	2.18–2.21
Labware (Pyrex, Kimax, Duran, and other borosilicate glasses)	2.23–2.72
Lead glass (lead-alkali-silicate)	2.65–6.00
Soda-lime-silicate (flat glass, bottle and window glass)	2.40–2.63
Zerodur (Schott trademark for a glass-ceramic)	2.53

PYREX VERSUS PYREX

Pyrex is a trademark rather than a type of glass. Originally, Pyrex items were all made from borosilicate glass. Nowadays, although Pyrex labware is still made of borosilicate glass, Pyrex kitchenware is made of ordinary soda-lime glass.

PROCEDURE III-1-2: DETERMINE DENSITY BY FLOTATION

1. Ask a friend or lab partner to make a questioned specimen for you, choosing among LDPE, HDPE, and PP.
2. Label four test tubes LDPE, HDPE, PP, and Q (for questioned), fill each tube about half full of 99% isopropanol or 100% acetone, and place them in the rack.
3. Drop each specimen into the corresponding test tube. It should sink to the bottom.
4. Using a pipette, gradually add water to the LDPE tube, with swirling, until the plastic specimen begins to show some buoyancy. You can ensure that the liquids are thoroughly mixed by covering the mouth of the test tube with your gloved thumb and inverting it several times.

5. As you approach neutral buoyancy, the plastic specimen sinks, but very slowly. Continue adding water dropwise until you just achieve neutral buoyancy, with the plastic specimen suspended in the tube and neither rising nor sinking, as shown in Figure III-1-2. If you go a bit too far and the specimen begins rising, add isopropanol dropwise until the specimen is just suspended.

Figure III-1-2: *When the density of the specimen and fluid are the same, the specimen neither rises nor sinks*

6. Place the 10 mL graduated cylinder on the balance pan and tare the balance (adjust it to read 0.00 grams with the cylinder in place).

7. Use a pipette to transfer liquid from the LDPE test tube to the graduated cylinder to fill the graduated cylinder as far as possible without exceeding the 10 mL line. In your lab notebook, record the mass of the liquid to the resolution of your balance and the volume of the liquid as accurately as possible. (Interpolate between the graduation lines on the cylinder.)

8. Calculate the density of the liquid in which the LDPE specimen had neutral buoyancy. For example, if the mass of the liquid is 7.64 g and the volume of that liquid is 8.30 mL, calculate the density of the liquid as (7.64 g / 8.30 mL) = 0.92 g/mL.

9. Repeat steps 4 through 8 for your HDPE, PP, and questioned specimens.

The procedure you just completed gives you accurate values for the densities of your plastic specimens. Sometimes, that information is needed. Other times, all you really need to determine is if the density of a questioned specimen matches the density of a known specimen. In that case, you can simply adjust the density of the liquid until one of the specimens has neutral buoyancy and then drop the other specimen into the liquid. If the second specimen also has neutral buoyancy, the densities of the two specimens are the same. If the second specimen is less dense, it floats. If it's denser, it sinks.

REVIEW QUESTIONS

Q1: **Why is the density of the liquid used to determine the density of a specimen by displacement immaterial?**

Q2: **Why would you use the flotation method rather than the displacement method to determine the density of a glass or plastic fragment?**

Q3: **Of what forensic significance is the density of a glass or plastic specimen?**

Compare Refractive Indices of Glass and Plastic Specimens

Lab III-2

EQUIPMENT AND MATERIALS

You'll need the following items to complete this lab session. (The standard kit for this book, available from *http://www.thehomescientist.com*, includes the items listed in the first group.)

MATERIALS FROM KIT

- Goggles
- Beaker, 100 mL
- Forceps
- Magnifier
- Pipettes
- Slides, deep-cavity

MATERIALS YOU PROVIDE

- Gloves
- Isopropanol or acetone
- Microscope
- RI matching fluids (from Lab II-7)
- Specimens: glass and plastic (see text)

WARNING

Glass and plastic fragments have sharp edges and may cut you. Handle them with forceps. Isopropanol is flammable. As a matter of good practice, you should always wear splash goggles, gloves, and protective clothing when working in the lab.

BACKGROUND

In Lab II-7, we used refractive index matching fluids to determine the refractive indices of various fiber specimens. In this lab session, we'll use those same fluids to determine the refractive indices of various glass and plastic specimens, and attempt to identify a questioned specimen by comparing its refractive index with the refractive indices of known specimens.

Most common glasses and plastics have RI values in the 1.47 to 1.6 range, although optical and other specialty glasses can range from less than 1.44 to more than 2. Table III-2-1 lists refractive indices for various common glasses, including a handful of the most common of the hundreds of types of optical glasses available.

Table III-2-1: *Typical refractive indices at the sodium D line and 20°C of some common glasses and plastics*

Material	n_D^{20} (typical)
Quartz	1.46
Pyrex	1.47
Automobile headlight glass	1.47–1.49
Crown glass	1.48–1.54
Poly(methylmethacrylate) (PMMA; acrylic glass)	1.49
Polypropylene (PP)	1.49
Soda-lime (windows, bottles)	1.51–1.52
Low-density polyethylene (LDPE)	1.51–1.53
High-density polyethylene (HDPE)	1.52–1.53
Borosilicate crown optical glass	1.52
Eyeglass lenses	1.52–1.53
Barium crown optical glass	1.57
Polyethylene terephthalate (PET)	1.57–1.58
Polycarbonate (Lexan, etc.)	1.58–1.59
Polystyrene (PS)	1.59
Flint optical glass	1.62

For forensic purposes, it's usually not necessary to determine the actual refractive indices of the known and questioned specimens, but merely to determine whether the refractive indices of the known and questioned specimens are the same. For example, a forensic examiner may be asked to determine whether tiny fragments of glass found embedded in the clothing of a hit-and-run victim are consistent with the glass of a shattered headlight in a suspect's automobile. By testing the known specimen directly against the questioned specimen, the examiner can determine whether or not those two specimens are consistent in terms of refractive index.

Although forensic glass examiners sometimes use refractometers (optical instruments that measure the RI of a specimen directly), most comparisons are done using refractive index matching liquids. Although using RI liquids may be more time-consuming, the advantage is that RI liquids can be used with even the tiniest specimens. As long as the specimen is large enough to be distinctly visible with a microscope, it's large enough to determine its RI and compare it with those of other specimens.

In this lab session, we'll determine refractive indices for small fragments of various known types of glasses and plastics and attempt to identify an unknown specimen by comparing its refractive index to those of our known specimens. You need only tiny known and "questioned" specimens of various glasses and plastics for this lab session. You can obtain those by pulverizing a larger specimen with a hammer (wear eye protection and wrap the larger specimen in several thicknesses of newspaper) or for soft plastics simply by cutting a small specimen. Have a friend or lab partner choose small fragments of at least one of your known glass and plastic specimens to serve as your questioned specimen(s).

One assumption may trip up even experienced forensic glass examiners. Modern glass is extremely uniform because it is made in huge, homogeneous batches. Old glass, particularly from the early 19th century and before, was made in much smaller batches that often lacked homogeneity, sometimes dramatically. For example, old leaded glass may show huge variations in refractive index from samples taken less than a centimeter apart from a poured or cast specimen. Back then, they simply didn't stir the mixture very well, and the lead compounds that cause such a profound difference in refractive index were not distributed uniformly throughout the melt.

When examining fragments from such a specimen, a forensic glass examiner may conclude that the specimens are inconsistent with each other even though they actually were a part of the same whole. And that's because the specimens *are* inconsistent with each other.

PROCEDURE III-2-1: COMPARE RI OF QUESTIONED AND KNOWN SPECIMENS

With fluids of closely spaced known refractive indices, it's easy enough to determine the RI of even the smallest glass or plastic fragment by comparing it with the various fluids. Of course, the actual comparisons can be tedious because each specimen may need to be matched against each of several fluids. Once you have some experience in matching, you should be able to find the closest match in two or three attempts. Once you have found a refractive index matching fluid that is a close match to your questioned specimen, you can use that fluid to test known specimens until you locate one whose refractive index is closely similar to that of the questioned specimen.

1. Label deep-cavity slides for each of your refractive index matching liquids and transfer several drops of each liquid to the well of the corresponding slide.
2. Place a small fragment of the first questioned specimen in a well that contains one of the RI fluids from the middle of the range, making sure the specimen is completely immersed in the fluid.
3. Place the slide on the stage and observe it at 40X or 100X.
4. Focus the specimen critically, and use the Becke line method described in Lab II-7 to determine whether the RI of your specimen is higher or lower than the fluid. By observing the degree of relief of the specimen, you can estimate whether the RIs of the specimen and fluid differ greatly, moderately, or minimally. If the specimen almost disappears in the fluid, you have a very close match. If the specimen stands out from the fluid, the RIs of the specimen and fluid are significantly different. You'll learn to judge the degree of difference after you make several observations.
5. Based on your observations in the preceding step, choose a second RI fluid that you believe should be a closer match for the specimen. Place a fragment of the specimen in that second fluid, and again use the Becke line method and the appearance of the specimen against the fluid background to judge how close the match is and in what direction the RI of the specimen differs from the RI of the fluid.

Either use a fresh specimen for each fluid or wash your specimen in isopropanol, acetone, or a similar solvent before transferring it from one fluid to the next.

6. If necessary, repeat step 5 with different fluids until you find the fluid that most closely matches the RI of the questioned specimen. Note the RI of that fluid and make your best estimate of the actual RI of the questioned specimen in your lab notebook.

7. Using the fluid that most closely matches the RI of the questioned specimen, test each of your known specimens until you find one, if any, that closely matches the RI of the questioned specimen.

8. When you have identified a known specimen that is consistent with the questioned specimen, place both of those specimens in the best-match RI fluid and compare the two specimens directly with each other using the Becke line method.

9. Repeat steps 2 through 8 for each of your other questioned specimens.

REVIEW QUESTION

Q1: **You are presented with one questioned glass specimen and a dozen known specimens. Propose a method for doing a quick preliminary screening to determine if the RI of the questioned specimen matches one or more of the known specimens.**

Observe Shatter Patterns

Lab III-3

EQUIPMENT AND MATERIALS

You'll need the following items to complete this lab session. (The standard kit for this book, available from *http://www.thehomescientist.com*, includes the items listed in the first group.)

MATERIALS FROM KIT

- Goggles
- Forceps
- Modeling clay
- Slides, flat

MATERIALS YOU PROVIDE

- Gloves
- Desk lamp or other top illuminator
- Hammer
- Microscope
- Newspapers or brown paper bags
- Tape, masking or duct

WARNING

The only real danger in this lab session is handling broken glass. Particularly when you are breaking glass, wear goggles to protect your eyes, and heavy gloves to prevent cuts.

BACKGROUND

Matching the density and the optical and chemical characteristics of glass or plastic specimens provides, at best, a class match from within what is often a huge class. For example, even if a questioned specimen of a broken beer bottle or windowpane matches known specimens exactly in density and in every optical and chemical respect, that questioned specimen may potentially also match millions of other beer bottles or windowpanes. Obviously, such positive matches are of very limited evidentiary value (although negative matches may be very useful indeed for elimination purposes).

But that doesn't mean it's impossible to individualize glass and plastic fragments. Like snowflakes or fingerprints, every glass or plastic fragment is unique in at least some small respect. Furthermore, two glass or plastic fragments that adjoined before being broken apart may be fitted back together like a three-dimensional jigsaw puzzle. This physical matching of the broken edges of shards is the only practical way to individualize questioned against known specimens of common glasses and plastics.

This physical matching is done first on a macro level, by fitting together relatively large fragments from the known and questioned specimens. But the nature of broken glass means that parts of that jigsaw puzzle are usually missing, leading to the possibility of the match being questioned by defense attorneys. For that reason, a successful macro match is followed by a micro match, in which the adjoining edges at the break between the two specimens are compared under a microscope. If those edges match in every respect, the match is conclusive.

> Because it is tedious and time-consuming (and therefore expensive), physical matching of glass or plastic shards is rarely done except for serious crimes. The most common of these crimes is hit-and-run with death or serious injury resulting. In such cases, the crime scene technicians make every effort to collect not just samples of broken glass and plastic, but every bit they can find, in the hope that one or more of these specimens can later be matched against a broken headlight or other glass or plastic components of a suspect vehicle.

In this lab session, we'll produce some glass shards and observe the physical characteristics that allow an individualized match to be established.

PROCEDURE III-3-1: PRODUCE GLASS SHARDS

1. If you have not already done so, put on your goggles and heavy gloves to protect your eyes and hands from broken glass.
2. Cover a hard, flat surface with several layers of newspaper, brown kraft paper, or a sheet of stiff card stock.
3. Place a microscope slide on the surface, and apply masking tape or duct tape to cover the entire top surface of the slide. Do not tape the slide to the work surface.
4. Rub the tape firmly with your finger to make sure the entire surface of the slide tightly adheres to the tape.
5. Strike the taped surface of the slide lightly with the hammer. Your goal is to break the slide into shards, not to pummel it into powder.

PROCEDURE III-3-2: OBSERVE AND COMPARE GLASS SHARDS

1. To begin, invert your specimen so that the glass side faces up and carefully place it on your microscope stage.

2. Illuminate the glass shards from above, and observe the matching broken edges at 40X and 100X. Note your observations in your lab notebook. If you have time and the necessary equipment, make sketches or shoot images of some of the significant details visible.

3. Place a small blob of modeling clay, about the size of a pencil eraser, in the center of a microscope slide.

4. Using your forceps, carefully remove two adjacent shards from the tape backing and embed them as close together as possible in the modeling clay, with the matching edges up and touching or nearly so. Choose two shards that are narrow enough that when they are mounted vertically will still allow your microscope to focus on the edges, and try to select an area where the fracture line is as close as possible to a straight line. When you embed the shards, place them with the same surface from each face to face. Try to keep the edges of both as close as possible to the same level so that both can be in focus at the same time.

5. Observe the fracture patterns on the edges of the two shards at 40X, 100X, and 400X, and note your observations in your lab notebook. If possible, make sketches or shoot images of some of the significant details visible.

Figure III-3-1 shows the matching edges of two glass shards longitudinally at 100X magnification.

Figure III-3-1: *Matching edges of two glass shards at 100X magnification*

REVIEW QUESTIONS

Q1: **Did you find longitudinal or cross-sectional comparisons more useful in matching your glass fragments? Why?**

Revealing Latent Fingerprints

Group IV

Even someone who knows nothing else about forensics knows about fingerprints. The individuality of fingerprints had been generally accepted as established by forensic scientists and courts by the early 20th century, and the billions of fingerprint specimens taken since then have confirmed fingerprints as unique individual characteristics. Figure IV-0-1 shows a full fingerprint that we made by pressing one of our fingers to a stamp pad and then rolling it against a sheet of paper.

Figure IV-0-1: *A typical full fingerprint taken under controlled conditions*

Unfortunately, prints found at a crime scene are usually partial, broken, smeared, or otherwise inferior to the perfect specimens found on fingerprint cards, so it's often impossible to do a full comparison. Fingerprint examiners use *points of comparison*, also called *points of identification*, to compare unknown fingerprints against known specimens. A point of comparison is a particular individual feature of a particular fingerprint, such as where a ridge ends or bifurcates or the shape and number of ridges present in a whorl. If sufficient points of comparison exist between two fingerprints, it's reasonable to assume that those two prints were produced by the same finger, even if parts of the questioned fingerprint are missing, smeared, or otherwise obscured. Figure IV-0-2 shows a typical questioned partial fingerprint found on a questioned document. Although some ridge detail is present, it's unlikely that this print could be identified against a known print.

Figure IV-0-2: *A typical questioned fingerprint*

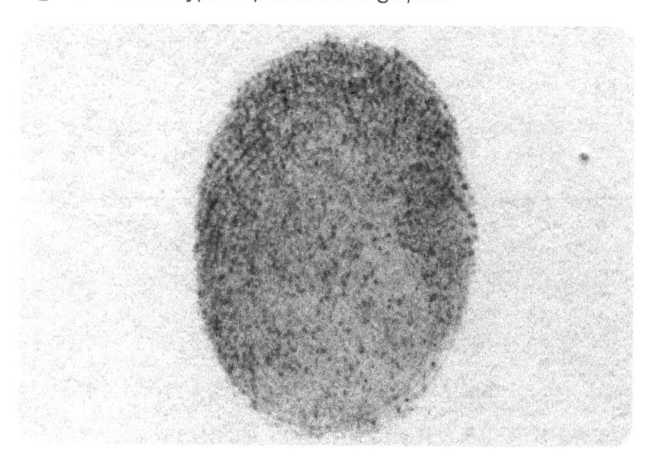

> **DENNIS HILLIARD COMMENTS**
>
> There are three levels of comparison:
>
> - **Level one** The overall pattern: loop, arch, whorl
> - **Level two** The dots, islands, bifurcations, ridge endings, etc.
> - **Level three** Poroscopy, the presence and spatial relationship of the pores on the print ridges
>
> With poor to moderate quality prints, no pores, smudging, or cross hatching, there needs to be more points of identification. With prints of excellent clarity and good poroscopy, an identification can be made with fewer points of identification.
>
> The FBI has become more conscious of verification of print identification after the Brandon Mayfield mishap. See: *http://en.wikipedia.org/wiki/Brandon_Mayfield.*

There are two types of fingerprints:

Patent fingerprints

Patent fingerprints are visible to the naked eye under ordinary light. *Visible fingerprints* are patent fingerprints made by fingers touching a surface after they have been in contact with ink, paint, grease, soot, blood, or some similar substance. *Plastic fingerprints* are patent fingerprints left on an impressionable material such as wet paint, modeling clay, tar, putty, wax, soap, and similar materials. Patent fingerprints of either type are ordinarily readily visible to crime scene investigators and may sometimes be photographed or lifted directly. In some situations, patent fingerprints may be treated to increase their visibility or contrast against the background surface.

Latent fingerprints

Latent fingerprints are invisible to the naked eye under ordinary light, but can be made visible by *dusting, chemical development*, or an *alternate light source*.

> In forensics, the term *alternate light source* (or ALS) is used generically to describe any bright light source that emits light at a single wavelength or a narrow band of wavelengths. An ALS may emit light at any wavelength from far ultraviolet through the visible spectrum and into the far infrared. For example, a standard "black light" fluorescent tube or UV LED flashlight is considered an ALS, as is a sodium- or mercury-vapor lamp. Beginning in the 1980s, lasers became widely used forensically as ALSs. Being expensive, bulky, and limited to a single wavelength made lasers less than ideal as ALSs, so in the 1990s lasers were gradually supplanted by portable ALSs that could be configured with filters or slits to emit narrow-band light over a wide variety of selectable wavelength ranges. Nowadays, lasers are seldom used in forensics labs or by crime-scene technicians.

After any processing needed to reveal or enhance the print, fingerprints of either type are preserved by photographing them or by *lifting* them, either by carefully applying transparent lift tape to the surface that contains the fingerprint, peeling the tape from the surface, and transferring it to a card, or by *electrostatic lifting*, which uses an electrostatically charged sheet of clear plastic to attract powder applied to the fingerprints. It's at this point that the forensic scientist's job ends and the fingerprint examiner's job begins.

DENNIS HILLIARD COMMENTS

Lifting is generally done to allow a print to be photographed. If the object is mobile, the print is not lifted but preserved in place with lifting tape. If the surface is irregular, a lift is made for photographic purposes. Although I have never seen a fingerprint lifted by "electrostatic lifting," it is often used to lift footprints in dust from flooring. In forensic laboratories or police departments that have officers trained in fingerprint examinations, it is my experience that the prints are processed and then examined by the same analyst. In Rhode Island and in many states throughout the northeastern US, the evidence suspected of having latent prints is collected by law enforcement officers. Our examiners train these officers to partially process certain types of evidence, since time is often of the essence.

Various methods are used to reveal latent prints. Some methods are non-destructive, which means if they are tried and fail to reveal prints, other methods may be used subsequently. Other methods—notably silver nitrate development and physical developer—are destructive, either in the sense that using them precludes using alternative methods to raise the prints or that using them may preclude testing the object for other types of forensic evidence, such as blood or DNA. The particular method or methods used, and the order in which they are applied, also depends upon the nature (porous, semiporous, or nonporous) and condition (e.g., wet, dry, dirty, sticky, etc.) of the surface that contains the prints, as well as the residue that constitutes the prints, such as perspiration, blood, oil, or dust.

Visual examination is always the first step in revealing latent fingerprints. Some latent prints are patent under strong, oblique lighting. Moving small objects to different angles under a fixed light source may reveal numerous prints, as may moving the light source itself when examining larger or fixed objects. Any latent prints that are revealed under oblique lighting are photographed before any subsequent treatment is attempted. Some prints revealed by visual examination may be undetectable by any other method. Done properly, visual examination is completely nondestructive. Done improperly, the handling required for visual examination may smudge or destroy prints that are invisible visually but potentially visible using other methods.

After visual examination is complete, the usual next step is to examine the specimen by using *inherent fluorescence*. Various components of the fingerprint residue including perspiration, fats and other organic components, and foreign materials present on the fingertips when the impression was made may fluoresce under laser, ultraviolet, or other alternate light sources. In a darkened room, the questioned surface is illuminated with the alternate light source and viewed through a filter of complementary color. For example, long-wave ultraviolet (black light) tubes emit some visible light in the deep violet part of the spectrum. Viewing a surface so illuminated through a deep yellow or orange filter blocks essentially all of the reflected incident violet light, making any inherent fluorescence emitted by the fingerprint residues in the yellow through red parts of the spectrum more clearly visible. The inherent fluorescence method is usable on any surface, including surfaces that cannot be treated with powders or chemical methods, and may reveal latent prints that are not revealed by any other method. Like visual examination, examination by inherent fluorescence is nondestructive.

After visual examination and inherent fluorescence examination are complete, other methods may be used to reveal additional latent fingerprints. Fingerprint powders, iodine fuming, and silver nitrate are considered the "classic" methods because they have been used since the 19th century. Despite their age and the availability of newer methods, all three of these methods, with some minor improvements, remain in use today.

Fingerprint powders

Fingerprint powders are used primarily for *dusting* nonporous surfaces such as glass and polished metal, most commonly to reveal latent fingerprints on immovable objects at crime scenes. Powders are often used in conjunction with super glue fuming to enable a lift to be made.

A very fine powder is applied to the area that contains the latent print. The powder adheres to the residues that make up the fingerprint. Excess powder is removed by gentle brushing or using puffs of air from a syringe. After the excess powder is removed, the fingerprint is revealed and can be photographed or lifted. Fingerprint powders are available in shades from white through black, which allows the fingerprint technician to choose a powder that contrasts with the background surface. Fluorescent fingerprint powders are useful for raising prints on printed or patterned surfaces, which might otherwise make it difficult to see the pattern of the print itself.

Magnetic fingerprint powders are used with magnetic brushes, which allow excess powder to be removed without actually touching the print. Magnetic powders are often used to raise latent fingerprints on paper surfaces, an exception to the general rule about powders being used only on non-porous surfaces.

Iodine fuming

Iodine fuming is used to reveal prints on porous and semiporous surfaces such as paper, cardboard, and unfinished wood. The object to be treated is placed in an enclosed chamber that contains a few crystals of iodine. Gently heating the iodine crystals causes them to sublime (go from solid phase to gas phase without passing through the liquid phase). The violet iodine vapor adheres selectively to fingerprint residues, turning them orange. These orange stains are fugitive, so they must be photographed immediately. After a period ranging from a few hours to a few days, the iodine stains disappear, leaving the specimen in its original state. The developed prints can be made semi-permanent by treating them with a starch solution, which turns the orange stains blue-black. These stains persist for weeks to months, depending on storage conditions.

Iodine spray reagent (ISR)

Iodine spray reagent (ISR) is a liquid analog to iodine fuming. Like iodine fuming, ISR is used to reveal prints on porous and semiporous surfaces such as paper, cardboard, and unfinished wood, but ISR can be used on specimens for which fuming is impractical. ISR is made up as two stock solutions that are combined to make the working solution. Solution A (iodine) is a 0.1% w/v solution of iodine crystals in cyclohexane. Solution B (fixer) is a 12.5% w/v solution of alpha-naphthoflavone in methylene chloride. The working solution is made up by combining A:B in a 100:2 ratio, mixing thoroughly, and filtering the working solution through a facial tissue or filter paper. The working solution is sprayed onto the questioned surface, using the finest mist possible. Latent prints develop immediately and should be photographed as soon as possible.

We didn't have any alpha-naphthoflavone on hand (or any cyclohexane, for that matter), so we decided to see what we could accomplish with what we did have available. Iodine has such a high affinity for the fats present in fingerprints that we thought almost any iodine solution should work, at least after a fashion. As it turned out, we were right.

We transferred a gram or so of iodine to a small spray bottle and added a few mL of lighter fluid, which formed a beautiful violet solution. Not all of the iodine dissolved, and we'd used the last of our lighter fluid, so we topped off the bottle with 70% ethanol. Iodine in ethanolic solution is brown, so we weren't surprised to see the solution turn a deep purple-brown color. One of us then pressed his fingers to a sheet of copy paper. We sprayed that area of the paper (in the sink; spraying iodine is very messy) and used a hair dryer to evaporate the solvent. Figure IV-0-3 shows the results. Not ideal, certainly, but much better than what we expected.

Figure IV-0-3: *Fingerprints revealed by spraying with an iodine solution*

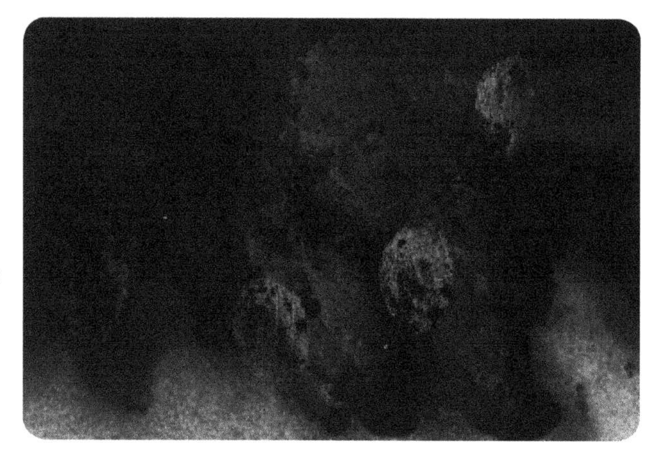

Silver nitrate

Silver nitrate is used to reveal prints on paper and similar surfaces. The surface is treated with a dilute solution of silver nitrate by spraying or immersion. The soluble silver nitrate reacts with the sodium chloride (salt) present in sweat to produce insoluble silver chloride. The surface may or may not be rinsed with water after treatment to remove excess silver nitrate. In either case, the treated surface is exposed to sunlight or an ultraviolet light source, which reduces the silver chloride to metallic silver, revealing the prints as gray-black stains. Careful observation is required to make sure the prints are not overdeveloped, particularly if the surface was not rinsed after treatment. In extreme cases of overdevelopment, the entire surface may turn black. Silver nitrate development is destructive, and so is used only after iodine fuming and other development methods. Three variants of silver nitrate solution are used: a 1% w/v aqueous solution, a 3% w/v aqueous solution, and a 3% w/v ethanolic solution. The alcohol solution is used on surfaces such as wax paper, coated cardboard, and polystyrene foam that repel water and so cause the aqueous solutions to bead.

Silver nitrate is used last, if it is used at all, because using it precludes subsequently using any other development method. Silver nitrate may succeed where other development methods fail because silver nitrate reacts with the nonvolatile sodium chloride present in fingerprint residues. Very old fingerprints may have lost all of their volatile residues, but the sodium chloride residue remains. Silver nitrate has been used successfully to develop latent prints that are years, decades, even centuries old.

Ninhydrin

Ninhydrin was introduced in 1954 as the first of the modern fingerprint development methods. In 1910, the English organic chemist Siegfried Ruhemann synthesized ninhydrin (triketohydrindene hydrate) and reported that it reacts with amino acids to form a violet dye that was subsequently named Ruhemann's Purple (RP). Forensic scientists must have been asleep at the switch, because it wasn't until 44 years later that Oden and von Hoffsten reported in the March 6, 1954 issue of *Nature* that ninhydrin could be used to develop latent fingerprints. Although amino acids are present in only tiny amounts in fingerprint residues, RP is so intensely colored that ninhydrin development produces stark visible images of latent prints.

Like iodine fuming and silver nitrate development, ninhydrin development is most useful for prints on porous surfaces. The questioned surface is simply sprayed with or dipped in a dilute solution of ninhydrin. After a period ranging from a few minutes to several hours, the prints self-develop as purple stains. In some cases, allowing development to continue for 24 to 48 hours reveals additional latent prints. Processing can be sped up by heating and humidifying the treated surface with a steam iron. After development with ninhydrin, prints may be sprayed with a 5% w/v solution of zinc chloride in a 25:1 mixture of MTBE (methyl tert-butyl ether) and anhydrous ethanol. This reagent causes a color shift from purple to yellow-orange and makes the developed prints fluoresce under an ALS.

Two variants of ninhydrin solution are used, depending on the surface to be treated. The standard formulation is a 0.5% w/v solution of ninhydrin in a 3:4:93 mixture of methanol:isopropanol:petroleum ether. The alternate formulation is a 0.6% w/v solution of ninhydrin in acetone.

DFO

DFO, also known by its chemical name of *1,8-diazafluoren-9-one*, operates by the same mechanism as ninhydrin, reacting with amino acids in fingerprint residues to form visible stains. DFO stains are much fainter than those produced by ninhydrin, but DFO stains fluoresce directly, without after-treatment. DFO was popularized by UK police forces, and is still more widely used in British Commonwealth countries than elsewhere. DFO is somewhat controversial. Many experts maintain that DFO is more sensitive than ninhydrin and provides superior detail. Other experts have questioned the reliability of DFO, and prefer to use ninhydrin alone.

If DFO is used, it must be used before ninhydrin, PD, or silver nitrate is applied. DFO reagent is a 0.05% solution of DFO crystals in a solution made up of methanol:ethyl acetate:acetic acid:petroleum ether in a 20:20:4:164 ratio. DFO is applied by spraying or immersion, followed by drying, retreating the surface, drying again, heating the treated surface to 50°C to 100°C for 10 to 20 minutes, and finally by viewing under an alternate light source at 495 nm to 550 nm. Raised prints are photographed through an orange filter. Because DFO reagent is expensive and the required procedure is complex and time-consuming, DFO treatment is used less often than it might otherwise be.

Sudan black

Sudan black is a dye that reacts with the sebaceous perspiration component of fingerprints to form a blue-black stain. Sudan black is used primarily for wet surfaces, including those contaminated with beverages, oil, grease, or foods, and is also useful for post-processing of cyanoacrylate-fumed prints, particularly those on the interior of latex or rubber gloves. The Sudan black reagent is simply a 1% w/v solution of Sudan black in 60% to 70% ethanol, although it is usually made up by dissolving the solid dye in 95% ethanol and then adding distilled water to reduce the ethanol concentration to 60% to 70%. In use, the questioned surface is immersed in the Sudan black solution for about two minutes and then rinsed gently with water. Raised prints are visible as blue-black stains.

Blood reagents

Blood reagents are used to develop latent prints and enhance visible prints that include blood in the fingerprint residues. These agents are also commonly used to reveal latent bloodstained footprints, hand marks, and so on.

Amido black, the oldest of these reagents, is a dye that stains proteins in blood residues blue-black. Several variants of amido black reagent are used. The most common is a 0.2% w/v solution of amido black in a 9:1 mixture of methanol:acetic acid. The questioned surface is sprayed with or immersed in this solution and allowed to soak for 30 seconds to 1 minute, after which it is rinsed with a 9:1 methanol:acetic acid solution. The dye/rinse procedure can be repeated to increase contrast. After a final rinse with water, the specimen is dried and photographed. Surfaces that are likely to be damaged by methanol can be treated with an alternate formulation that contains 0.3% amido black w/v, 0.3% sodium carbonate w/v, and 2% 5-sulfosalicylic acid w/v in an 89:5:5:1.25 mixture of water:acetic acid:formic acid:Kodak Photo-Flo 600. Questioned surfaces are sprayed with or immersed in this solution for 3 to 5 minutes and then rinsed with water. Again, the treatment can be repeated to increase contrast. An alternate water-based amido black reagent can be made up that is 0.2% w/v with respect to amido black and 1.9% w/v with respect to citric acid in a 998:2 mixture of water:Kodak Photo-Flo 600. The questioned surface is sprayed with or immersed in this solution and allowed to soak for 30 seconds to 1 minute, after which it is rinsed with water. Repeated treatments may be used to increase contrast.

Leucocrystal violet (LCV) is an alternative to amido black, and provides similar results. LCV reagent is a solution in 3% hydrogen peroxide that is 0.2% w/v with respect to LCV, 2% w/v with respect to 5-sulfosalicylic acid, and 0.74% w/v with respect to sodium acetate. When this solution is sprayed on a questioned surface, the prints develop in about 30 seconds, after which the surface is blotted dry and photographed.

Coomassie brilliant blue is another alternative to amido black that provides similar results. Coomassie brilliant blue reagent is a 0.1% solution of Coomassie brilliant blue R dye in a 2:9:9 mixture of acetic acid:methanol:water. When this solution is sprayed on a questioned surface, the prints develop in 30 to 90 seconds, after which the surface is rinsed with a 2:9:9 solution of acetic acid:methanol:water. Repeated treatments may be used to increase contrast. After the final treatment, the surface is rinsed with distilled water, dried, and photographed.

Crowle's double stain is still another alternative to amido black that also provides similar results. As you might expect, the developer solution uses two dyes. It contains 0.015% w/v Coomassie brilliant blue R and 0.25% crocein scarlet 7B in a 3:5:92 mixture of trichloroacetic acid:acetic acid:water. The questioned surface is sprayed with or immersed in this solution, allowed to soak for 30 to 90 seconds, and then rinsed with a 3:97 mixture of acetic acid:water. After the final treatment, the surface is rinsed with water, dried, and photographed.

DAB, also known by its chemical name of *3,3'-diaminobenzidine tetrahydrochloride*, is the newest of the blood reagents. The DAB method is relatively complicated and expensive, but it sometimes provides usable results where no other method works. The DAB process requires four reagents. solution A, the fixer, is a 2% w/v solution of 5-sulfosalicylic acid in distilled water. solution B, the buffer, is a 1:8 mixture of 1 M pH 7.4

phosphate buffer solution in distilled water. Solution C is a 1% w/v solution of DAB in distilled water. Developer is made up by combining 180 parts by volume of Solution B with 20 parts Solution C and one part 30% hydrogen peroxide.

Prints may be developed by the *DAB submersion method* or the *DAB tissue method*, depending on which is better suited for the specimen. For the submersion method, the specimen is soaked for 3 to 5 minutes in a tray of solution A to fix the prints, followed by rinsing for 30 seconds to 1 minute in a tray of distilled water. After the first rinse, the specimen is soaked for up to five minutes in a tray of developer, until maximum contrast is achieved. After a final rinse in distilled water, the specimen is air-dried or dried with a hair dryer, after which it is photographed. For the tissue method, the surface is covered with an unscented facial tissue or a thin paper towel, which is then sprayed with solution A and allowed to soak for three to five minutes. The tissue is removed and the area is rinsed for 30 seconds to 1 minute with distilled water. A new tissue is placed over the subject area, saturated with developer, and allowed to soak for up to five minutes. When development is complete, the area is rinsed thoroughly with distilled water, dried, and photographed.

Small particle reagent (SPR)

Small particle reagent (SPR) is a liquid suspension of solid particles of dark gray molybdenum disulfide, applied to the questioned surface by spraying or dipping. SPR works in the same way as fingerprint powders—by physical adhesion of particles to fatty fingerprint residues—but unlike dry fingerprint powders, SPR can be used for processing wet surfaces, including surfaces soaked in liquid accelerants and other organic solvents. SPR is also used on glossy nonporous surfaces, such as glass and plastic, coated glossy papers, and surfaces covered with glossy paint. Prints raised by SPR are extremely fragile, and should be photographed before any attempt is made to lift them. Modified versions of SPR are available in white, gray, and fluorescent forms, all of which are based on chemicals other than molybdenum disulfide.

> Although SPR is readily available from forensic suppliers, you may want to try making your own. To do so, add about 5 g of molybdenum disulfide (dry powder lubricant sold as Moly Lube and similar trade names) to about 100 mL of water to which you've added 1 mL of liquid dish washing detergent. Agitate this suspension thoroughly before use, and apply it by spraying or dipping. Allow the SPR to act for one minute, and then rinse the surface gently with water. We tried this ad hoc method, and it actually kind of worked.

Superglue fuming

Superglue fuming, also called *cyanoacrylate fuming* from the primary component of super glue, was discovered by accident in 1976 when Masao Soba noticed white fingerprints on the surface of a super glue container. Frank Kendall improved the process and adapted it to latent fingerprint development, reporting his findings in a 1980 paper. Since that time, superglue-fuming has become one of the most frequently used latent print development processes.

Super glue fuming is used to develop latent prints on nonporous glossy surfaces such as glass, plastic, and polished metal. Like dusting, cyanoacrylate fuming is a physical process. Cyanoacrylate vapor is selectively attracted to fingerprint residues, where it builds up as a crystalline white deposit. The developed latent prints may be photographed as is, or may be dusted or treated with various dyes that enhance the visibility and contrast of the prints. Although the exact mechanism remains unknown, it's suspected that the super glue fumes are catalyzed by the tiny amount of moisture attracted by the sodium chloride residues in latent fingerprints.

The standard method for super glue fuming is to place the object to be fumed in an enclosed chamber (aquariums are often used) that contains a small electric heater. An aluminum weighing boat is placed on the heater, and the temperature is set to high. When the boat is hot, a few mL of cyanoacrylate are added to the boat. Fuming commences immediately and is ordinarily complete after 30 seconds to 10 or 15 minutes. Alternatively, a cotton ball can be soaked in 0.5 M sodium hydroxide and allowed to dry. Once dry, the cotton ball is placed in the chamber and moistened with a few drops of cyanoacrylate. Fuming begins within a few seconds and is allowed to continue until the latent prints become visible.

Superglue-fumed prints can be photographed directly, or treated with dyes to increase the visibility and contrast of the prints and makes them easier to see against a patterned background surface. With the exception of Sudan black, most of these dyes are fluorescent, with absorption wavelengths that match to the emission wavelengths of commonly used forensic alternate light sources. For example, *rhodamine 6G* may be excited with a light source from 495 nm (blue) to 540 nm (green-yellow), with maximum absorption at 525 nm (green). Rhodamine 6G fluoresces in the range of 555 nm (yellow-green) to 585 nm (orange), with maximum emission at 566 nm (yellow). By viewing the treated surface through a filter that blocks wavelengths below about 555 nm but passes longer wavelengths, the treated fingerprints are visible as a yellow glow against a dark background.

Various fluorescent dyes are used alone or in combination to post-process superglue-fumed prints. Standard individual dyes include rhodamine 6G, *Ardrox*, 7-(p-methoxybenzylamino)-4-nitrobenz-2-oxa-1,3-diazole (MBD for short), *basic yellow 40*, *safranin O*, and *thenoyl europium chelate*. The most commonly used combination is RAM, a mixture of rhodamine 6G, Ardrox P133D, and MBD. Other mixtures are also used, including RAY (rhodamine 6G, Ardrox, and basic yellow 40), and MRM 10 (MBD, rhodamine 6G, and basic yellow 40).

Physical developer (PD)

Physical developer (PD) is useful for developing latent fingerprints on most porous surfaces and some nonporous surfaces. It is particularly useful for revealing latent prints on paper currency, paper bags, and porous surfaces that have been wet. PD is a destructive process, and so is always used last, if at all. PD is an alternative to the silver nitrate method. You can use one or the other, but not both. Whichever you use must be the last method you apply. PD is normally used after DFO and/or ninhydrin, and often reveals latent prints that neither of these methods revealed.

The "physical" part of the name is a misnomer. PD is not a physical process (like dusting), but a chemical one. It depends on a redox reaction that reduces silver ions to metallic silver, which stain the latent fingerprints a gray-black color. The PD working solution is unstable in the sense that it must be used immediately after it is made up, but it is this very instability that allows PD to work as well as it does. PD is expensive, complex, finicky, destructive, and requires a great deal of experience to get good results. Despite these criticisms, PD is used because it often gets results when no other method works. For this reason, many forensics labs routinely use PD as the final step in processing latent prints.

The PD process requires four solutions, with a fifth solution optional. Solution A is a 2.5% w/v solution of maleic acid in distilled water. Solution B (redox solution) is an aqueous solution that is 3% w/v with respect to ferric nitrate, 8% w/v with respect to ferrous ammonium sulfate, and 2% w/v with respect to citric acid. Solution C (detergent) is an aqueous solution that is 0.3% w/v with respect to n-dodecylamine acetate and 0.4% w/v with respect to Synperonic-N. Solution D is a 20% w/v solution of silver nitrate. Solution E (bleach) is a 1:1 mixture of standard chlorine laundry bleach with water.

The specimen to be treated is first placed in a tray of solution A and agitated for five minutes or until any bubbling has ceased, whichever is longer. The specimen is then transferred to a second tray that contains the working PD redox solution, made by combining solutions B:C:D in a 100:4:5 ratio and mixing thoroughly. The specimen is soaked in the working PD redox solution for 5 to 15 minutes, with constant agitation, after which it is rinsed thoroughly with water, air-dried or dried with a hair dryer, and photographed.

Bleach solution may be applied at the operator's discretion as a final step. This solution darkens the developed prints, lightens the background, and removes ninhydrin stains, but it may also eliminate detail that is visible before bleaching. The specimen is simply dipped in the bleach solution for about 15 seconds and then rinsed, dried, and photographed. It's important that the final rinse be thorough, because otherwise the specimen may degrade very quickly.

Adhesive surface techniques

The FBI uses the phrase *adhesive surface techniques* to describe four processing methods used to raise latent prints on sticky surfaces such as the adhesive side of sticky tapes, sticky labels, peel-and-stick plastics, and so on. Surprisingly, three of the four methods—*alternate black powder, ash gray powder,* and *sticky-side powder*—are powder based. (One might think that these powders would adhere equally well to the adhesive and the fingerprint residues, but they do not.) All three powders are applied in the same way—they are made into a thin paste, brushed onto the questioned surface, and rinsed off with cold water—and provide similar results. The primary differences among these three powders are their colors, chosen to provide contrast against different surface colors. The final method employs a 0.1% w/v solution of *gentian violet* (also called crystal violet) in water. The questioned surface is sprayed with or immersed in the gentian violet solution for one to two minutes, after which it is removed and rinsed with cold water. The gentian violet stains the latent prints, which can then be viewed and photographed under ordinary light.

Vacuum metal deposition (VMD)

Vacuum metal deposition (VMD) is similar conceptually to cyanoacrylate fuming, but substitutes metal vapor for cyanoacrylate vapor. The questioned surface is placed in a chamber from which the air is evacuated. The chamber also contains small pieces of gold and zinc that can be heated electrically until they vaporize. The specimen is exposed first to gold vapor, and then to zinc vapor. The metal vapors adhere selectively to fingerprint residues, revealing latent prints as metal-plated traces on a pristine substrate. Because VMD requires expensive equipment and materials, its use is limited to well-equipped and well-funded forensics labs.

The FBI categorizes these processes as Standard (used routinely) or Optional (used only in special situations or as supplements to Standard processes), as follows:

Standard processes

Adhesive surface techniques (alternate black powder, ash gray powder, gentian violet, and sticky-side powder), amido black (methanol base), amido black (water base—Fischer 98), DAB, DFO, fingerprint powders, iodine fuming, ISR, LCV, ninhydrin (petroleum ether base), PD, RAM, silver nitrate, Sudan black, super glue fuming, and VMD.

Optional processes

Amido black (water base), Coomassie brilliant blue, Crowle's double stain, fluorescent super glue dyes (Ardrox, MBD, MRM 10, rhodamine 6G, safranin O, and thenoyl europium chelate), Liqui-Drox, and ninhydrin (acetone base).

Obviously, from this plethora of techniques—and these are only the most popular of a larger group—we needed to choose a subset for the lab sessions in this group. We settled on iodine fuming, ninhydrin, and super glue fuming, which happen to be the Big Three in real forensic labs and also have the advantage of being inexpensive and using materials that are relatively easy to acquire. We'll also use gentian violet to develop prints on cellophane tape, and we'll dust to develop prints on glass. Finally, we'll use two liquids found in most homes to reveal latent fingerprints on brass cartridge cases.

Dusting and Lifting Latent Fingerprints

Lab IV-1

EQUIPMENT AND MATERIALS

You'll need the following items to complete this lab session. (The standard kit for this book, available from *http://www.thehomescientist.com*, includes the items listed in the first group.)

MATERIALS FROM KIT

- Goggles
- Fingerprint brush
- Fingerprint powder, black
- Fingerprint powder, white

MATERIALS YOU PROVIDE

- Gloves
- Camera with macro capability (optional)
- Desk lamp
- Index cards or paper (white and black)
- Scanner (optional)
- Tape, packing or similar transparent
- Ultraviolet (UV) light source (optional)
- Specimens: objects with fingerprints

BACKGROUND

We'll begin with the oldest fingerprint development method, dusting. With the exception of using magnetic powders to treat recently touched paper, dusting is used almost exclusively on nonporous surfaces, and can provide excellent results if it's done skillfully. If not done skillfully, dusting can easily damage or destroy any latent fingerprints present, as we found out and you probably will, too. In this lab session, we'll dust various specimens, using dark or light dusting powder as appropriate for the color of the surface. We'll then perform a tape lift to preserve the fingerprints we've developed.

Before you get started, you need to create some specimens to be tested. Choose several nonporous items, such as a drinking glass, beverage can, and so on. Use a soft cloth to polish the item clean, removing any contamination. For your first attempts at dusting, you may find it difficult enough to get usable prints even under perfect conditions. You can increase the likelihood of having usable prints on your specimens by rubbing your forefinger against your nose or forehead and then carefully pressing your finger into contact with the surface, making sure not to smear the prints. Once you're comfortable with this training-wheels version of developing prints by dusting, you can try dusting random objects from around the house to get a better idea of the highly variable quality of real latent fingerprint specimens.

In addition to creating latent fingerprints on several non-porous objects, create at least one set of latent fingerprints on an ordinary sheet of paper. You'll use that specimen to determine how suitable dusting is for developing prints on non-porous surfaces.

PROCEDURE IV-1-1: DUSTING LATENT FINGERPRINTS

1. Wearing gloves and handling the questioned object by the edges or otherwise as required to avoid damaging any latent prints, observe the object by oblique lighting from the desk lamp or other directional light source. Record your observations in your lab notebook and note the approximate location of any latent prints that are made visible by the oblique lighting.

2. If you have a black light or other UV light source, repeat step 1 using that light source.

3. Place the object on a clean, flat surface with the suspected location of the latent prints accessible. (Don't forget to wear gloves.)

4. Choose the dark or light fingerprint dusting powder, according to which will provide better contrast with the color of the surface.

5. Transfer a small amount of the dusting powder into the lid or work directly from the jar that contains the powder. Dip just the tips of the bristles of the brush into the powder so that a small amount of powder is retained by the bristles. Tap the brush gently to return excess powder to the container.

6. Under a strong light, use a circular, twirling motion to sweep the brush gently over the area to be treated, as shown in Figure IV-1-1, allowing the bristles to just barely contact the surface. Continue depositing powder lightly until the latent fingerprint begins to develop, concentrating on that area as it becomes clearer where the latent prints are on the surface. If necessary, add more powder to the brush using the procedure in step 5. When the ridges begin to appear, change the direction of motion to follow the direction of the ridges. Once the fingerprint is developed clearly, stop dusting immediately. Beginners tend to overdevelop prints, which almost invariably causes loss of detail if not loss of the entire print.

Figure IV-1-1: *Barbara dusting a specimen for fingerprints*

7. Use the brush or a puffer bulb gently to remove any excess powder. (Okay, we admit it; we used our mouths to puff off excess powder, but that's a horrible practice.) You can also use canned air if you do so *extremely* carefully, keeping the canned air nozzle far enough away from the dusted print to avoid blowing away everything, including the print. If you have a camera, shoot an image of the revealed print.

8. Repeat steps 1 through 7 for each of your other specimens, including at least one set of latent fingerprints on a sheet of paper.

You'll probably find that your first efforts are poor, but you will improve rapidly with practice. Of course, getting really good at dusting prints requires lots of practice. Professional fingerprint technicians can work wonders when dusting latent prints, but then they have years of experience in doing it. Figure IV-1-2 shows a specimen after dusting but before lifting the fingerprints.

Figure IV-1-2: *Dusted prints on the surface of the specimen*

PROCEDURE IV-1-2: LIFTING DEVELOPED FINGERPRINTS

1. Choose one of your better developed prints. (If possible, shoot an image of a print before attempting to lift it. Accidents happen.)

2. Wearing gloves, lift the free end from the roll of lifting tape and smoothly pull out about 6 to 7.5 cm (2.5 to 3") of tape from the roll. Don't touch the sticky surface of the tape, and do not cut the tape from the roll.

3. Press the free end of the tape into contact with the surface, starting 5 to 6 cm from the nearest part of the dusted print. Make sure the tape adheres firmly to the surface.

4. Beginning at the free end, use your fingers to carefully press the tape down onto the surface, making sure that no air bubbles are trapped.

5. Continue pressing tape onto the surface, unrolling more as necessary, until you have covered the entire print with tape and continued for a couple of centimeters past the print.

6. Using the roll as a handle, peel the tape from the surface using one smooth motion, as shown in Figure IV-1-3. It helps to put one finger on the free end of the tape to make sure the tape doesn't curl back on itself.

Figure IV-1-3: *Barbara lifting a dusted fingerprint from a specimen*

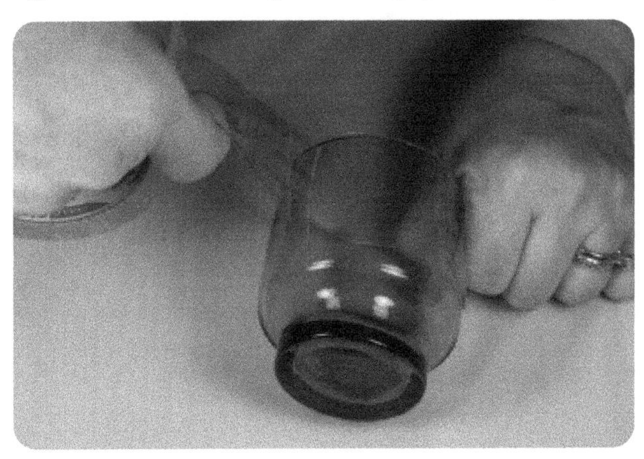

7. Stick the free end of the tape near one edge of a transfer card of a color that contrasts with powder you used to dust the print. Make sure the free end adheres tightly to the transfer card, and then carefully press the tape into contact with the transfer card, making sure to avoid air bubbles.

8. Cut the used tape from the roll and press the free end into contact with the transfer card. Label the transfer card with your initials, the date and time, and the object from which the print was lifted.

9. Repeat steps 2 through 8 for your other specimens.

> This procedure works well for most lifts, but what if the specimen isn't flat? If the contour of the surface makes it impossible to follow these directions exactly, modify the procedure as necessary, for example, by removing the tape from the roll before pressing it into place.

Figure IV-1-4 shows a transferred print.

Figure IV-1-4: *A partial fingerprint after transfer*

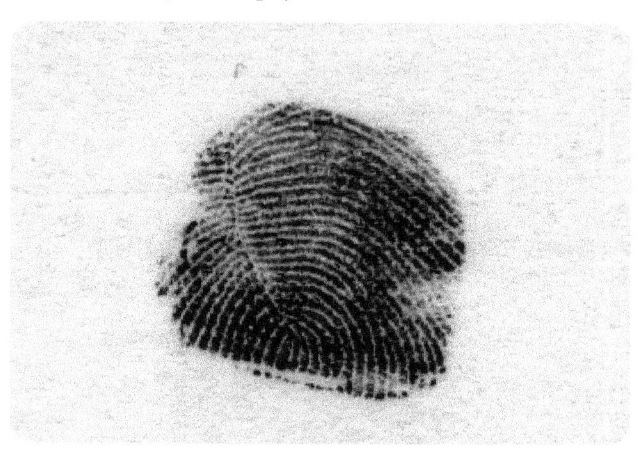

REVIEW QUESTIONS

Q1: Should dusting be the first or last method attempted to raise latent fingerprints? Why?

Q2: Is dusting better suited for porous or nonporous surfaces?

Revealing Latent Fingerprints Using Iodine Fuming

Lab IV-2

EQUIPMENT AND MATERIALS

You'll need the following items to complete this lab session. (The standard kit for this book, available from *http://www.thehomescientist.com*, includes the items listed in the first group.)

MATERIALS FROM KIT

- Goggles
- Magnifier
- Iodine crystals

MATERIALS YOU PROVIDE

- Gloves
- Iodine fuming chamber (see text)
- Scanner and software (optional)
- Starch water (see text)
- Specimens: paper with latent fingerprints

WARNING

Iodine crystals are toxic and corrosive and stain skin and clothing. (Stains can be removed with a solution made by dissolving a vitamin C tablet in water.) Iodine vapors are toxic and irritating. Perform this experiment only outdoors or in a well-ventilated area. Wear splash goggles, gloves, and protective clothing.

BACKGROUND

Iodine fuming has been used since the turn of the 20th century to develop latent fingerprints on porous surfaces, particularly paper. Iodine fuming is still widely used because it is inexpensive and easy, sensitive, and is nondestructive (the stains it produces are ephemeral). If it is used at all, iodine fuming is normally the first processing method attempted. Some forensics texts state that iodine fuming is used less often nowadays than formerly. That may be true in the limited sense that there are now many alternatives, but iodine fuming is still used frequently by many forensics labs. It's cheap, fast, effective, and completely reversible. What's not to like?

In the lab, iodine fuming is done in a chamber, but the process was adapted to field use quite early. The first *iodine fuming wands* were simple tubes with a small reservoir for iodine crystals. The operator warmed the tube in his hand and blew gently into one end of the wand. His breath vaporized iodine and expelled iodine vapor from the other end of the tube, which was aimed at the surface to be treated. Modern versions of the iodine fuming wand substitute battery power for body heat and the operator's breath to avoid the risk of inhaling iodine vapor, but the principle remains the same.

DENNIS HILLIARD COMMENTS

I have seen iodine fuming done in a plastic zip-lock bag. The paper is placed in the bag with several iodine crystals and allowed to sit for a period of time while the crystals sublimate inside the zip-lock bag.

You can make an iodine fuming chamber from any airtight container large enough to contain your specimens. As Dennis mentioned, zip-lock bags are a popular choice, as are plastic kitchen storage containers with snap-on lids. You don't need much iodine: an iodine crystal the size of a pinhead is sufficient to fill a large container with iodine vapor. The small amount of iodine crystals included with the FK01 Forensic Science Kit is sufficient to fume many, many specimens. When you finish fuming a specimen, remove it quickly from the chamber, making sure as little as possible of the iodine vapor escapes. You can use that same chamber several times without replenishing the iodine.

You can fume at room temperature, but heating the iodine crystals even slightly vaporizes them, making fuming proceed faster. Forensic technicians often place their hands on the outside of the bag to use their own body heat to vaporize the iodine crystals. If you're using a plastic kitchen storage container, just immerse its base in a dish of warm tap water.

The fugitive nature of iodine-developed prints is a two-edged sword. On the one hand, it's nice that iodine fuming is easily reversible. Developed prints simply disappear within a few hours or days as the iodine gradually sublimes from the prints, and that process can be accelerated by warming the prints slightly. On the other hand, there are times when it would be nice if the developed prints were a bit more permanent. The first method used to stabilize iodine-developed prints was to treat them with a starch solution. Iodine and starch combine to form a deep blue-black complex, which persists for weeks to months, depending on storage conditions. Later, benzoflavone was introduced as an after-treatment for iodine-developed prints. Prints treated with benzoflavone are effectively permanently fixed.

Rather than making up a starch solution, you can substitute water in which potatoes, pasta, rice, or another starchy food has been boiled. Starch solution spoils quickly, even when refrigerated, so make up your starch solution no more than a day before using it and discard any unused solution.

In this lab session, we'll use iodine fuming to develop latent prints on paper. We'll then treat those developed prints with a starch solution to fix them, at least temporarily.

Before you get started, you need to create some fingerprint specimens to be fumed. Since iodine fuming is used almost exclusively for paper specimens, you should produce specimens on various types of paper, including ordinary copy paper, cotton bond, kraft or wrapping paper, thin card stock, the paper side of photographs, glossy magazine paper, and so on. You can also test random paper items you find around the house. (Remember, iodine fuming is reversible, so you won't damage a specimen by fuming it.)

We found that latent fingerprints were difficult or impossible to develop by iodine fuming if we'd washed our hands soon before touching the paper. Apparently, soap removes all or most of the skin oils and other residues to which iodine fumes adhere. We obtained good results with specimens that we'd touched when our hands hadn't been washed recently, and the best results occurred when we intentionally rubbed our fingers against our noses or foreheads before making the latent prints.

PROCEDURE IV-2-1: FUMING LATENT FINGERPRINTS WITH IODINE

In this procedure, we'll that assume you're using a zip-lock plastic bag as your fuming chamber.

1. If you have not already done so, put on your splash goggles, gloves, and protective clothing.

2. Open the zip-lock bag and transfer one or two tiny iodine crystals to it. An amount the size of a pinhead is sufficient.

3. Place your first specimen inside the bag, expand the bag so that it contains some air space, and then zip the bag closed.

4. The crystals begin sublimating immediately, filling the bag with iodine vapor. Depending on the size of the bag and specimen, the amount of iodine you use, and how you vaporize it, latent prints should start becoming visible within anywhere from a few seconds to a few minutes as faint orange smudges on the specimen.

5. Allow the fuming to continue until the ridge detail is evident in the developing prints or until no further change occurs. Depending on the specimen, the amount of iodine vapor present, and other factors, this may require anything from just a minute or so to several hours or more.

6. When development appears to be complete, quickly remove the first specimen, insert the second specimen, and re-zip the bag. (Be careful not to inhale the iodine vapor, which is irritating and has a strong chlorine-like odor.) You can use the iodine vapor already present in the chamber to develop additional specimens, adding a few more iodine crystals as required to keep the chamber filled with iodine vapor.

7. Place the specimen on a clean, flat surface and examine it carefully under strong light with the magnifier. Depending on the specimen, you may see only faint orange smudges with little or no ridge detail, as shown in Figure IV-2-1, or you may see well-developed orange prints with considerable ridge detail visible.

> The dark orange area at the top-left corner of Figure IV-2-1 was nearest the iodine crystals in the fuming bag. If a porous specimen is exposed long enough to concentrated iodine fumes, even areas that have no fingerprints will eventually assume an overall orange color.

Figure IV-2-1: *Fingerprints immediately after iodine fuming*

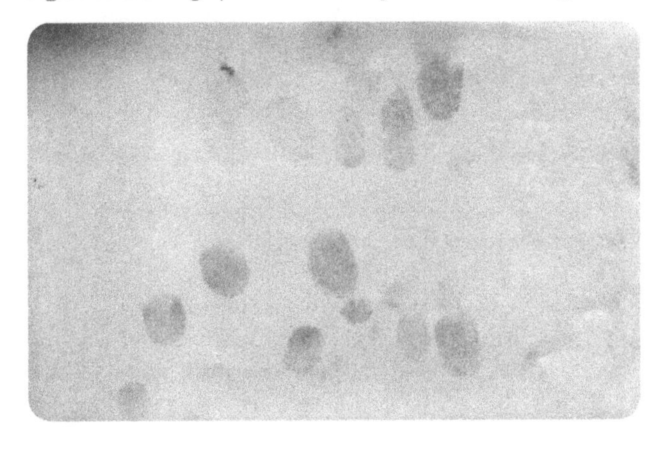

8. Regardless of how much or how little ridge detail is visible, forensic technicians always make an image of the specimen immediately after fuming it. If you have a camera

or a scanner, capture an image of the fumed fingerprints for your records. Record the pertinent details for the specimen in your lab notebook.

> In addition to serving as a permanent record of the fumed fingerprints, the scanned image is the departure point for making enhanced images that may show detail that's invisible in the original image. By using Photoshop, The GIMP, ShowFoto, or a similar image-manipulation program to adjust contrast, gamma, and other image characteristics, it is often possible to reveal ridge detail sufficient to allow a match to be made.
>
> Figure IV-2-2 is a gamma-enhanced version of Figure IV-2-1. In the original scanned image, essentially no ridge detail was visible. In the enhanced image, sufficient ridge detail is present to possibly allow a match to be made.

Figure IV-2-2: *Figure IV-2-1 after gamma enhancement*

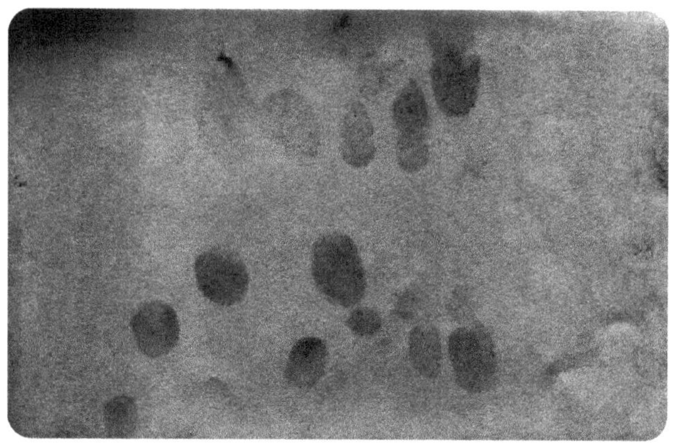

9. Use your image-manipulation software to enhance the image to reveal as much detail as possible.

10. After you have a good image of the original fumed specimen, use a starch solution to develop the iodine stains. You can spray the specimen with the starch solution or gently swab the solution onto the fumed prints using a cotton ball, cotton swab, or the corner of a paper towel. You needn't drench the specimen; just dampening it is sufficient. The iodine reacts with the starch instantly, changing the color of the fumed prints to blue-black, as shown in Figure IV-2-3. Fumed prints treated with starch solution often show contrast enhancement that reveals additional detail.

11. Record your observations in your lab notebook and tape the developed specimen into your lab notebook, as well as printed copies of your initial scan and any enhanced images that you created from that scan.

12. Repeat steps 3 through 11 for your other specimens. (Retain at least two iodine-fumed specimens that you have not treated with starch solution for use in the following lab session.)

Figure IV-2-3: *Iodine-fumed fingerprint after development with starch solution*

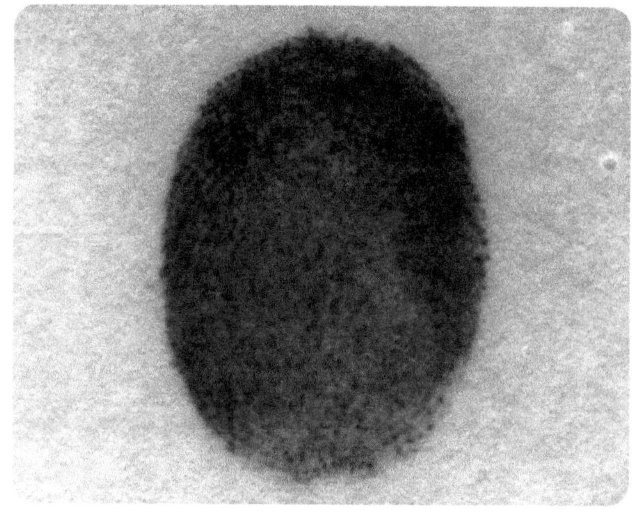

REVIEW QUESTIONS

Q1: **For what type(s) of specimen is iodine fuming best suited?**

Q2: **Is iodine fuming normally the first method attempted for revealing latent prints? Why?**

Q3: **What are two after-treatments used to preserve iodine-fumed fingerprints?**

Q4: **Why did we shoot an image of the iodine-fumed prints before we treated them with starch?**

Q5: **How much variation in quality did you observe among the prints you fumed on different types of paper? Were there major variations in quality among prints fumed on the same type of paper?**

Q6: **What other types of forensically significant prints might iodine fuming be used to develop?**

Revealing Latent Fingerprints Using Ninhydrin

Lab IV-3

EQUIPMENT AND MATERIALS

You'll need the following items to complete this lab session. (The standard kit for this book, available from *http://www.thehomescientist.com*, includes the items listed in the first group.)

MATERIALS FROM KIT

- Goggles
- Magnifier
- Ninhydrin powder
- Zinc chloride solution

MATERIALS YOU PROVIDE

- Gloves
- Acetone
- Bottle, small storage
- Paper towels
- Scanner and software (optional)
- Spray bottle, fingertip (optional)
- Steam iron or oven
- UV light source (optional)
- Specimens: paper with latent fingerprints
- Specimens: paper with fumed fingerprints

WARNING

Acetone is flammable. Ninhydrin is irritating. Zinc chloride is toxic and corrosive and emits hydrochloric acid fumes when wet. Wear splash goggles, gloves, and protective clothing, and use the ninhydrin spray outdoors or under an exhaust fan. Note that ninhydrin reacts with any amino acids, including those on your skin, to form Ruhemann's Purple. If you get ninhydrin solution on your skin, the purple stains, although harmless, will persist for several days.

BACKGROUND

Since it was first used for developing latent fingerprints in 1954, ninhydrin has become the most common method used to reveal prints on porous surfaces. Nearly all forensic labs use ninhydrin for this purpose, and some seldom use anything other than ninhydrin. Ninhydrin is cheap, sensitive, and commercially available in disposable spray cans. The developed prints are a high-contrast purple that's readily visible on most paper backgrounds. If iodine fuming or DFO is to be used, either or both must be used before ninhydrin, in that order. If it is to be used, ninhydrin must be used before silver nitrate or PD.

Ninhydrin solution can be applied by various methods. Most professional forensic labs use ninhydrin supplied in commercial spray cans, but many use manual spray bottles such as the fingertip sprayers sold in drugstores. (Using manual spray bottles allows the lab to choose a particular ninhydrin formulation according to its own preferences.) Some technicians prefer to use tray development, immersing the specimen in a shallow dish of ninhydrin solution. Still others prefer to use swab development, applying the ninhydrin solution to the specimen by gently swabbing with a cotton ball that is saturated with the ninhydrin solution. You can use any or all of these methods for your own testing.

Because the solvent does not take part in the reaction that forms Ruhemann's Purple, nearly any organic solvent can be used successfully. We've used ordinary rubbing alcohol (ethanol or isopropanol), acetone, petroleum ether, and the mixed alcohols recommended by the FBI; all appear to work identically. Ninhydrin development occurs slowly at room temperature and humidity. Although some stains may appear within seconds to minutes of applying the ninhydrin solution, complete development may take 24 to 48 hours. After the ninhydrin solution dries, the development process can be accelerated by increasing the temperature and humidity. We used an ordinary steam iron with the specimen sandwiched between paper towels, and found that full development occurs within a few minutes under those conditions.

Professional forensic labs often use heat and humidity to develop ninhydrin-treated latent fingerprints because they usually need them quickly. When time is not of the essence, allowing ninhydrin-treated latent fingerprints to develop naturally at room temperature for a day or two often provides better detail in the developed prints.

Various after-treatments are sometimes used to enhance prints developed by ninhydrin, including treating the developed prints with nickel nitrate solution, zinc chloride reagent, or other metal-based reagents, which may intensify the prints and/or make them fluorescent under an ALS. One of the more effective after-treatments is simply to saturate the developed prints with blank solvent (e.g., if the specimen was treated with ninhydrin in acetone, saturating the developed prints with plain acetone) and then expose the developed prints to heat and humidity. The exact mechanism of this enhancement is uncertain, but it seems likely that the use of blank solvent allows unreacted ninhydrin still present on the specimen to combine with unreacted amino acids from the latent prints.

In this lab session, we'll use ninhydrin solution to develop both untreated latent prints and prints that have already been processed by iodine fuming. For the former, create some fresh fingerprint specimens by using the procedure described in the preceding lab session. Ninhydrin is used to develop prints on various nonporous surfaces, but it is used primarily to develop latent prints on paper. Select various types of paper from around the house to use for your specimens. Use only paper items you are willing to discard after the experiment; ninhydrin stains are persistent. For the latter, use specimens that you processed in the preceding lab session.

FORMULARY

If you are using the FK01 Forensic Science Kit, dissolve the ninhydrin powder supplied with the kit in 100 mL of acetone. (Acetone purchased at a hardware store or paint store is fine.) If you do not have the kit, make up ninhydrin solution by dissolving about 0.5 gram of ninhydrin powder in about 100 mL of acetone. The concentration is not critical. You can substitute ethanol, isopropanol, petroleum ether, or another organic solvent if you don't have acetone. We've tried numerous solvents, including lighter fluid and paint thinner, with similar results. We've found ninhydrin solutions to be stable for at least several months if stored in a cool, dark place.

If you are using the FK01 Forensic Science Kit, add 5 mL of the zinc chloride solution supplied with the kit to 90 mL of acetone. If you do not have the kit, make up the zinc chloride solution by dissolving about 0.25 g of zinc chloride in about 5 mL of acetone, adding 0.75 mL of glacial acetic acid, and adding acetone to make up the solution to 100 mL. The concentration is not critical.

PROCEDURE IV-3-1: DEVELOPING LATENT FINGERPRINTS WITH NINHYDRIN

1. If you have not already done so, put on your splash goggles, gloves, and protective clothing. Work outdoors or in a well-ventilated area with no flame or other ignition source nearby.

2. If you are spraying or swabbing, place the specimen print-side up on paper towels or old newspaper to protect the work surface. If you're using tray development, place the specimen print-side up in the dish or tray.

3. Apply ninhydrin solution sufficient to dampen the surface of the specimen, as shown in Figure IV-3-1. Don't drench it, but make sure the entire surface is dampened with ninhydrin solution.

4. Repeat steps 2 and 3 for your other specimens.

5. Allow the specimens to air dry for a few minutes. Some bluish or purplish ninhydrin stains may be faintly visible at this point, but do not be concerned if no stains are evident.

6. Place one of the treated specimens aside for a day or two to allow development to occur at room temperature.

7. Make a sandwich with two thicknesses of paper towels, followed by the specimen (print side up), and then two more layers of paper towels.

Figure IV-3-1: *Spraying a specimen with ninhydrin solution*

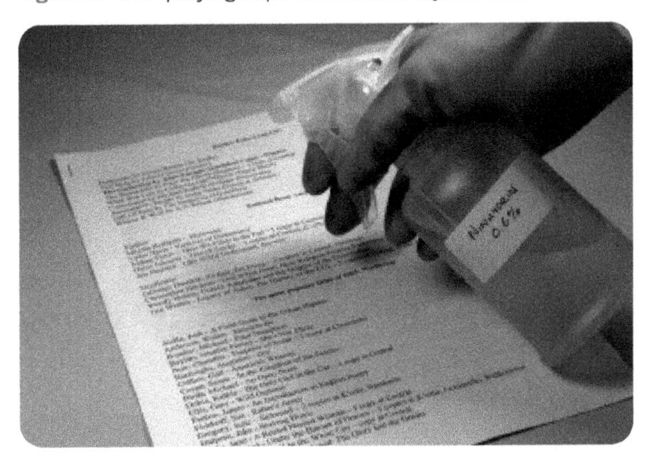

8. Set the steam iron to low heat and iron the sandwich for several minutes, using steam occasionally, as shown in Figure IV-3-2. You can check development progress periodically by peeling back the top layer of paper towels.

If you don't have a steam iron, you can achieve similar results by soaking the top layer of paper towels in water and then placing the sandwich in an oven set to about 110°C (230°F). Heat the sandwich for 15 to 20 minutes, or until the paper towels have dried.

Figure IV-3-2: *Applying moist heat to develop the fingerprints treated with ninhydrin solution*

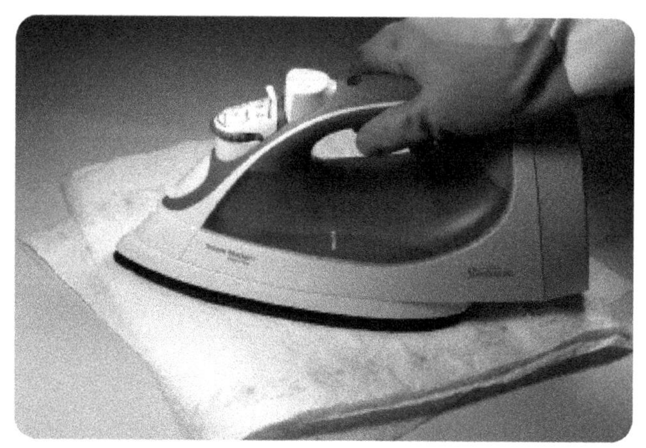

9. When development appears to be complete, place the specimen on a clean, flat surface and examine it carefully under strong light with the magnifier. You should see fingerprints, which may vary from featureless smudges to prints showing considerable ridge detail. If you have a camera or scanner, capture an image of the developed fingerprints for your records, and use your image-manipulation software to attempt to increase the visible detail. Record the pertinent details for each specimen in your lab notebook.

Figure IV-3-3 shows the latent fingerprints revealed by ninhydrin development.

Figure IV-3-3: *Latent fingerprints revealed by ninhydrin development*

PROCEDURE IV-3-2: NINHYDRIN AFTER-TREATMENTS

Forensic labs use many different after-treatments for prints that have been developed by ninhydrin. These supplementary procedures may or may not increase the contrast and level of detail present in the prints. In this procedure, we'll test three of the most common after-treatments: retreating with ninhydrin solution, retreating with blank solvent, and retreating with a zinc chloride solution.

To be able to directly compare the results of these three after-treatments, produce a fingerprint specimen with many latent prints on one sheet of paper. Develop the prints with ninhydrin, as described in the preceding procedure, and then cut the sheet into four pieces, each of which contains approximately the same number of similar-appearing prints. Label one of these pieces "Control" and keep it for comparison to the other three pieces, each of which will receive a different after-treatment.

1. If you have not already done so, put on your splash goggles, gloves, and protective clothing. Work outdoors or in a well-ventilated area with no flame or other ignition source nearby.

2. Label the first test specimen "Ninhydrin," and subject it to the ninhydrin development process, including heat/humidity development, described in the preceding procedure.

3. Label the second test specimen "Blank." Subject this piece to the same procedure, but substitute plain acetone for the ninhydrin solution.

4. Label the third test specimen "Zinc." Spray this specimen with the diluted zinc chloride solution to dampen the surface, allow it to air dry, and then dampen the specimen again with zinc chloride solution and allow it to air dry. Finally, use the heat/humidity development process described in the preceding procedure to develop the zinc-treated specimen.

5. After drying all of the specimens, examine and compare them with one another to determine whether any or all of them show additional detail versus the specimen that has not undergone any after-treatment. Shoot or scan images of each specimen and record the details in your lab notebook. Use your image-manipulation software to attempt to increase visible detail.

REVIEW QUESTIONS

Q1: **With which component of fingerprint residues does ninhydrin react to form Ruhemann's Purple?**

Q2: **Which two common fingerprint development methods must be used before ninhydrin, if they are to be used at all?**

Q3: **What four after-treatment methods are commonly used to enhance fingerprints developed with ninhydrin?**

Q4: **Did the iodine-fumed prints show any additional detail after treatment with ninhydrin?**

Q5: **Which, if any, of the after-treatment methods increased the detail visible in your ninhydrin-treated prints?**

Q6: **If you had developed a real questioned latent fingerprint with ninhydrin, which after-treatment(s) would you attempt, in what order, and why?**

Revealing Latent Fingerprints Using Superglue Fuming

Lab IV-4

EQUIPMENT AND MATERIALS

You'll need the following items to complete this lab session. (The standard kit for this book, available from *http://www.thehomescientist.com*, includes the items listed in the first group.)

MATERIALS FROM KIT

- Goggles
- Fingerprint brush
- Fingerprint powder, black
- Magnifier
- Slide, flat

MATERIALS YOU PROVIDE

- Gloves
- Acetone
- Aluminum foil
- Baking soda
- Camera (optional)
- Cotton balls (real cotton)
- Fuming chamber (see text)
- Index cards or paper (white)
- Paper towels
- Scanner (optional)
- Superglue, cyanoacrylate (bottle or tube)
- Tape, packing or similar transparent
- Specimens: nonporous with latent fingerprints

WARNING

Acetone is flammable. Handle superglue carefully, or you might find you've glued yourself to yourself. (If you have a mishap, use acetone to remove the superglue.) Superglue fumes are not particularly toxic, but they are irritating. Work outdoors or under an exhaust fan. Although it didn't happen to us, it's possible that the cotton ball will catch fire if you use too much superglue. Have a container of water handy to flood the chamber if necessary. Wear splash goggles, gloves, and protective clothing.

BACKGROUND

For 80 years or more, dusting was the only widely used method for raising latent fingerprints on nonporous surfaces. Various incremental improvements were made over the decades—including the introduction of superior powder formulations, fluorescent powders, and magnetic powders—but dusting for fingerprints in 1980 was essentially unchanged from dusting for fingerprints in 1900.

That all changed with the introduction of superglue fuming, which quickly became the default method for raising latent fingerprints on nonporous surfaces. The method is simple: the object to be treated is placed in a chamber with a source of humidity present and then superglue fumes are introduced into the chamber. Development is usually complete within a few minutes.

DENNIS HILLIARD COMMENTS

For fuming, we have used a 10-gallon fish tank with Lexan covers. A small beaker of hot water is placed in the tank to provide humidity. An aluminum boat holds the superglue, with a cup warmer to warm it. You can use wire to hang plastic bags, etc. To ease clean up, cover the inside of the tank with aluminum foil that can be discarded and replaced.

Superglue-developed prints are often dusted with black powder to allow for a lift, especially on irregular surfaces, for preservation by photography. DFO can be mixed with superglue to give a fluorescent property to the print.

Superglue fuming works even at room temperature, although it may take hours to a day or more for development to be complete. Superglue fuming is often done by using special packs that emit superglue fumes as soon as they're opened or by using heat to vaporize a puddle of superglue, but an alternative method is available that depends on an interesting chemical reaction between cyanoacrylate ester and cotton (as well as wool and other fabrics). The reaction between superglue and cotton is extremely exothermic (heat producing), which is why it's a very bad idea to wear cotton gloves when you work with superglue. In fact, if you saturate a cotton ball with superglue, the cotton ball may eventually burst into flames. This reaction is base-catalyzed, which means the reaction can be sped up if the cotton balls are first soaked in a solution of sodium hydroxide, sodium carbonate, or sodium bicarbonate (baking soda) and then dried. If you add a few drops of superglue to a treated cotton ball, the superglue heats up quickly and begins to emit fumes within a few seconds to a minute or so.

Prints developed with superglue may be viewed and photographed directly with an oblique light source, but they are often treated with special fluorescent dyes to improve contrast and detail. Examined under an appropriate ALS with proper filtration between the eye or camera and the specimen, dye-treated superglue prints stand out starkly against the background. We'll forego dye treatment in this lab session, because most of these dyes are expensive and difficult to obtain. What we will do is fume various nonporous objects with superglue and observe the results, and then dust some of the fumed specimens to increase the contrast of the developed prints.

Choose a variety of small, nonporous, disposable objects as test specimens. We used microscope slides, which are cheap and fit easily into our small chamber. You can also try coins, small glass or plastic bottles, and similar nonporous objects. Also obtain at least one latent fingerprint specimen on glossy magazine paper.

PROCEDURE IV-4-1: PREPARING FOR SUPERGLUE FUMING

There's some work to be done before you begin the superglue fuming process. First, soak several cotton balls (at least one for each specimen you intend to fume, and it's a good idea to have a few extras) in a paste made of baking soda and tap water and place them aside to dry completely. Make your specimens by pressing your fingers onto the nonporous objects you selected. Again, you'll get the best results by rubbing your finger against your nose or forehead before making the impression on the objects. (Yes, we know that criminals aren't normally this cooperative, but we want to make things as easy as possible for beginners.)

The fuming chamber can be almost any disposable glass or plastic-lidded, wide-mouth container that's large enough to contain your specimens. (Professional forensic labs often use bespoke fuming chambers, but they're just as likely to use aquariums or fishbowls.) For our chamber, we used a plastic tub that had contained grocery store potato salad. We've also used the one-quart plastic tubs that Chinese restaurants use for take-out wonton and egg-drop soup orders and the wide-mouth jars in which Costco sells nuts.

Fold some aluminum foil to make a boat large enough to hold the cotton ball, put one of the dry treated cotton balls in the boat, and then place it in the bottom of the chamber. To provide the humidity necessary for superglue fuming, place a crumpled paper towel soaked in hot tap water or a small container of hot tap water in the bottom of the chamber. Replace the lid on the chamber and allow it to sit for a few minutes to saturate the air in the chamber with water vapor.

PROCEDURE IV-4-2: FUMING LATENT FINGERPRINTS WITH SUPERGLUE

1. If you have not already done so, put on your splash goggles, gloves, and protective clothing.

2. Place the specimen in the fuming chamber, as shown in Figure IV-4-1, orienting it so that the surface that contains the latent prints is open to the atmosphere in the chamber.

3. Add a few drops of superglue to the cotton ball, and immediately replace the lid on the fuming chamber. The cotton ball should begin to emit fumes within a few seconds. Allow development to continue for several minutes. You can check development progress by quickly lifting the lid and examining the specimen, although this allows fumes to escape. If necessary, you can add a few more drops of superglue to the cotton ball to increase fume output, or replace the original cotton ball with a new one and add a few drops of superglue.

Figure IV-4-1: *Placing the specimen in the superglue fuming chamber*

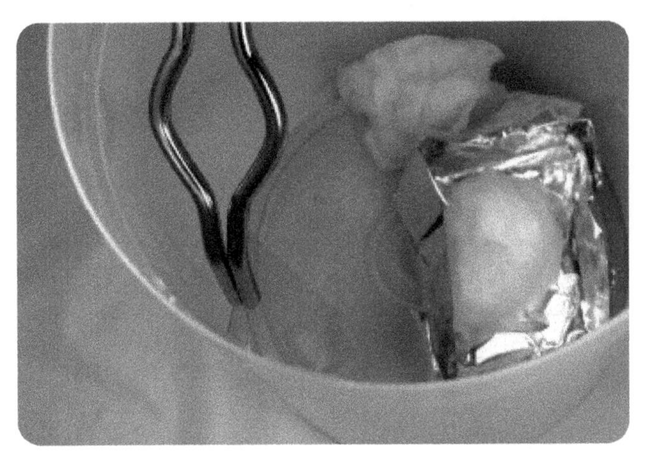

4. When development appears to be complete, open the container (be careful of the fumes) and remove the specimen. You can dispose of the paper towel, boat, and cotton ball in your household trash.

5. Place the specimen on a clean, flat surface and examine it carefully under strong oblique light with the magnifier or loupe. You should see fingerprints revealed in considerable detail as crystalline white traces. If you have a camera, shoot an image of the developed fingerprints for your records. Record the relevant details for the specimen in your lab notebook. If the specimen is small enough, tape it to the pertinent page of your lab notebook and label it.

6. Repeat steps 2 through 5 for your other specimens.

Figure IV-4-2 shows a latent fingerprint revealed by superglue fuming.

Figure IV-4-2: *Latent fingerprint revealed by superglue fuming*

PROCEDURE IV-4-3: DUSTING AND LIFTING SUPERGLUE-FUMED FINGERPRINTS

Latent fingerprints that have been fumed with superglue appear as white or colorless traces that may be difficult to photograph, depending on the background. Several after-treatments are used with superglue-fumed prints, including various standard dyes and fluorescent dyes. But among the most common after-treatments is simple dusting using ordinary dusting powder of a color that contrasts with the background. (Black powder is most commonly used because untreated superglue-fumed prints generally contrast well with dark backgrounds.)

The other advantage of dusting superglue-fumed prints is that it allows the prints to be lifted. Because the superglue traces are much more robust than the skin oils of the original latent print, a superglue-fumed print can often be dusted and lifted repeatedly until a good transfer is obtained.

1. Choose one of your fumed specimens that shows poor contrast between the superglue traces and the background.

2. Dust, lift, and transfer the print(s) with black powder by using the procedure described in Lab IV-1.

3. Repeat step 2 at least once to obtain a second transfer of the same print(s).

4. If you have a scanner, scan each set of transferred prints and use your image-manipulation software to enhance the images to reveal as much detail as possible.

5. Record your observations in your lab notebook. Attach the original transfers and copies of the enhanced scanned prints.

REVIEW QUESTIONS

Q1: For which specific types of surfaces is superglue fuming appropriate?

Q2: Is superglue fuming more akin to physical development processes such as dusting, or to chemical development processes, such as ninhydrin? Why?

Q3: What other treatments are commonly used after superglue fuming?

Revealing Latent Fingerprints On Sticky Surfaces

Lab IV-5

EQUIPMENT AND MATERIALS

You'll need the following items to complete this lab session. (The standard kit for this book, available from *http://www.thehomescientist.com*, includes the items listed in the first group.)

MATERIALS FROM KIT

- Goggles
- Forceps
- Gentian violet solution, 0.1% aqueous
- Magnifier
- Pipette
- Slides, flat

MATERIALS YOU PROVIDE

- Gloves
- Camera (optional)
- Paper towels
- Plastic sheet, transparent (optional; see text)
- Scanner (optional)
- Specimens: sticky with latent fingerprints

WARNING

Gentian violet solution stains anything it contacts. You can remove stains from your skin and some materials using ethanol or isopropanol. Some fabrics may be stained indelibly. Wear splash goggles, gloves, and protective clothing.

BACKGROUND

Gentian violet stain has been used for decades to develop latent fingerprints on nonporous surfaces, particularly the adhesive side of sticky tapes. The specimen to be treated is simply immersed in or floated on a 0.1% w/v aqueous solution of gentian violet for one to two minutes and then rinsed with water. Repeated stain/rinse cycles may be used to intensify the color of the stains. With some types of tape, a clearing solution of 1 M hydrochloric acid is used as the final rinse, to remove background staining without affecting the developed prints. When the print is fully developed, it can be viewed and photographed under ordinary light. Gentian violet is non-destructive, so it is often used first on sticky tape specimens. If gentian violet fails to develop the print, other powder-based methods can be attempted. Obviously, the aqueous solution of gentian violet should not be used on tapes that use water-soluble adhesives. In this lab session, we'll use gentian violet stain to develop latent fingerprints on various types of tape and other sticky surfaces.

Obtain as many different types of sticky tape and other adhesive materials as possible. In addition to ordinary cellophane tape, you can try masking tape, packaging tape, electrician's tape, medical adhesive tape, duct tape, sticky computer labels, and so on. For ease of handling, we used microscope slides to secure our specimens and prevent them from curling. We placed each specimen sticky-side up on a slide and then used additional tape to secure the two ends.

Preserving developed prints on sticky tape can be problematic because of the nature of the surface. Transparent tape can be pressed into contact with a white sheet of paper, but that reverses the prints to a mirror image and is not usable for opaque tapes. Because the adhesive often remains sticky after treatment, you can't simply leave the surface exposed. One method is to mount the developed prints by placing the sticky side of the tape in contact with a thin sheet of stiff transparent plastic. (We used a sheet of plastic that was used to protect the screen of a new LCD display we'd purchased. Acetate theater gels or similar sheets also work.) The drawback to mounting the developed prints is that it may make it difficult to see the prints on some surfaces. Overall, the best method of preserving gentian violet developed prints is to photograph or scan them.

FORMULARY

If you do not have the FK01 Forensic Science Kit, you can make up a 0.1% w/v solution of gentian violet by dissolving 0.1 g of gentian violet crystals in 100 mL of distilled or deionized water. Gentian violet is available from lab supply vendors and some drugstores in solid form, and often as a 1% aqueous solution, which can be diluted one part solution to nine parts water to make up a 0.1% solution. Gentian violet is sold under many names, including crystal violet, methyl violet 10B, basic violet 3, brilliant violet 58, Gram stain, and many others. (Gentian violet is the hexamethyl form of methyl violet, which is also available in tetramethyl [methyl violet 2B] and pentamethyl [methyl violet 6B] forms.)

PROCEDURE IV-5-1: PREPARING SPECIMENS FOR GENTIAN VIOLET DEVELOPMENT

1. Label a flat microscope slide for your first specimen. Record the details in your lab notebook.

2. Put a latent fingerprint on your first tape specimen by pressing the tape against your fingertip and then peeling it away.

3. Touching only the edges of the tape, place the specimen sticky side up on a microscope slide and carefully secure both ends of the tape to the slide using additional tape. Make sure the non-sticky side of the specimen is in close contact with the slide and as flat as possible.

> With many types of tapes, this step is harder than it sounds. The tape wants to curl in on itself, stick to itself, and stick to you. It is, after all, sticky tape. You can make things easier by using the forceps or the tip of a pencil to position the tape and hold it in place while you secure it.

4. Repeat steps 1 through 3 with your other tape specimens.

PROCEDURE IV-5-2: DEVELOPING SPECIMENS WITH GENTIAN VIOLET

1. If you have not already done so, put on your splash goggles, gloves, and protective clothing.

2. Place your first specimen sticky-side up on several layers of paper towels. (Be careful: gentian violet stains many surfaces indelibly, including most kitchen counter materials.)

3. Use a pipette to transfer a few drops of gentian violet solution to the exposed area of the tape. If the solution beads, use the tip of the pipette to spread the solution over the entire exposed surface.

4. Allow the stain to work for a minute or so, and then draw it back up into the pipette. You can use the same few drops of stain to treat many specimens.

5. Grasp the specimen with the forceps and rinse it under cold running water for several seconds, as shown in Figure IV-5-1. (You can begin staining the next specimen while you rinse the current specimen.)

Figure IV-5-1: *Rinsing a specimen after treating it with gentian violet solution*

6. Examine the rinsed specimen for visible fingerprints. If the fingerprints are distinct, set the specimen aside to drain and dry. If the prints are stained only lightly, repeat the stain/rinse cycle several times, until no further improvement is evident.

7. Place the specimen on a clean, flat surface and examine it carefully under strong oblique light with the magnifier. You should see fingerprints revealed in considerable detail as violet stains. If you have a camera, shoot an image of the developed fingerprints for your records. Record the pertinent details for the specimen in your lab notebook.

8. Repeat steps 3 through 9 for your other specimens.

Figure IV-5-2 shows a latent fingerprint revealed by gentian violet staining on a dark tape.

Figure IV-5-2: *Latent fingerprint revealed by gentian violet staining on dark tape*

Figure IV-5-3 is a raw scan of a similar latent print on a white sticky label. The "ghost" images like the reversed "49" at the lower-left corner of the image are bleed-through from the printing on the other side of this sticky label. Figure IV-5-4 shows the same image after we ran it through our image-manipulation software to increase contrast and visible ridge detail. Amazingly, in addition to the expected ridge detail, there is actually considerable pore detail visible in this image.

> Incidentally, the images in Figures IV-5-3 and IV-5-4 are of a "real" specimen, not one we dummied up for this lab session. Robert was packing science kits for shipment, applying warning labels required by the US Postal Service. One of the labels stuck to his finger, so he decided to use it as a real live specimen.

Figure IV-5-3: *Raw scan of latent fingerprint revealed by gentian violet staining on a white sticky label*

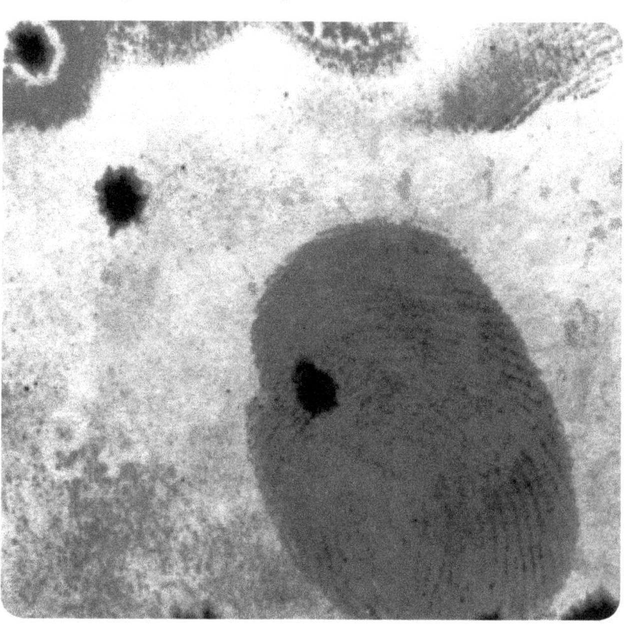

Figure IV-5-4: *An enhanced version of Figure IV-5-3*

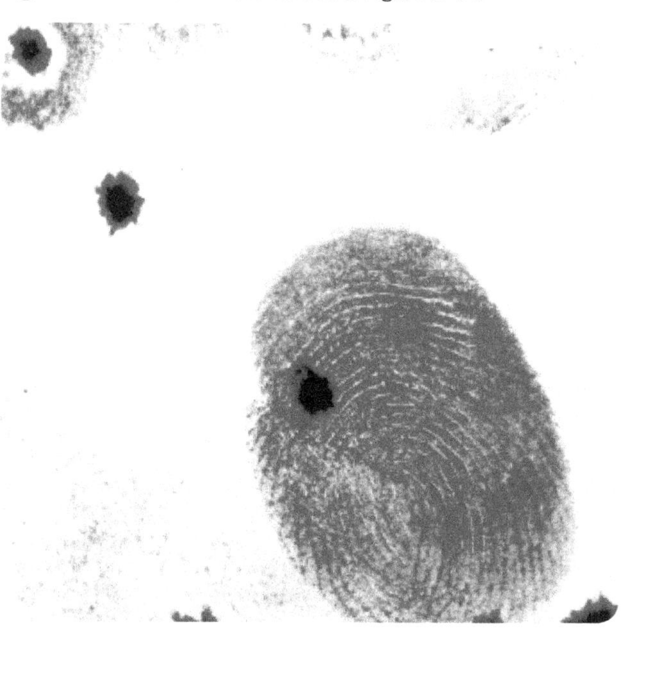

REVIEW QUESTIONS

Q1: **For which specific types of specimens is gentian violet solution best suited?**

Q2: **You are presented with a cellophane tape specimen that may contain latent fingerprints on either or both surfaces. How would you proceed, and why?**

Revealing Latent Fingerprints On Brass Cartridge Cases

Lab IV-6

EQUIPMENT AND MATERIALS

You'll need the following items to complete this lab session. (The standard kit for this book, available from *http://www.thehomescientist.com*, includes the items listed in the first group.)

MATERIALS FROM KIT

- Goggles
- Beaker, 100 mL
- Cartridge case, brass, fired
- Centrifuge tube, 1.5 mL
- Forceps
- Graduated cylinder, 100 mL
- Magnifier
- Stirring rod

MATERIALS YOU PROVIDE

- Gloves
- Camera (optional)
- Paper towels
- Hydrogen peroxide, 3%
- Vinegar, distilled white
- Specimens: uncoated brass (optional)

WARNING

None of the materials used in this lab session present any serious hazard. The 3% hydrogen peroxide solution is irritating and may bleach clothing or other items. As a matter of good practice, always wear goggles, gloves, and protective clothing when doing lab work.

BACKGROUND

Many latent fingerprint development methods are useful on a broad range of surfaces. Others, such gentian violet development of prints on adhesive surfaces, are optimum for one particular special type of surface or a narrow range of surfaces. In this lab session, we'll look at another specialized development technique, this one used for raising latent fingerprints on brass cartridge cases.

There are actually two common methods for developing latent fingerprints on brass cartridge cases, both of which depend on the fact that the oils and fats in fingerprint residues coat the brass and prevent aqueous solutions from contacting it. One of those methods uses gun bluing solution, which is readily available from sporting goods stores. This solution darkens exposed brass, leaving the brass covered by the fingerprint residues bright and shiny. The second method uses even more readily available solutions, ordinary vinegar and 3% hydrogen peroxide, available from any drugstore. That's what we'll use.

PROCEDURE IV-6-1: TREAT SPECIMENS WITH ACIDIFIED HYDROGEN PEROXIDE

1. If you have not already done so, put on your splash goggles, gloves, and protective clothing.

2. Make up the acidified hydrogen peroxide reagent in the beaker by adding 30 mL of 3% hydrogen peroxide to the beaker, followed by 21 mL of distilled white vinegar. Swirl or stir the solution to mix it.

3. Using the forceps, transfer a brass cartridge case upon which you have placed a latent fingerprint to the beaker. Make sure the cartridge case is fully immersed in the solution and that the solution is in contact with all of the exterior surface of the case. If air bubbles adhere to the case, tap it gently with the stirring rod to dislodge them.

4. Swirl the beaker frequently to expose fresh solution to the latent print, or use the forceps to swirl the cartridge case in the solution, as shown in Figure IV-6-1. As development proceeds, the brass that is not protected by the fingerprint residues begins to assume a dull yellowish color. The solution gradually assumes a greenish tinge, caused by copper salts leached from the brass case. Continue development until the latent print is clearly visible against the background.

Figure IV-6-1: *Treating cartridge cases with acidified hydrogen peroxide*

5. When development is complete, use the forceps to remove the case from the beaker, rinse it thoroughly with running water, and then allow it to dry.

> Transfer a small amount of the greenish solution to a 1.5 mL microcentrifuge tube, and retain it for Lab V-1.

6. Examine the case carefully under strong oblique light with the magnifier. You should see the fingerprint revealed in considerable detail. If you have a camera, shoot an image of the developed fingerprints for your records (or images; on the round cartridge case, the entire print may not be visible in a single image). Record the pertinent details for the specimen in your lab notebook.

Figure IV-6-2: *Partial latent fingerprint on a cartridge case revealed by acidified hydrogen peroxide development*

> **DENNIS HILLIARD COMMENTS**
>
> We have had no success in developing prints of value on fired cartridge casings. We theorize that any residue is vaporized when the cartridge is fired. Also, the surface is round, and partial prints have limited value. Whenever possible, it is more useful to develop prints on the weapon.

Figure IV-6-2 shows a partial latent fingerprint on a cartridge case, revealed by development with acidified hydrogen peroxide.

> In September 2008, a new method for revealing fingerprints on cartridge cases was published. John Bond, a forensic scientist with the Northamptonshire Police in the UK, discovered that the copper and brass alloys used in many cartridge cases are corroded very slightly by fingerprint residues, leaving a very faint fingerprint impression. Even when such impressions are much too faint to be revealed by chemical methods or an ALS, Bond discovered that they could be revealed by subjecting the cartridge case to high voltage and treating it with fine metal-based particles similar to laser printer toner. For more information about Bond's technique, visit *http://www.technologyreview.com/communications/21331/*.

REVIEW QUESTION

Q1: **Latent fingerprints on cartridge cases are sometimes the only way to tie a criminal to the weapon he used. Propose an explanation.**

Detecting Blood

Group V

Detecting blood, in the field and in the lab, is an important part of forensic work, but it is surprisingly difficult to establish unambiguously that a suspect stain is in fact blood. Even if obvious splatters or pools of blood-like material are found at a crime scene, it can't be assumed that they are blood. More than one investigator has been fooled by paint, hydraulic fluid, or other liquids that resemble blood, sometimes quite closely. Furthermore, bloodstains are by no means always obvious. Bloodstains may have been washed away by the criminal, leaving only invisible traces, or the blood may have been deposited on foliage or other materials that make it difficult to see.

If the crime scene covers a large area, it may not even be immediately obvious where to begin looking for latent bloodstains (also called occult bloodstains, where occult is used in the sense of hidden). After taking specimens of any patent bloodstains, forensic technicians use one, two, or all three of the following methods to "scan" the crime scene for possible latent bloodstains. All three of these methods are presumptive. A positive result does not establish the presence of blood, but provides a reasonable supposition that blood may be present.

Alternate light sources

Unlike most body fluids, blood does not fluoresce at visible wavelengths, but that doesn't mean an alternate light source is entirely useless in locating blood stains. Blood strongly absorbs light in the near UV and violet ranges, so illuminating a suspect area with a black light may render bloodstains more visible because they are darkened relative to the background.

Luminol

In a 1937 paper, the German chemist Walter Specht was the first to suggest the use of *luminol* for forensic blood detection. An aqueous or alcoholic solution of luminol and an oxidizer is catalyzed by the iron present in the hemoglobin component of blood to produce *3-aminophthalate* (3-APA) in an excited state. Excited 3-APA molecules quickly return to their base state, emitting photons that are visible as a characteristic weak blue luminescence (not fluorescence; in the presence of iron or another catalyst, luminol emits light directly).

Luminol is extraordinarily sensitive to the presence of blood, yielding positive results at dilutions as high as 100,000,000:1, and by some reports 1,000,000,000:1. Washing a surface thoroughly or even painting over a bloodstained surface often leaves sufficient traces of blood to yield a positive test with luminol. Unfortunately, this high sensitivity is accompanied by low selectivity. Many materials other than blood—including copper and other metals, laundry bleach, and many food items—yield positive luminol tests that are indistinguishable from positive results caused by actual blood. Interestingly, most body fluids other than blood do not react with luminol.

When luminol is used at a crime scene, cameras and luminescent markers are set up first, the crime scene is darkened, and then luminol solution is sprayed liberally on all surfaces that may have latent bloodstains. (It's not uncommon for forensic technicians to use what amounts to garden sprayers to spray luminol solution literally by the liter.) Walls are sprayed first to detect spatter patterns, followed by the ceiling to detect cast-off patterns, followed by the floor to detect footprints, drag marks, and so on.

Luminol, at least in aqueous solution, is considered non-destructive. Although it may interfere with some older serological tests, using aqueous luminol does not interfere with subsequent PCR DNA analysis.

Fluorescein

Aqueous or alcoholic solutions of *fluorescein* are used in the same way as luminol to detect latent bloodstains. The chief differences are that fluorescein is much less sensitive than luminol—10,000:1 versus 10,000,000:1 or more—and must be illuminated by an ALS (alternate light source) at about 445 nm (indigo) to induce fluorescence at 520 nm (green). The deep violet visible light produced by most "black light" bulbs may produce faint fluorescence with fluorescein; more intense forensic ALS units produce much brighter fluorescence. Like luminol, fluorescein is subject to false positives caused by many common materials. With all of those drawbacks, you might reasonably wonder why any forensic scientist would use fluorescein instead of luminol. The answer is, they don't. Fluorescein is used only if luminol fails to yield usable results. At times, fluorescein may reveal latent bloodstains when luminol fails, because materials are present that either yield false positives with luminol or suppress the luminescence of luminol. In particular, fluorescein is used when the presence of bleach interferes with luminol.

If patent possible bloodstains are present, or if a positive result with any of the preceding tests indicates that latent bloodstains may be present, the next step is to run a color-change test to verify the result from the preliminary screening test. Color-change tests are less sensitive than luminol, but more selective. Although a positive result with the color-change test is not confirmatory in a legal sense, it adds weight to the presumption of the presence of blood and indicates that legally confirmatory tests should be done in the lab. If the color-change test is negative, investigators know not to waste scarce lab resources to investigate further.

The first such color-change test, the guaiacum test, was introduced in 1865. In the presence of blood, guaiacum and hydrogen peroxide react to form the dye guaiacum blue. Although it is relatively sensitive, about 50,000:1, the guaiacum test is extremely unselective, yielding false positives with numerous materials. Despite its lack of selectivity, the guaiacum test was commonly used until 1904, when the benzidine test was introduced. In the presence of acetic acid and hydrogen peroxide, benzidine reacts with blood to form the dye benzidine blue. The benzidine test is no more selective than the guaiacum test, but is considerably more sensitive at about 250,000:1. The next color-change test reagent, o-toluidine, was introduced in 1912. In the presence of acetic acid and ethanol, o-toluidine reacts with blood to form a characteristic blue-green stain. The o-toluidine test is no more selective than the guaiacum or benzidine tests, but is considerably more sensitive, at about 5,000,000:1.

The benzidine and o-toluidine tests remained in use for decades, until it was learned that both of these compounds are strongly carcinogenic. Fortunately, alternative methods are available, including the following:

Kastle-Meyer test

The *Kastle-Meyer* (KM) test, introduced in 1901 by Kastle and improved in 1903 by Meyer, uses an alkaline phenolphthalin solution (the reduced form of the phenolphthalein commonly used as an acid-base indicator). Phenolphthalin in alkaline solution is colorless or a very pale straw yellow, as opposed to phenolphthalein, which is bright pink in alkaline solution. KM reagent reacts with the heme component of the hemoglobin in blood and hydrogen peroxide to produce the oxidized phenolphthalein form, which turns bright pink.

The KM test can be run in two ways. If the phenolphthalin reagent and the hydrogen peroxide are combined first, the test is extremely sensitive (10,000,000:1 or better) but is also extremely unselective. Conversely, if the reagents are applied sequentially—phenolphthalin reagent first, followed by hydrogen peroxide—the test is much less sensitive (~20,000:1) but extremely selective. For that reason, if KM is to be used as a secondary test, the sequential method is used for its higher selectivity. For that method, a small amount of the suspect stain is transferred to a cotton swab moistened with water. A drop or two of alcohol is placed on the swab to lyse the blood cells and release the catalase enzyme present in hemoglobin. A drop or two of the alkaline phenolphthalin reagent is placed on the swab and observed for a few seconds. An immediate pink stain is inconclusive, indicating the presence of an inorganic oxidizer or a base. (Blood may also be present, but is screened by the oxidizer or base.) If the swab remains colorless, a drop or two of drugstore 3% hydrogen peroxide is placed on the swab. If blood is present, the catalase enzyme catalyzes the breakdown of the hydrogen peroxide into water and oxygen, and the swab immediately assumes a pink color. (Atmospheric oxygen causes the swab to turn pink within a minute or so even in the absence of blood.) False positives can be produced by vegetable peroxidases, including those present in horseradish, potato, and other vegetable matter.

The KM test is still widely used because it combines high sensitivity and selectivity with low cost. Well-funded departments often use the more convenient (and more expensive) tetramethylbenzidine test (described in just a moment), but the KM test remains a mainstream procedure.

Although, like all catalytic tests, the KM test is subject to false positives, those false positives may be at least somewhat distinguishable from true positives. That is, in the presence of interfering materials, many blood reagents yield color changes that are indistinguishable from the color change that occurs in the presence of actual blood. KM reagent reacts with many interfering materials to yield a color change, but that color often has a yellowish, orangish, or reddish tint rather than the pure bright pink of a true positive.

Leucomalachite green

The *leucomalachite green* (LMG) test works on the same principle as the KM test. In its reduced form, LMG is colorless. When it reacts with the oxygen released by catalase catalysis of hydrogen peroxide, LMG is oxidized to the blue-green dye malachite green. In use, a small amount of the suspect stain is transferred to a cotton swab moistened with water. A drop or two of LMG solution is placed on the swab and observed for a few seconds. An immediate blue-green stain is inconclusive, indicating the presence of an inorganic oxidizer. If the swab remains colorless, a drop or two of drugstore 3% hydrogen peroxide is placed on the swab. If blood is present, the catalase enzyme catalyzes the breakdown of the hydrogen peroxide into water and oxygen, and the swab immediately assumes a blue-green color. Although the LMG test is less sensitive than the KM test and is subject to many of the same interferences, including vegetable peroxidase enzymes, it is sometimes still used because it is less affected by the presence of reducing agents like ascorbic acid (vitamin C), which produce false negatives with the KM test.

Tetramethylbenzidene

The *tetramethylbenzidine* (TMB) test was introduced in 1976 as a safer alternative to the benzidine and o-toluidine tests. At about 1,000,000:1, the sensitivity of the TMB test is superior to the sequential KM test, and its selectivity is similar. TMB is the most commonly used color test because, in addition to its high sensitivity and selectivity, it is very convenient. A packaged TMB test is available in the form of *Hemastix*, which are plastic test strips originally designed for detecting blood in urine. The treated pad on these strips contains 3,3',5,5'-tetramethylbenzidine and diisopropylbenzene dihydroperoxide (an oxidizer similar to hydrogen peroxide). When moistened with distilled water, these strips react with blood within a minute to form a colored stain that typically ranges from orange through green. If blood is present in high concentration, the stain may assume a blue color. Some forensics departments use Hematest tablets rather than Hemastix strips. These tablets work similarly and provide essentially identical results.

In this group, we'll explore the sensitivity and selectivity of KM reagent and use it to detect latent bloodstains.

Testing the Sensitivity and Selectivity of Kastle-Meyer Reagent

EQUIPMENT AND MATERIALS

You'll need the following items to complete this lab session. (The standard kit for this book, available from *http://www.thehomescientist.com*, includes the items listed in the first group.)

MATERIALS FROM KIT

- Goggles
- Blood, synthetic
- Centrifuge tube, 1.5 mL
- Kastle-Meyer reagent
- Pipettes
- Reaction plate, 24-well

MATERIALS YOU PROVIDE

- Gloves
- Cotton swabs
- Hair dryer (optional)
- Hydrogen peroxide, 3%
- Nickel (US coin)
- Paper (white copy or printer)
- Pencil solution from Lab IV-6
- Ultraviolet (UV) light source (optional)
- Vinegar, distilled white
- Watch or clock with second hand
- Water, distilled or deionized
- Specimens, food (see Procedure V-1-4)

WARNING

Kastle-Meyer reagent is corrosive. Ethanol is flammable, so avoid open flame. Hydrogen peroxide is an oxidizer and irritant, and may bleach clothing and other materials with which it comes into contact. Zinc dust or powder is pyrophoric (spontaneously combustible) in the presence of moisture. Wear splash goggles, gloves, and protective clothing.

BACKGROUND

Soon after its introduction more than a century ago, the Kastle-Meyer (KM) test became the most commonly used color test for patent and latent bloodstains. Before the introduction of luminol, KM was sometimes used in the same way luminol is used today—to scan large areas for latent bloodstains. Combined KM reagent (phenolphthalin and peroxide mixed together) was used for that purpose, often in a spray bottle. The high sensitivity of combined KM reagent—1:1,000,000 or more—was ideal for such preliminary scans. After the combined KM test revealed any latent potential bloodstains, individual stains were tested with the more selective KM sequential method for preliminary confirmation.

In this lab session, we'll test KM reagent against known concentrations of blood to determine its sensitivity (both combined and sequential), and then use it to test latent bloodstains and potential false-positive and false-negative materials.

FORMULARY

If you don't have the FK01 Forensic Science Kit, you can purchase Kastle-Meyer reagent from any forensic supply vendor or make it up yourself as follows:

1. If you have not already done so, put on your goggles, gloves, and protective clothing.
2. Transfer 50 mL of distilled or deionized water to a 250 mL Erlenmeyer flask.
3. Weigh out 20 g of potassium hydroxide and add it to the flask in small portions, swirling until the solid dissolves. (**Caution: this process is very exothermic, and the resulting solution is extremely corrosive**.)
4. Weigh out 2 g of phenolphthalein powder, add it to the flask, and swirl until the powder dissolves. The solution turns bright pink.
5. Weigh out 20 g of zinc powder and add it to the flask.
6. Add a boiling chip, and stopper the flask loosely with a cotton ball to minimize evaporation.
7. Place the flask on a hotplate, and heat the solution until it approaches a boil.
8. Reduce the heat and allow the solution to simmer until the bright pink solution turns colorless (or a very pale straw yellow), which may require an hour or two. Add distilled water as necessary during this process to replace water lost to evaporation and keep the volume of the solution near the original level.
9. After the solution turns colorless, remove the flask from the heat and allow it to cool to room temperature.
10. Make up the solution to 100 mL using 70% ethanol.
11. Decant the solution into a brown glass storage bottle labeled "Kastle-Meyer reagent." Add about 0.5 g of zinc granules to the storage bottle to prevent oxidation.

Caution: wet zinc powder is pyrophoric (catches fire spontaneously). React the excess zinc powder in the flask with hydrochloric acid to form zinc chloride solution. Flush that solution down the drain with copious water.

This solution remains usable for several months if stored at room temperature in a tightly stoppered bottle, and for years if refrigerated.

PROCEDURE V-1-1: PREPARE KNOWN DILUTIONS OF BLOOD

To test the sensitivity of KM reagent, we need to produce known dilutions of blood that cover a wide range of concentrations. The easiest way to do this quickly is to use the following procedure, which is called *serial dilution*.

1. If you have not already done so, put on your splash goggles, gloves, and protective clothing.

2. Transfer 0.5 mL of distilled or deionized water to each well of the 24-well reaction plate. (The second line up from the tip on a disposable pipette is the 0.5 mL line.)

3. Transfer 0.5 mL of the synthetic blood to the well A1 of the reaction plate. Use the tip of the pipette to mix the water and blood. Draw up and expel the contents of the well several times to ensure complete mixing. Record the contents of that well in your lab notebook as blood:water 1:1.

4. Draw up 0.5 mL of the blood-water mix from well A1 and expel it into well A2. Mix the contents of that well thoroughly, and record them in your lab notebook as blood:water 1:3.

5. Draw up 0.5 mL of the blood-water mix from well A2 and expel it into well A3. Mix the contents of that well thoroughly, and record them in your lab notebook as blood:water 1:7.

6. Repeat this serial dilution procedure for the remaining wells, producing the following nominal concentrations:

   A1 – 1:1

   A2 – 1:3

   A3 – 1:7

   A4 – 1:15

   A5 – 1:31

   A6 – 1:63

   B1 – 1:127

   B2 – 1:255

   B3 – 1:511

   B4 – 1:1,023

   B5 – 1:2,047

   B6 – 1:4,095

   C1 – 1:8,191

   C2 – 1:16,383

   C3 – 1:32,767

   C4 – 1:65,535

   C5 – 1:131,071

   C6 – 1:262,143

   D1 – 1:524,287

   D2 – 1:1,048,575

   D3 – 1:2,097,151

   D4 – 1:4,194,303

   D5 – 1:8,388,607

   D6 – 1:16,777,215

Of course, there are far too many significant figures in these dilutions for our actual level of accuracy. We'd actually record them to many fewer significant figures. For example, for wells B3 and following, we'd record values of "~1:500," "~1:1,000," "~1:2,000," "~1:4,000," and so on.

You'll use these serially diluted blood specimens in the following procedures. If you complete this lab session in one day, you can simply leave the specimens uncovered. If you do the procedures over a period of days, replace the lid on the reaction plate to prevent evaporation. You needn't refrigerate the specimens between sessions. After all, real bloodstains at crime scenes are not refrigerated.

PROCEDURE V-1-2: SPOT KNOWN DILUTIONS OF BLOOD

Although there are obviously exceptions, most latent bloodstains that are processed by crime-scene investigators are dry. To determine the sensitivity of KM reagent to such dried bloodstains, we'll produce simulated dried bloodstains by transferring our various concentrations of blood to paper and allowing them to dry. We'll need two such specimens: one each to test sensitivity of KM reagent using the combined and sequential procedures. We'll later prepare additional sheets to test the effects of various contaminants that produce false negatives with KM reagent.

1. If you have not already done so, put on your splash goggles, gloves, and protective clothing.
2. Using a pencil, label two sheets of copy or printer paper, each with an array of the approximate concentrations of the 24 wells in the reaction plate. Make a 2 to 3 cm circle next to each label.
3. Place the sheets side-by-side on a layer of paper towels to absorb any soak-through.
4. Use a clean pipette to withdraw a tiny amount of the solution in well D6. Put one drop of that solution in the centers of the 1:16,000,000 circles on both sheets of paper, as shown in Figure V-1-1. Expel any remaining solution from the pipette back into well D6.
5. Repeat step 4 using solution from well D5 to spot the 1:8,000,000 circles on both sheets of paper.
6. Repeat step 5 until you have spotted all 24 circles on both sheets of paper with the corresponding blood dilutions.

Figure V-1-1. Spotting diluted blood specimens

7. Allow the sheets of paper to dry naturally, or use a hair dryer to speed the process.

> When the sheets have dried, it will be obvious why we made a circle for each dilution. Most of the spots will be invisible once dried.

PROCEDURE V-1-3: TEST SENSITIVITY OF KASTLE-MEYER REAGENT

In this procedure, we'll test the sensitivity of KM reagent against dried bloodstains, using both the combined and sequential protocols. We know that, under ideal conditions, the sensitivity of the combined KM reagent can be as high as 1:10,000,000. We don't really expect to see a positive result at our 1:8,000,000 concentration. A very small amount of that specimen is spread across a relative large area of paper, and a very faint positive result may be too pale to see. What we do expect is to see a positive result at at least moderately low to very low blood concentrations.

1. If you have not already done so, put on your splash goggles, gloves, and protective clothing.

2. Use a clean pipette to draw up 0.5 mL (the second line up from the tip of the pipette) of Kastle-Meyer reagent and transfer it to a 1.5 mL microcentrifuge tube.

3. Use another clean pipette to draw up 0.5 mL of 3% hydrogen peroxide and transfer it to the microcentrifuge tube. Gently draw up and expel the solutions several times to mix them thoroughly.

4. Working quickly, add one drop of the combined Kastle-Meyer reagent to each of the 24 circles on the first specimen sheet. Also add one drop of the combined KM reagent to a corner of the sheet that has not been exposed to any of the blood specimens.

5. Observe all 24 circles and the corner for any color change. Record your observations in your lab notebook.

> Retain this combined Kastel-Meyer reagent for Procedure V-1-4, if you intend to do it immediately following this procedure; otherwise, mix fresh reagent for it.

We'll use the second specimen sheet to determine the sensitivity of sequential Kastle-Meyer testing. Even in the absence of blood or a material that yields false positives, Kastel-Meyer reagent is oxidized by air to yield a pink color, so the first task is to determine how long it takes for atmospheric oxygen to produce a false positive reaction.

6. Add one drop of Kastel-Meyer reagent to an area of the paper where no blood is present. Wait a couple seconds, note the time, and then add one drop of 3% hydrogen peroxide to that same spot. (If a pink color develops in the short time between making the KM reagent spot and adding the hydrogen peroxide, something is causing a false positive. It may be the paper itself or some other environmental factor.)

7. Observe the spot carefully. When the first hint of a pink color appears, note the elapsed time. The exact time required varies with test conditions, the type of paper, and other factors, but is typically in the range of 30 seconds to one minute. Record the actual time required in your lab notebook.

While the sensitivity of combined Kastle-Meyer testing is often stated as 1:1,000,000, that of sequential testing is usually stated as more like 1:20,000. Accordingly, we can start with a concentration near that level—such as the spot made with solution C3—rather than testing every one of the very dilute spots.

8. Add one drop of KM reagent to spot C3 on the second specimen sheet. Observe the spot for a couple of seconds to make sure that no reaction occurs.

9. Add one drop of 3% hydrogen peroxide to spot C3 and observe the spot. A color change to pink within about 15 to 20 seconds is a positive test.

10. If a clear positive reaction occurred with spot C3, repeat the test for C4 and other more dilute spots until no reaction occurs. The most dilute spot that showed a positive reaction tells you the sensitivity of sequential KM testing under your conditions. Record that information in your lab notebook.

11. If no reaction occurred with spot C3, repeat the test for spot C2. Continue testing additional spots if necessary until you locate the spot with the minimum concentration that provides a clear positive reaction. Record that information in your lab notebook.

PROCEDURE V-1-4: TEST SELECTIVITY OF KASTLE-MEYER REAGENT

1. If you have not already done so, put on your splash goggles, gloves, and protective clothing.

2. Produce two spotted specimen sheets as you did in Procedure V-1-2, but rather than diluted blood specimens, use dilute solutions of various materials that might cause false positives.

   Use as many different materials as you have available and have time to test. In particular, examine materials that a criminal might use to attempt to clean up blood stains, such as chlorine laundry bleach, hand soap, laundry detergent, and Formula 409 or a similar cleaner. Also test dilute solutions of potentially-interfering materials that might be found as low-level environmental pollutants: copper-zinc (the solution from Lab IV-6); copper-nickel (soak a US or Canadian nickel coin in distilled white vinegar overnight); iron (soak a nail or steel wool in vinegar overnight). Finally, test various food items and other common household materials, including horseradish (or Horsey Sauce from Arby's), mashed or ground potato, broccoli, cauliflower, or other vegetables. You can also try various liquids that are insoluble in water, such as oils, glues, cosmetics, and paints and varnishes.

3. Test the spots on the first sheet by using the steps described in Procedure V-1-3 for combined Kastle-Meyer testing. Record your observations in your lab notebook, including any departures from the pure pink color of a true positive.

4. Test the spots on the second sheet by using the steps described in Procedure V-1-3 for sequential Kastle-Meyer testing. Record your observations in your lab notebook, again, including any departures from the pure pink color of a true positive.

PROCEDURE V-1-5: FIELD TESTING WITH KASTLE-MEYER REAGENT

Until now, we've been testing KM reagent under controlled conditions. Of course, criminals seldom leave bloodstain specimens conveniently arranged on sheets of paper. In the field, KM reagent is normally used to detect bloodstains by swabbing those stains with a damp cotton swab and then testing the tip of the swab. Under those circumstances, KM reagent is less sensitive than under controlled conditions.

Because sequential KM testing is normally used to confirm suspicious stains revealed by luminol, we'll use sequential KM to test swabs taken from various objects that contain one or more of the dilute blood specimens we prepared in Procedure V-1-1. This method closely simulates actual field testing.

1. If you have not already done so, put on your splash goggles, gloves, and protective clothing.

2. Contaminate various surfaces with one drop each of the 1:1,023 blood dilution, mark the locations of the bloodstains (which in real life would have been revealed by luminol), and allow the specimens to dry.

The 1:1,023 blood specimen is much too dilute to leave visible stains, so use your imagination to choose various surfaces to test. You might, for example, choose the kitchen counter, hardwood, tile, and carpeted floors, a piece of clothing, the bathroom sink or shower, a painted wall, and so on.

3. Moisten the tip of a cotton swab with a drop or two of distilled or deionized water. You want the cotton damp, but not dripping wet. Swab the area of the stain with the tip of the swab and place the swab in a labeled bag or other container. (It does no harm to allow the swab to dry before performing the test.)

4. Apply one drop of KM reagent to the tip of the first swab and observe it for a couple seconds. There should be no color change. A color change suggests a false positive.

5. Apply one drop of 3% hydrogen peroxide to the tip of the swab. If blood is present on the swab at detectable levels, the swab should assume a pink color within 15 seconds or so. Again, atmospheric oxygen creates false positives after (typically) 30 seconds to one minute of exposure, so a pink color that doesn't appear until 30 seconds or more into the test should be recorded as a negative test. Record your observations in your lab notebook.

Figure V-1-2 shows negative (left) and positive KM swab tests. A negative swab remains white or has only an extremely pale yellow coloration caused by the KM reagent itself. A positive swab has an unmistakable pink coloration.

Figure V-1-2: *Swabs showing negative (left) and positive KM tests for blood*

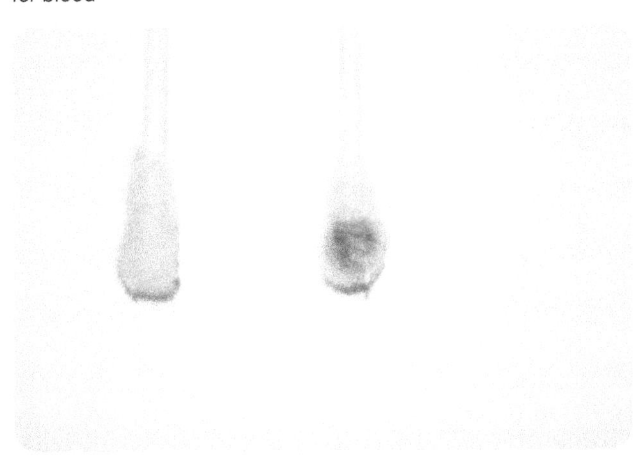

6. Repeat steps 4 and 5 for each of your other exposed swabs.

REVIEW QUESTIONS

Q1: **What are the advantages and disadvantages of the combined KM method versus the sequential KM method?**

Q2: **When spotting the diluted blood specimens, why did we start with well D6 rather than well A1?**

Q3: **Why did we apply hydrogen peroxide and KM reagent to a corner of the paper specimen that had not been exposed to any of the blood specimens?**

Q4: **What color changes did you observe while testing the sensitivity of the KM reagent? What conclusions did you make?**

Q5: **At what minimum blood concentration did combined KM reagent yield a positive result?**

Q6: **At what minimum blood concentration did sequential KM reagent yield a positive result?**

Q7: **Why did we spot the specimen sheet to be tested with sequential KM testing with all concentrations, rather than just those of at least 1:20,000?**

Q8: **After completing Procedure V-1-4, what did you conclude about the selectivity of KM reagent? Did you observe any differences in selectivity between combined testing and sequential testing?**

Impression Analysis Group VI

Impression analysis makes up a significant part of the workload of any forensic laboratory. Any time an object is pressed into contact with another object, an impression may result. For example, the tread of a shoe may be impressed into snow, mud, soft dirt, or a carpet. A hit-and-run driver's vehicle bumper may leave characteristic impressions in an object he strikes, or even on the body of his victim. A burglar's crowbar may leave characteristic impressions on a door frame or striker plate, and the pliers used by a terrorist bomb maker to cut wires may leave characteristic impressions on the wire.

Impressions are useful primarily to the extent that they are individualizable. Primarily, but not exclusively. Consider a shoe imprint found at a crime scene. The amount of information that can be gained from that shoe imprint depends on the quality of the imprint and the condition of the shoe that made it. If, for example, the shoe that made the imprint was relatively new and the quality of the imprint low, it may be that the most that can be determined is the size, manufacturer, and model of the shoe. While it may be useful to know that the imprint was made by, say, a men's size 10 Adidas Uraha running shoe, there are tens of thousands of individual shoes that might have made that impression, relegating this particular impression to class evidence. Conversely, a good impression made by a well-worn shoe may be individualizable because it contains characteristic wear patterns, cracks, and other individual elements.

Individual characteristics may arise from manufacturing differences or from subsequent wear during use, or both. For example, every rifle and pistol leaves impressions on every bullet it fires, as well as every cartridge case. The relatively soft bullet is pressed into tight contact with the rifling present in the barrel, shown in Figure VI-0-1, leaving an impression of the rifling marks on the bullet itself. General rifling characteristics—the number, size, separation, handedness, and degree of twist—are class evidence that allow a bullet to be placed in a class with other bullets that have similar general rifling characteristics. Every Ruger .357 Magnum Police Service Six revolver, for example, marks bullets with similar rifling marks, so examining a fired bullet for these general rifling characteristics can at best identify the manufacturer and model of the weapon.

Figure VI-0-1: *Looking down the muzzle of a .44 revolver at the spiral rifling lands and grooves*

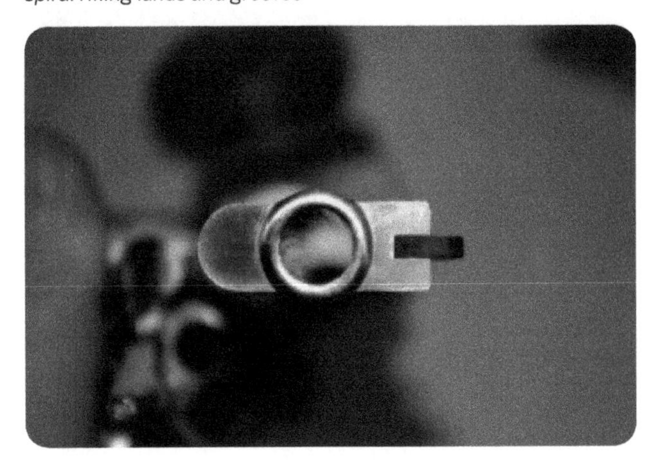

However, we can match a particular fired bullet or cartridge case to a specific weapon by looking beyond the class characteristics of the particular weapon to its individual characteristics. For example, the rifling marks produced by one particular Ruger .357 Magnum Police Service Six revolver differ microscopically from those produced by any other Ruger .357 Magnum Police Service Six revolver. This is true because cutting rifling into a barrel produces microscopic striations that differ for each barrel. The tooling used to cut the rifling into the barrels differs slightly from one set to the next, and the individual tooling wears each time it is used to cut rifling in another barrel. As a result, no barrel is exactly like any other.

The same is true for impressions made on cartridge cases by firing pins, extractors, ejectors, and the chamber itself. Figure VI-0-2 shows the impressions made by the firing pin of the same Ruger .357 Magnum Police Service Six revolver on the primers of two Remington-Peters .38 Special cartridges. Cartridge cases and primers are typically made from brass, aluminum, or other soft metals that are ideal for receiving impressions from the hardened steel components of the firearm. All of these impressions can be individualized, and, particularly in combination, they can be used to match a particular fired cartridge case to the specific weapon that fired it.

Figure VI-0-2: *The firing pin impressions made by a revolver on the primers of two .38 Special cartridges*

Hand tools have similar microscopic individual characteristics. For example, the hardened steel cutting edge of a wire cutter has microscopic nicks that are impressed into the relatively soft metal of a wire as striations each time a wire is cut. The working edges of screwdrivers, chisels, crowbars, bolt cutters, scissors, and similar tools have similar individualizable microscopic variations, all of which may be matched against the impressions they leave.

Professional forensics labs have many tools at their disposal for impression analysis, including expensive instruments such as comparison microscopes and scanning electron microscopes. But expensive equipment isn't necessarily needed to compare and match the types of impressions commonly found at crime scenes. In this group, we'll compare and match various types of impressions using only a standard compound microscope.

Tool Mark Analysis

EQUIPMENT AND MATERIALS

You'll need the following items to complete this lab session. (The standard kit for this book, available from *http://www.thehomescientist.com*, includes the items listed in the first group.)

MATERIALS FROM KIT

- Goggles
- Forceps
- Magnifier
- Slides, flat

MATERIALS YOU PROVIDE

- Gloves
- Camera with microscope adapter (optional)
- Desk lamp, booklight or other top illuminator
- Marking pen
- Microscope
- Pie pans, disposable aluminum
- Scissors
- Tape, masking
- Specimens, screwdrivers (see text)
- Specimens, cutting pliers (see text)

WARNING

None of the activities in this lab session present any serious hazard. The cut edges of aluminum sheets may be razor sharp, so use care when handling them.

BACKGROUND

Any tool that is used to cut, pry, or otherwise manipulate objects almost unavoidably leaves characteristic impressions on those objects. For durability, most tools—and, in particular, their working surfaces—are made from hardened steel or other even harder alloys. Tools are normally used to work softer materials such as wood, plastics, and softer metals, so the tool invariably leaves an impression on the material being worked rather than vice versa. The most common types of tool mark impressions analyzed by forensic labs are compression impressions—such as those made when a plier cuts a wire or a bolt cutter cuts a padlock shackle—and scoring impressions—such as those produced when a screwdriver, chisel, or crowbar is used to pry open a window, door, safe, or similar object.

Ordinarily, a forensic examiner uses a *comparison microscope* to compare impressions. A comparison microscope is essentially two identical standard microscopes whose optical trains have been physically linked to provide an eyepiece image that shows the image from one microscope on the left side of the eyepiece field and the image from the second microscope on the right side of the eyepiece field. Because each of the microscopes has an independent mechanical stage, the two specimens can be positioned in the eyepiece side by side, allowing direct visual comparison and imaging.

Physical comparison microscopes are extremely expensive. Basic Chinese models cost $2,000 or more, and a professional model from a German or Japanese manufacturer with all the bells and whistles can easily cost $50,000 or more. Although physical comparison microscopes are still sold and remain in common use, the trend is toward virtual comparison microscopes, which allow comparisons to be made from two standard microscopes equipped with digital cameras. The images from the two microscopes are combined in software and displayed on a computer monitor.

In this lab session, we'll make compression impressions and scoring impressions on soft metal objects and compare and attempt to match impressions made by known tools with questioned impressions.

To make the compression impressions, you'll need at least two cutting pliers—wire cutters, needle-nose pliers, diagonal cutters, and so on. If your own toolbox has only as limited selection of cutting pliers, borrow examples of these items from friends and neighbors. To make the scoring impressions, you'll need at least two flat-blade screwdrivers, both or all of the same size. Again, friends and neighbors are a good source of diverse specimens.

We'll make the scoring impressions on aluminum sheets. Even heavy-duty aluminum foil isn't strong enough to prevent tearing when making scoring impressions, so we used disposable aluminum pie plates, which are available inexpensively in any supermarket. For the compression impressions, we'll use the cutting pliers to cut small widths of aluminum sheet.

Digital images made through your microscope are extremely useful for comparison purposes, particularly if your microscope is set up next to a computer. We shot images as we worked and transferred them to the computer. That allowed us to make direct side-by-side comparisons between what we were observing through the microscope and what we'd seen in earlier observations. If you do not have the equipment needed to shoot images through your microscope, make accurate sketches, instead.

PROCEDURE VI-1-1: PRODUCE AND COMPARE COMPRESSION SPECIMENS

To compare and match a questioned cut wire specimen against known wire cutters, it's necessary to produce at least one and often many known exemplars with the known wire cutter. Many, because the cutting blades of most cutters are usually much wider than the wire cut with them. For example, the cutting blades of a diagonal cutter may be 1.5 cm wide. If that cutter is used to cut a 24-gauge copper wire, which is 0.511 mm in diameter, the blades are more than 29 times the width of the wire. The forensic examiner has no way of knowing what part of the blades was used to cut the wire, so the only option is to produce many overlapping exemplars that cover the entire width of the blades. That may mean making 50 or more exemplars, to allow for some overlap, and then making 200 or more separate comparisons.

> Why are 200 comparisons needed for 50 exemplars? Remember, the examiner has no way of knowing which of the two sides of the blades made the cut, and each cut has a top side and a bottom side. If the examiner is comparing one side of the questioned cut wire specimen against exemplars produced with known cutting tools, to rule out a match, she must compare both sides of both cut ends of each known exemplar. (Although any wire specimen starts out as round, when it's cut, it ends up with two distinct sides because as the cut is made, the cutting tool crushes the wire to conform to the flat blades.)
>
> You might think it would be possible to reduce the required number of known exemplars by using heavier gauge wire. For example, if the questioned specimen is 24 gauge (0.511 mm), you might instead use heavier 10-gauge wire, which is 2.588 mm in diameter, or five times wider than the thinner wire. And in fact this is often done in less serious cases (see below), either by using heavier-gauge wire or by using thin, flat, wide, soft metal plates. But striation patterns can differ if the specimens are of different thickness, so although such shortcuts might be used in a serious case to establish a quick presumptive match, for evidentiary purposes, the specimens are made with wire that is consistent with the questioned specimen.
>
> You might also think it should be possible simply to image the cutting blade and compare the striations on it with those in the cut wire specimen. Unfortunately, this doesn't work in practice. The cutting process is sequential. The edges of the blade make first contact, impressing their microscopic imperfections on the surface of the wire. As the cut continues, the wire is crushed and deformed and subsequently comes into contact with portions of the blades farther and farther from the edges. The imperfections in these areas of the blades are impressed upon those made earlier, changing their appearance.

Obviously, that's a great deal of work and occupies a lot of skilled technician time, so such complete comparisons are normally done only in very serious cases, such as a terrorist bombing. In such cases, the forensic lab will attempt to match the questioned wire specimen as closely as possible to the wire they use to make the exemplars—including gauge and material—and will take as much time as necessary to either confirm or rule out a match.

In less serious cases, the forensic examiner may take a shortcut by using a wider material to produce the known exemplars. To continue the example from above, clearly there's no way a typical diagonal cutter can cut a wire that's 1.5 cm in diameter. The solution is to substitute a heavier gauge wire or a flat, thin plate made of a soft metal (such as aluminum, copper, or lead) that is as wide as the jaws of the cutter, or nearly so. Or, more commonly, two pieces of soft metal plate are used, each perhaps two-thirds the width of the blades. This provides sufficient overlap for the two known exemplars to be compared quickly with the questioned wire specimen.

To simplify this procedure and save time, we'll use widths of aluminum sheet rather than wires. Supply a volunteer with an aluminum strip 7 or 8 cm long and a bit narrower than the narrowest width of the cutting blades of any of your specimens. Ask your volunteer to chose one of your cutting plier specimens, use it to cut that strip in half with a single, continuous cut, label one of the cut halves with an arrow pointing to the edge cut with the questioned plier, and give you that cut half as your questioned specimen.

1. Prepare two aluminum strips for each of your known specimens, each about 7 or 8 cm long and roughly two-thirds to three-quarters the width of the cutting blade of that plier specimen. One one side of each strip, label both ends with the number of specimen and an arrow pointing toward the center of the strip. For example, you might label the first pair of strips:

   K1A ⟶ ⟵ K1A

   K1B ⟶ ⟵ K1B

   Labeling each strip in this fashion allows you to identify unambiguously which plier was used to make the cut, whether the cut was made by the left or right side of the blades, and whether the cut was made by the top or bottom blade.

2. Position the K1A strip in the cutting blades of the K1 plier specimen, with one edge of the strip as close as possible to the edge of the cutting blades nearest the tip of the plier. Make the cut in a single, smooth motion.

3. Position the K1B strip in the cutting blades of the K1 plier specimen, with one edge of the strip as close as possible to the edge of the cutting blades nearest the handles of the plier. Make the cut in a single, smooth motion.

4. Repeat steps 2 and 3 with the strip pairs for each of your other questioned specimens.

5. Place a flat slide on the stage of your microscope. Set the microscope to the lowest magnification (usually 40X), place your questioned specimen on the slide, and position it until the cut edge roughly divides the eyepiece field and one end of that cut edge is in view, as shown in Figure VI-1-1. Focus on the edge of the specimen.

Figure VI-1-1: *Compression striation patterns at 40X magnification*

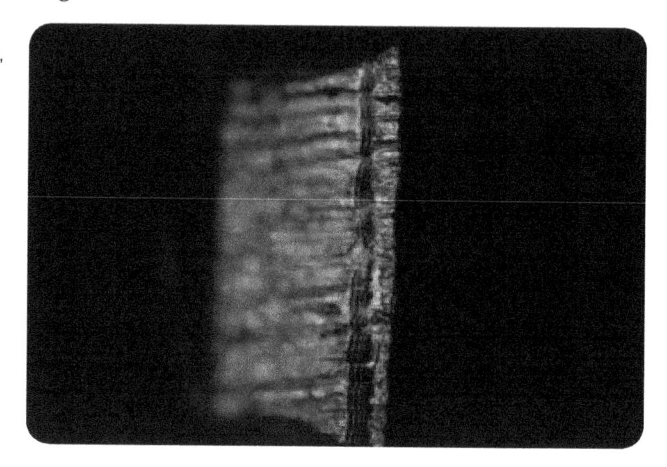

6. Place your first K1 specimen on top of the questioned specimen, with the cut edge of the known specimen almost but not quite overlapping the cut edge of the questioned specimen. Refocusing as necessary, compare the two cut edges to determine if both have a similar pattern of striations. (At 40X, a typical microscope has a field of view of about 4 mm, so depending on the width of the specimens, you may have to reposition them several times to compare the entire edge of the known specimen against the entire edge of the questioned specimen.)

> Keeping the edges of two specimens flat, aligned, and in good focus is rather finicky. Keep at it, and you'll soon get the hang of it.

7. If you don't find a match, turn the known specimen over to compare the edge of its other side against the questioned specimen.

8. If you don't find a match, repeat steps 6 and 7 with the second K1 strip.

9. If you don't find a match, repeat steps 6, 7, and 8 with the K2 specimens, K3 specimens, and so on, until you find a match.

10. Record your observations in your lab notebook. If you have the equipment necessary, shoot images of the matching specimens.

PROCEDURE VI-1-2: PRODUCE AND COMPARE SCORING SPECIMENS

Producing and comparing scoring specimens is similar to producing and comparing compression specimens. Scoring specimens are actually easier and require less work because scoring instruments have only two sides to take into account, and scoring specimens have only one.

Provide your volunteer with a flat section of an aluminum pie plate and your selection of screwdrivers. Ask the volunteer to choose one of your cutting plier specimens, hold the screwdriver at a 30° to 45° angle to the aluminum, as shown in Figure VI-1-2, and, while pressing down heavily enough to score but not tear the sheet, drag the flat tip of the screwdriver across the aluminum to score it.

Figure VI-1-2: *Producing a scored specimen*

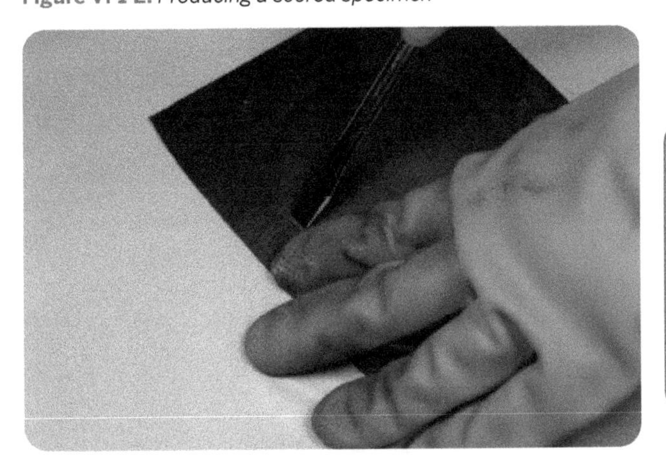

> Place the thin aluminum sheets on a hard, flat surface such as a workbench or a piece of scrap wood. If you use a counter top or similar surface, be very careful not to score it as well as the specimens.

1. Label sections of aluminum sheet for each of your known screwdriver specimens.

2. Use the K1 screwdriver to score the corresponding aluminum sheet twice, once with each side of the blade. Make the scoring impressions as close together as possible without overlapping.

3. Repeat step 2 with each of your known screwdriver specimens.

4. Use the scissors to cut a section from the questioned specimen roughly 3 cm long and wide enough to contain the full width of the scoring trace.

5. Place the questioned section on a microscope slide with the cut edge roughly centered on the slide, and tape the specimen in place. Set the microscope to its lowest magnification, turn on your top illuminator, and position the questioned specimen until the cut edge roughly divides the eyepiece field and one edge of the scoring is in view.

> Depending on the width of the screwdriver blades you used, you may or may not be able to see the entire scored section in one field of view. At 40X, the field of view of most microscopes is roughly 4 mm wide, which is slightly wider than a 1/8" blade and slightly narrower than a 3/16" blade. If the blade width is greater than the field of view, simply compare just that portion of the scoring that is visible in the field of view.

6. Use the scissors to cut a section from the K1 specimen long enough to grasp easily and wide enough to contain both scoring traces.

7. If your scoring impressions are narrower than the field of view, with one edge of the scored portion of the questioned specimen in the field of view, while looking through the microscope, place one short edge of the K1 specimen as close as possible to the edge of the questioned specimen and center the two scoring impressions against each other.

If the scoring impressions match, it should be immediately obvious. (Ignore any impressions visible on the cut edges of the specimens; what we're interested in is only the parallel scored impressions.)

> If the scoring impressions are wider than the field of view, you'll need to slowly slide the edge of the K1 specimen along the edge of the questioned specimen, looking for a match as you move the questioned specimen.

8. If the first scoring impression on the K1 specimen does not match the scoring on the questioned specimen, compare the second scoring impression (from the other side of the blade) on the K1 specimen against the questioned specimen. If that scoring impression doesn't match the questioned specimen, reverse the K1 specimen and compare the other ends of the two scoring impressions again the questioned specimen.

9. If necessary, repeat steps 6 through 8 to compare the questioned scoring impression against both ends of both scoring impressions on the K2 known specimen, then on the K3 known specimen, and so on until you locate a match.

10. When you locate a match, record the details in your lab notebook. Make a sketch or shoot an image of the matching striation patterns.

REVIEW QUESTIONS

Q1: **What are two common types of tool impression marks?**

Q2: **What issue must be taken into account when comparing a cut wire sample against a plier suspected of having made the cut, and why?**

Q3: **If you are presented with a questioned cut wire specimen and a wire cutter suspected of having made the cut, how many comparisons must you make to rule out a match, and why?**

Q4: **When we compared scoring impressions, why did it not matter if we compared only a portion of the questioned specimen scoring against our known specimens?**

Matching Images to Cameras

Lab VI-2

EQUIPMENT AND MATERIALS

You'll need the following items to complete this lab session. (The standard kit for this book, available from *http://www.thehomescientist.com*, includes the items listed in the first group.)

MATERIALS FROM KIT

- Goggles
- Forceps
- Slides, flat

MATERIALS YOU PROVIDE

- Gloves
- Camera with microscope adapter (optional)
- Microscope
- Specimens, film (see text)
- Specimens, digital images (see text)

WARNING

None of the activities in this lab session present any real hazard. It is still good practice to wear goggles and gloves to prevent contaminating specimens.

BACKGROUND

Although digital photography has now rendered film photography essentially obsolete in all but a few specialized niches, matching a film negative or slide to the camera that made it has been decisive in many investigations over the years and will continue to be so for a long time to come. Why? Because matches made between years- or decades-old negatives or slides and a camera still in the possession of a suspect may still be useful in establishing a link between a questioned film specimen and a suspect.

A film/camera match can be made in either or both of two ways. First, if a questioned film and the questioned camera are both available, a forensic scientist can shoot new images with the camera and compare those negatives or slides to the questioned film. Second, a questioned film not in the possession of the suspect can be compared against another film found in the possession of the suspect to establish whether the two films were shot with the same camera.

But compared how? Every film camera has a film chamber with a mask that delimits the area of the film that is exposed to light when an image is made. Figure VI-2-1 shows the film chamber of a typical 35mm SLR camera. (We shot this image with strong oblique lighting to reveal the cruft that accumulates unless the film chamber is blown out frequently.) To the naked eye, the edges of the mask may appear perfectly smooth and regular but, like all manufactured objects, these edges have microscopic imperfections. Because those imperfections are recorded on the edge of each negative or slide made with the camera, each frame has a built-in "fingerprint" that can be used to match that frame to the camera that made it—in theory, anyway.

Figure VI-2-1: *The film chamber of a typical 35mm SLR camera*

In practice, it may or may not be possible to make such a match. Why? Because the resolution of a particular type of film may be too low to record the microscopic imperfections for a particular camera. As you might expect, more expensive cameras are better finished and show less variability than cheaper cameras.

In particular, it's impossible to match film to camera if the film was shot with a disposable camera, which in recent years have been used to shoot billions of frames of color negative film. The only comparison possible for images shot with such cameras is within frames on the one roll of film that came in the camera.

The type of film is also significant. Fast, grainy films are more likely to obscure any microscopic roughness in the edges of the film chamber than are slower, finer-grained films. Most color negative films and most color slide films have relatively thick emulsion layers, which makes it more difficult to discriminate edge roughness from image grain. Black-and-white negative films and thin-emulsion slide films such as Kodachrome are generally easier to match to a specific camera because the edge details are easier to discriminate from image grain and random grain outside the image area. Even factors such as the focal length of the lens used to make the image and its aperture setting play a role in determining whether a particular frame of film can be matched to a particular camera.

So, at most we can say that it is sometimes possible to establish a match between a frame of film and the camera that made it. In this lab session, we'll examine the edges of film frames at 40X and 100X to determine if we can link those frames to the cameras that made them. Although you can of course examine the entire perimeter of each film frame, the best starting point is usually a frame corner, where microscopic imperfections are both more likely to occur and easier to compare.

Making sure that you're comparing the same corner between different frames is another matter. If the film frame is square, such as the 6 × 6 cm (2¼ × 2¼") format used by many cameras that accept 120, 220, or 620 roll film, things are pretty straightforward. Such cameras, whether they have eye-level or waist-level finders, are almost always held in the same orientation. (There's no point in rotating the camera when the image is square.) Oblong formats such as full-frame 35mm are a different story. If the frame is in landscape format, you can usually safely assume the camera was held normally in a horizontal position (people almost never turn a 35mm camera upside down to shoot an image). But if the frame is in portrait format, you may need to make two comparisons for that frame because some people rotate a camera left-side upward to take a portrait shot, and others right-side upward. Some may do it either way, depending on the situation. And, of course, for close-up shots and some others, it's not always obvious from the subject matter whether the camera was held in the portrait or landscape orientation.

For this lab session, you'll need a selection of black-and-white negatives, color negatives, and/or color slides shot with various cameras that use the same film format (ideally, full-frame 35 mm). Black-and-white or color negative strips can be used directly. Most color slides are mounted, with the mount obscuring the edges of the image, so you'll need to disassemble the mount before examining the image. Borrow negatives and slides from relatives, friends, and neighbors to increase the variety of your specimens. Wear gloves to avoid leaving fingerprints on the film specimens. Designate one frame—or, better yet, a strip—as your "questioned" specimen, which you'll attempt to match against your other specimens.

Obviously, it's difficult to do accurate comparisons without a comparison microscope, because you have to remove one specimen before viewing the next. Lacking a comparison microscope, the next-best way to do comparisons is to shoot images through your microscope and compare images rather than comparing specimens directly. It's ideal to have your microscope set up next to your computer. You can then shoot images as you work, transfer them to the computer, and make direct comparisons between images you shot earlier and what you're currently viewing through the microscope.

PROCEDURE VI-2-1: MATCHING FILMS TO CAMERAS

1. Set your microscope to low (40X) magnification, place an empty flat slide on the stage, place your questioned film specimen emulsion (dull) side up on the slide, and turn on the illuminator.

2. Position the film frame to center one corner in the field of view and focus on the edge of the frame, looking for roughness of any sort on the edge of the image.

> You'll probably find that you need to refocus constantly as you examine film frames. Part of the problem is that a film emulsion (and the image recorded on it) is relatively thick in microscopic terms, and the depth of focus of a microscope even at low magnification is insufficient to bring all parts of the emulsion into focus simultaneously. Also, rather than lying flat, film tends to curl slightly, so when the image is focused on one part of the frame edge, other parts will not be focused. Finally, the nature of the film itself may lead to fuzzy images. Black-and-white films use a very thin emulsion, and the image is made up of tiny grains of silver, which can be focused sharply. Color films are a different matter. In addition to having much thicker emulsions than black-and-white films, the image in color negative and slide films is made up of tiny fuzzy clouds of dye rather than sharp-edged grains of silver.

3. Scan the entire perimeter of the film frame at low magnification to determine if there is a major flaw, such as a nick in the film chamber edge. Such gross flaws are surprisingly common, and the presence of such a flaw makes it easy to confirm or eliminate other specimens quickly.

4. After you've scanned the perimeter of the film frame and returned to the initial corner, center that corner in the field of view and increase magnification to 100X. You're looking for any roughness in the lines that join to make the corner. Small raised bumps along the edge of the film chamber mask intrude into the image area, causing unexposed areas within the frame. Small depressions along the edge of the film chamber allow the image area to extend outside the general line of the edge.

> Even at 100X, you'll find that film grain becomes intrusive, making it difficult to discriminate between random film grains and actual roughness in the edge mask. Any magnification higher than about 100X is useless because the trees (edge details) disappear into the forest (film grain).

5. After you complete your examination of the first corner, switch to low magnification and center another corner in the field of view. Repeat step 4 for that corner, noting any visible imperfections.

6. Repeat step 5 for the third and fourth corners of the frame.

7. If none of the four corners has characteristic imperfections, it's unlikely that you can match this film frame to a particular camera. (It's possible that characteristic imperfections could exist other than at the corners, but in practical terms, it's difficult or impossible with a standard microscope—particularly if you don't have a calibrated mechanical stage—to identify and match imperfections other than those near corners.) Choose a negative or slide from a different camera as your questioned film.

8. If one or more of the corners on the questioned film frame has characteristic imperfections, compare each of your known specimens against the questioned specimen to identify which known specimen(s) are consistent with the questioned specimen. Record your observations in your lab notebook. If you have the necessary equipment, shoot images of the flawed areas that correspond in the questioned and known specimens.

Figure VI-2-2 shows a 40X view of a color negative frame taken in a camera that has no obvious edge imperfections (at least in this part of the frame). Figure VI-2-3 shows a 40X view of a color slide frame taken in a camera that has obvious mask roughness at the top-left corner. Other frames taken by this camera show the same pattern, and so can be matched to this specific camera.

Figure VI-2-2: *Color negative film at 40X showing no individualizable mask roughness*

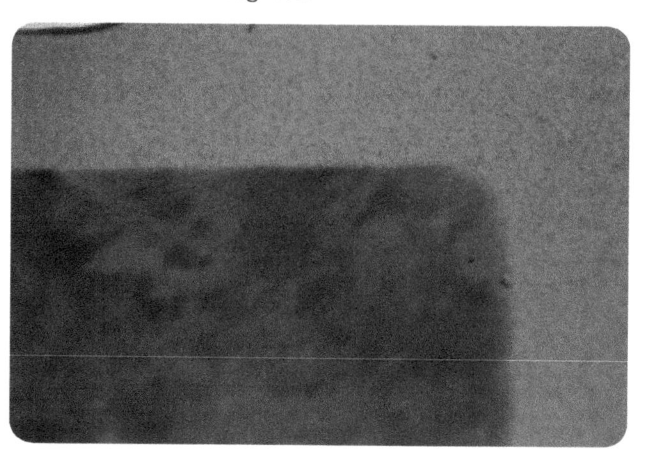

Figure VI-2-3: *Color slide film at 40X showing individualizable mask roughness on the top left edge*

Imperfections on the edge of the film chamber may be permanent or transient. Permanent imperfections are those caused by painting anomalies, machining marks, casting or mold marks, scratches, and similar features. Transient imperfections may be caused by tiny dust particles or similar contaminants and may manifest for anything from a single frame to an entire roll of film or occasionally over several rolls of film. If the imperfections on two frames appear to match very closely but one frame has imperfections that are not present on the second frame, they may be caused by dust particles on the first frame.

PROCEDURE VI-2-1: FORENSIC EXAMINATION OF DIGITAL IMAGE FILES

Nowadays, of course, most images are captured as digital files. Nearly all digital cameras and some scanners store extra data (called metadata) with their image files, usually in exchangeable image file format (EXIF). The camera manufacturer decides exactly what information is stored as metadata, but the make and model of camera is always stored, and often its serial number. Figure VI-2-4 shows some of the EXIF metadata for an image taken with one of our Pentax digital SLR cameras.

Figure VI-2-4: *EXIF metadata stored with an image*

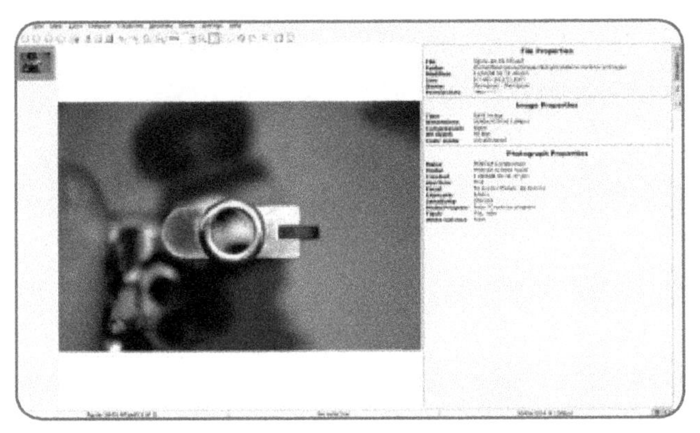

The EXIF metadata from an unmodified digital camera image file can almost always identify at least the make and model of camera that captured the image, usually the lens focal length used, and sometimes the individual camera by serial number. But post-processing with Photoshop or a similar graphics program can inadvertently (or advertently) corrupt or remove metadata, and a criminal who is aware of this issue may intentionally remove metadata from an image.

Obtain digital camera images for analysis from several sources. If you own or have owned multiple digital cameras, you can examine your own image files to determine which camera produced which images. Otherwise, obtain image files from friends and family, or simply download image files from Flickr or another source.

Attempt to obtain as much information as possible about each image and the camera that produced it. If the image software on your computer doesn't display full EXIF information, download and install a suitable program that will do so. Many free image viewers are available for Windows, OS X, and Linux. Some display only a subset of the EXIF data—for example, make and model of camera but not serial number—while others display all of the EXIF data present in the image file.

Until a few years ago, an image file that lacked metadata could not be matched to the individual camera that made the image, nor even to the brand and model of camera. In November 2008, *New Scientist* magazine (issue 2682, page 30) reported that forensic scientists have developed a new method of computer analysis that allows a JPEG image without metadata to be matched with 90% accuracy to the brand and model of camera that made the image. This can be done because every digital camera produces JPEG images by processing raw sensor data using an algorithm specific to that brand and model of camera (or, sometimes, to a related line of camera models). By examining a particular pixel and determining the characteristics of the pixels that surround it, the analysis program can determine the algorithm that was used to produce the image and therefore the brand and model of camera that produced the image file. Although this procedure provides only a class match rather than an individual match, it may still be very helpful to investigators.

REVIEW QUESTIONS

Q1: **What major elements determine whether or not a particular negative or slide can be identified as having been made by a particular camera? Why?**

Q2: **In general, are you more likely to be successful comparing a color or black-and-white film specimen? Why?**

Perforation and Tear Analysis

Lab VI-3

EQUIPMENT AND MATERIALS

You'll need the following items to complete this lab session. (The standard kit for this book, available from *http://www.thehomescientist.com*, includes the items listed in the first group.)

MATERIALS FROM KIT

- Goggles
- Slides, flat

MATERIALS YOU PROVIDE

- Gloves
- Camera with microscope adapter (optional)
- Microscope
- Specimens, tape dispensers (see text)

WARNING

None of the activities in this lab session present any real hazard. It is still good practice to wear goggles and gloves to prevent contaminating specimens.

BACKGROUND

Many products are packaged in rolls. Some rolls—paper towels, toilet paper, some plastic bags, and so on—are perforated to deliver portions of fixed length. Other rolls—waxed paper, plastic wrap, aluminum foil, packaging tape, and so on—include a tear bar to allow cutting portions of any arbitrary length. In either case, microscopic analysis of the perforations and tear patterns as well as the material on either side of the cut may allow a match to be established between two adjacent portions of the material.

In some cases, only a class match can be obtained. For example, if cellophane tape is used to seal a letter bomb, subsequent examination of a tape dispenser found in the possession of a suspect may establish only that the suspect tape specimen was torn from a dispenser that was similar or identical to the dispenser found in the possession of the suspect. Although that may be useful information in itself, it remains class evidence because one roll of cellophane tape is difficult or impossible to discriminate from another similar roll, and similar tape dispensers are manufactured in large numbers.

In other cases, an individual match can be obtained. For example, plastic lawn bags are made from recycled plastic, and often show streaks and striations where different types of plastic were incorporated into the bulk mass of plastic from which the bags were extruded. In addition to perforation/tear pattern matching, forensic examination may be able to match these streaks and striations to establish that the questioned plastic bag was immediately adjacent to the next bag in a roll found in the possession of a suspect.

In this lab session, we'll compare the perforated/torn edges of sticky tape specimens obtained from various dispensers to determine which of our specimens are consistent with originating from the same tape dispenser. The ubiquitous 3/4" (19 mm) Scotch Magic Tape in various dispensers is ideal for this lab session, although you can substitute masking tape, medical adhesive tape, packaging tape, or any other type of tape for which you can obtain specimens from a variety of dispensers. The important things are that the tape specimens should not be easily discriminated by naked eye and that you have a variety of different dispensers. (Make sure that you can identify each dispenser uniquely. If necessary, label each of them temporarily.)

PROCEDURE VI-3-1: PRODUCE AND EXAMINE TAPE SPECIMENS

1. Ask a friend or lab partner to choose one of the dispensers, tear a short piece of tape from it, and apply that piece of tape to a microscope slide, with one of the ends near the center of slide, as shown in Figure VI-3-1.

Figure VI-3-1: *Mount the questioned specimen on a microscope slide*

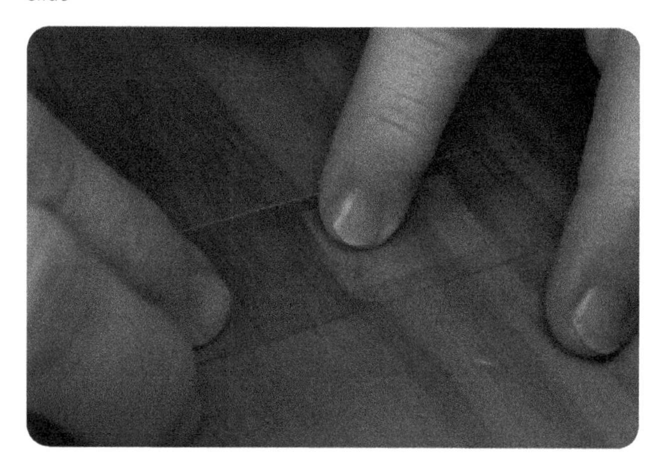

2. Tear off a short piece of tape from one of the dispensers and place it on the unused side of the slide so that its perforated edge closely abuts but is not in contact with or overlapping the perforated edge of the questioned specimen.

3. Examine the edges of the two specimens at low magnification and note similarities and differences, including the number of points present in the tear bar and the angle of the cuts made by the tear bar, as shown in Figure VI-3-2.

4. Repeat steps 2 and 3 with known tape specimens from each of the dispensers. Determine which of the tape dispensers may have been the source of the questioned specimen and which (if any) can be ruled out on the basis of the tool marks they leave on the cut specimens.

Figure VI-3-2: *Two tape specimens whose edges are dissimilar*

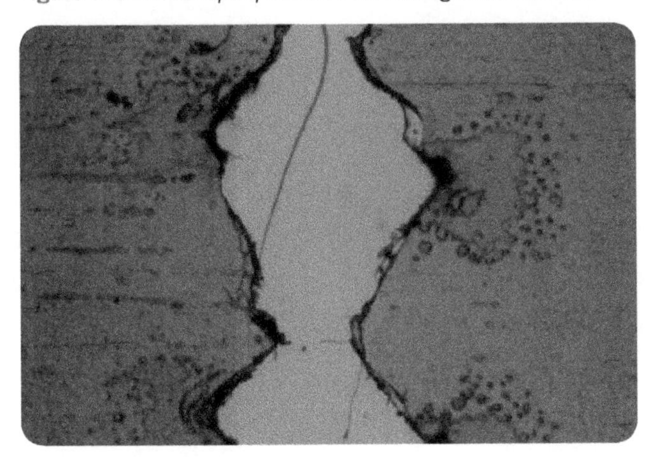

REVIEW QUESTION

Q1: You have established to your satisfaction that a questioned tape specimen is consistent in every respect with a specimen from a known dispenser, and must therefore have come from that or a similar dispenser. What other unrelated tests might you do on the questioned tape specimen to attempt to establish a link between the questioned tape specimen and the suspect? Why?

Forensic Drug Testing Group VII

Drug tests are a major part of the workload of most forensics labs. The vast majority of forensic drug testing focuses on recreational drugs rather than pharmaceutical drugs, although, of course, many pharmaceutical drugs are misused for recreational purposes. Recreational drugs include those—such as cocaine, methamphetamine, and heroin—which have few or no legitimate medical uses as well as drugs such as oxycodone, hydrocodone, methylphenidate (Ritalin), numerous opiates, and various steroids, which have legitimate medical uses but are often diverted for recreational use. Finally, recreational drugs include those sometimes referred to as "designer drugs," street drugs that are manufactured exclusively in underground laboratories and have never been manufactured by legitimate pharmaceutical companies.

In this group of lab sessions, we'll perform a variety of drug tests using techniques similar or identical to those used in professional forensic labs. Obviously, it's illegal to possess real controlled substances even in the spirit of scientific inquiry, so, with the exception of one lab session, we'll make do with legal substitutes. For that lab session, we'll test for the presence of actual cocaine and methamphetamine. How can we legally obtain specimens of cocaine and methamphetamine? The answer might surprise you. Read on to find out.

Presumptive Drug Testing

Lab VII-1

EQUIPMENT AND MATERIALS

You'll need the following items to complete this lab session. (The FK01 Forensic Science kit for this book, available from *http://www.thehomescientist.com*, includes the items listed in the first group.)

MATERIALS FROM KIT

- Goggles
- Hydrochloric acid
- Mandelin reagent
- Marquis reagent
- Pipettes
- Scott reagent
- Spot plate, 12-well
- Stirring rod
- Specimen: acetaminophen
- Specimen: aspirin
- Specimen: chlorpheniramine
- Specimen: diphenhydramine
- Specimen: ibuprofen
- Specimen: poppy seeds
- Specimen: two questioned (see text)

MATERIALS YOU PROVIDE

- Gloves
- Salt, table (use as known specimen)
- Specimens: drug (optional; see text)
- Toothpicks (flat)
- Watch, clock, or other timer

WARNING

Perform this lab session only in a well-ventilated area. Most of the chemicals used to make up the presumptive test reagents are irritating, toxic, corrosive, mutagenic, carcinogenic, or all of those. Concentrated hydrochloric and sulfuric acids are corrosive. Formaldehyde and ammonium metavanadate are toxic. Read the MSDS for each chemical you use and follow the listed precautions. Wear splash goggles, gloves, and protective clothing.

In addition to the specimens included in the kit, you may be able to obtain legal specimens of some controlled substances. For example, some prescription cough medicines contain codeine, as does Tylenol with Codeine (Tylenol 1, 2, 3, or 4, in different dosages). In some jurisdictions, some OTC drugs include codeine in very small amounts, which are nonetheless more than sufficient for these tests.

Ask anyone you know who uses a prescription pain reliever if you can have a tiny specimen. The smallest scraping from a tablet or even the dust from an empty bottle is sufficient. If you have a pet, try testing tiny specimens of any medication your veterinarian has prescribed.

You can also test specimens of whatever prescription and OTC medications you have in your bathroom cabinet. In particular, try testing other pain relievers (such as naproxen sodium) and allergy medications (including pseudoephedrine, brompheniramine, and so on). The kitchen spice rack is also a fertile source of spices such as mace, thyme, sage, oregano, and so on. Use your imagination, and test as many substances as you have time for.

FORMULARY

If you're not using the kit or presumptive test reagents purchased elsewhere, you'll need to make up your own presumptive test reagents. The amount of each reagent you need depends on the number of tests you intend to run and whether you run the tests on a spot plate or in test tubes. For a spot plate, 5 or 10 mL of each reagent is sufficient to run many tests. If you use test tubes, we recommend preparing at least 10 to 25 mL of each reagent, depending on how many tests you intend to run.

Wear full protective equipment (gloves, splash goggles, and protective clothing) and work in a well-ventilated area when making up these reagents.

A.1 Scott reagent Dissolve 0.5 g of reagent-grade cobalt thiocyanate in 25 mL of distilled or deionized water. If you do not have cobalt thiocyanate, substitute 0.68 g of cobalt chloride hexahydrate and 0.56 g of potassium thiocyanate for the 0.5 g of cobalt thiocyanate. In solution, these two chemicals dissociate to form cobalt ions and thiocyanate ions. The spectator chloride ions and potassium ions do not interfere with the test.

A.4 Mandelin reagent Dissolve 0.1 g of reagent-grade ammonium metavanadate in 10 mL of reagent-grade concentrated (96% to 98%) sulfuric acid.

A.5 Marquis reagent Add 10 mL of reagent-grade concentrated sulfuric acid to 0.25 mL of reagent-grade 37% formaldehyde solution (formalin).

These reagents remain usable indefinitely if they are stored in the dark in airtight containers, but if you have any doubt, use freshly prepared reagents to ensure that your results are reproducible.

If you want to experiment with other presumptive drug test reagents, you can make them up as follows:

A.2 Dille-Koppanyi reagent (modified) Dissolve 0.1 g of cobalt(II) acetate dihydrate in 100 mL of methanol, add 0.2 mL of glacial acetic acid and mix to make solution A. Add 5 mL of isopropylamine to 95 mL of methanol and mix to make solution B. This is a primary test reagent for pentobarbital, phenobarbital, and secobarbital, and a secondary test reagent for amobarbital. A light purple color is a positive test. To run the test, add two volumes of solution A to the specimen, either as a solid or dissolved in chloroform, followed by one volume of solution B.

A.3 Duquenois-Levine reagent (modified) Dissolve 2.5 mL of acetaldehyde and 2 g of vanillin in 100 mL of 95% ethanol. This is a primary test reagent for tetrahydrocannibinol (THC), found in marijuana. A color blue through a light purplish blue

FORMULARY (CONTINUED)

to a deep purple is a positive test. To run the test, add one volume of the reagent to an ethanol extract of the specimen and shake for one minute. Add one volume of concentrated (37%) hydrochloric acid, agitate gently, and observe the color produced. Add three volumes of chloroform and determine whether the color is extracted from the mixture.

A.6 Nitric acid reagent Use reagent-grade concentrated (68%) nitric acid. This is a primary test reagent for (except as noted) chloroform solutions of: codeine (light greenish yellow is positive), heroin (pale yellow), mescaline (dark red), morphine (brilliant orange-yellow), and opium (as powder; dark orange-yellow). This reagent is also a secondary test for dimethoxymethamphetamine hydrochloride (very yellow), LSD (strong brown), MDA hydrochloride (light greenish yellow), and oxycodone hydrochloride (brilliant yellow). To run the test, add the reagent dropwise to the specimen as a powder or in chloroform.

A.7 p-DMAB reagent Dissolve 2 g of para-dimethylaminobenzaldehyde in a mixture of 50 mL of 95% ethanol and 50 mL of concentrated (37%) hydrochloric acid. This is a primary test reagent for LSD. To run the test, dissolve the specimen in chloroform and add the reagent dropwise. A deep purple color is positive.

A.8 Ferric chloride reagent Dissolve 2 g of anhydrous ferric chloride or 3.3 g of the hexahydrate salt in 100 mL of distilled water. This is a primary test reagent for morphine monohydrate. To run the test, dissolve the specimen in methanol and add the reagent dropwise. A dark green color is postive.

A.9 Froede reagent Dissolve 0.5 g of molybdic acid or sodium molybdate in 100 mL of hot concentrated (98%) sulfuric acid. This is a primary test reagent for (except as noted) chloroform solutions of: codeine (very dark green is positive), heroin (deep purplish red), MDA hydrochloride (greenish black), morphine monohydrate (deep purplish red), and opium (as powder, brownish black). This reagent is also a secondary test for dimethoxymethamphetamine hydrochloride (very yellow-green), LSD (moderate yellow-green), and oxycodone hydrochloride (strong yellow). To run the test, add the reagent dropwise to the specimen as a powder or in chloroform.

A.10 Mecke reagent Dissolve 1 g of selenious acid in 100 mL of hot concentrated (98%) sulfuric acid. This is a primary test reagent for (except as noted) chloroform solutions of: codeine (very dark bluish green is positive), heroin (deep bluish green), MDA hydrochloride (very dark bluish green), mescaline hydrochloride (moderate olive), morphine monohydrate (very dark bluish green), and opium (as powder, olive black). This reagent is also a secondary test for dimethoxymethamphetamine hydrochloride (dark brown), hydrocodone tartrate (dark bluish green), LSD (greenish black), and oxycodone hydrochloride (moderate olive). To run the test, add the reagent dropwise to the specimen as a powder or in chloroform.

A.11 Zwikker reagent Dissolve 0.5 g of copper(II) sulfate pentahydrate in 100 mL of distilled water to make solution A. Add 5 mL of pyridine to 95 mL of chloroform and mix to make solution B. This is a primary test reagent for the barbiturates pentobarbital, phenobarbital, and secobarbital, with which it yields a light purple color as a positive test. To run the test, dissolve the specimen in chloroform and add one volume of solution A, followed by one volume of solution B.

A.12 Simon's reagent Dissolve 1 g of sodium nitroprusside in 50 mL of distilled water, add 2 mL of acetaldehyde, and then mix thoroughly to make solution A. Dissolve 2 g of anhydrous sodium carbonate in 100 mL of distilled water to make solution B. This is a primary test reagent for d-methamphetamine hydrochloride (dark blue is positive) and dimethoxymethamphetamine hydrochloride (deep blue). It is also a secondary test reagent for MDMA hydrochloride (dark blue) and methylphenidate hydrochloride (pale violet). To run the test, dissolve the specimen in chloroform and add one volume of solution A followed by two volumes of solution B.

WARNING: If you make up any of these reagents, read the MSDS for each chemical first, and observe all safety precautions. Several of these reagents are extremely toxic and/or extremely corrosive.

BACKGROUND

Presumptive drug tests, also called *spot-color drug tests*, are quick pass/fail tests that are used to determine the presence or absence of various controlled substances in a specimen. In a typical presumptive drug test, a tiny amount of the suspect specimen is transferred to a spot plate, and a drop or two of a presumptive test reagent that is specific to that substance is added. If the expected, specific color change occurs, that substance is presumed to be present in the specimen. If some other color change—or no color change at all—occurs, that specimen is presumed not to contain the substance for which the test was undertaken.

> Although we make presumptive color spot tests for drugs sound very straightforward, in the real world, it's often more complicated. At best, these color spot tests can indicate the probable presence of a major component of the suspect specimen. If an illicit drug is in fact present in the specimen but has been cut with another substance that also reacts with the presumptive test reagent, the color changes that result from the reagent's reaction with the illicit drug may be obscured by the color changes that occur when the test reagent reacts with the substance used to cut the illicit drug.

Presumptive drug tests are used in the field for quick screening of suspect substances, but their use extends into the laboratory, as well. Although presumptive tests are neither as sensitive nor as selective as the instrumental tests that are used to unambiguously identify specific drugs, presumptive tests are much faster and cheaper than instrumental tests. For example, although a $500,000 spectrophotometer may ultimately be used to identify a drug unambiguously for the purpose of courtroom testimony, that machine is almost certainly in high demand, and time on it is strictly rationed. It's not unusual for such instruments to be booked solid 24 hours a day for weeks or even months in advance. Using presumptive drug tests to screen questioned specimens avoids wasting rare and expensive machine time slots on specimens that are known not to contain drugs.

A positive presumptive drug test is not admissible in court, because presumptive drug tests are subject to *false positives*. For example, a presumptive test reagent that produces a blue-green color if cocaine is present may produce the same or a similar blue-green color if any one of several over-the-counter antihistamines is present, even if no cocaine is present in the specimen. In theory, then, presumptive drug tests are of very limited use.

In practice, the converse is true. Why? Because although a positive result from one presumptive test reagent may leave room for doubt, testing that same suspect specimen with additional presumptive test reagents can eliminate most or all of the substances that produce false positives with the first reagent. For example, if the reagent described in the preceding paragraph—Scott reagent, which we'll use in the next lab session—produces a blue-green color change with a suspect substance, that substance *may* contain cocaine, but it may instead contain only a legal antihistamine drug that is known to produce a false positive with Scott reagent. Retesting the suspect substance with Mandelin reagent allows the forensic scientist to eliminate most of the false positives provided by the Scott reagent test. If Mandelin reagent produces a deep orange-yellow color change, that indicates that cocaine may be present in the specimen. Most of the substances that produce false positives with Scott reagent produce either a different color change or no color change at all with Mandelin reagent, allowing them to be discriminated from cocaine by using both test reagents.

You'll note we say "most substances," and that's the reason why presumptive tests are not accepted as conclusive by courts. For example, brompheniramine maleate—which is commonly used in over-the-counter allergy, cold, and flu medications—cannot in a legal sense be definitively discriminated from cocaine by either reagent. Table VII-1-1 makes the problem clear. Mandelin reagent yields an orange color with both substances, and Scott reagent yields a blue-green color with both. On first glance, it might seem difficult to discriminate these two compounds with either reagent.

Table VII-1-1. Color changes with Mandelin reagent and Scott reagent with brompheniramine maleate and cocaine

Compound	Mandelin reagent	Scott reagent
Brompheniramine maleate	Strong orange (7.5YR 7/14)	Brilliant greenish blue (5B 6/10)
Cocaine HCl	Deep orange yellow (10YR 7/14)	Strong greenish blue (5B 5/10)

But professional forensic labs don't depend on verbal descriptions like "strong greenish blue." Instead, they compare the test results against standard Munsell color charts to determine if an exact match exists. For example, the strong orange color designated Munsell 7.5YR 7/14 produced by Mandelin reagent with brompheniramine maleate differs subtly but distinctly from the deep orange-yellow color 10YR 7/14 produced by Mandelin reagent with cocaine. So, although the subtlety of those color differences may raise reasonable doubt in court, forensic scientists who do the test are sure in their own minds that a 10YR 7/14 result with Mandelin reagent in fact indicates that cocaine is almost certainly present in the specimen. Sure enough that that they'll schedule time on that $500,000 machine to prove it.

> No one completely understands the exact mechanisms that cause the characteristic color changes in presumptive drug tests, but in general terms, these changes stem from the reactive natures of both the drugs and the test reagents. Drugs are complex organic compounds. Presumptive test reagents include strong acids and other very reactive chemicals. When a drug reacts with a presumptive test reagent, the drug undergoes significant changes and may even be ripped apart into smaller fragments. The altered drug or its fragments react with one or more of the chemical compounds present in the presumptive test reagent to yield new compounds that are strongly colored.

In *Color Test Reagents/Kits for Preliminary Identification of Drugs of Abuse* (*http://www.ncjrs.gov/pdffiles1/nij/183258.pdf*), which we call 183258 for short, the National Institute of Justice (NIJ) specifies a dozen different reagents that can be used to perform presumptive drug tests. For this lab session, we chose the three presumptive test reagents that are most often used for field tests.

Mandelin reagent

Mandelin reagent, which the NIJ designates A.4, is one of the most frequently used presumptive drug tests. The Mandelin test is sensitive and detects a very wide range of illicit drugs, including amphetamines, some alkaloids (including cocaine and mescaline), and opioids like morphine and codeine. Specimens are tested in powder form or with chloroform as a solvent.

Marquis reagent

Marquis reagent, which the NIJ designates A.5, is another of the most frequently used presumptive drug tests. The Marquis test is very sensitive and detects a wide range of illicit drugs, including amphetamines, mescaline, and opioids like morphine and codeine. Specimens are tested in powder form or with chloroform as a solvent.

Scott reagent

Scott reagent, which the NIJ designates A.1, is another frequently used presumptive drug test. The Scott test yields positive results for cocaine, many synthetic anesthetics and antihistamines, and some opioids. Unlike Mandelin reagent and Marquis reagent, both of which yield a rainbow of colors specific to different substances, a positive Scott reagent test is invariably within a narrow range of greenish-blue colors. That makes it problematic to discriminate among different substances that yield positive tests, although it is quite useful for detecting a range of compounds that are not detected by other presumptive tests. Specimens are tested in powder form or with chloroform as a solvent. A positive test with this reagent yields a bluish-green color. If the test is positive, adding concentrated hydrochloric acid dropwise causes the solution to assume a pink color, which confirms the initial positive result. Also, if you mix the blue-green aqueous solution with a few drops of chloroform and allow the liquids to separate, the blue-green tint is extracted into the chloroform layer.

Table VII-1-2 shows the color changes expected when the compound shown in the left column is treated with Mandelin reagent, Marquis reagent, and Scott reagent. Note that the color changes are often not as straightforward as Table VII-1-2 implies. For example, when Marquis reagent is added to a specimen that contains MDA, the solution does not turn black immediately. Instead, it assumes a pale greenish-blue tint for the first several seconds, which gradually begins to darken. After 10 to 15 seconds, the color becomes nearly black. Other compounds and test reagents may produce many different colors as the compound reacts with the reagent over several seconds. The colors listed in Table VII-1-2 are the final color, after the reagent has reacted for one full minute with the compound.

Bolded text in a color change column indicates that this is a preferred presumptive test for this compound. A preferred test is usually one that is more sensitive to the presence of the compound in question than other presumptive tests, although there can be a wide range of sensitivities even among preferred tests. For example, Mandelin reagent and Marquis reagent are both preferred tests for codeine, but the threshold detection limit for Mandelin reagent is 20 μg (micrograms) of codeine, versus only 1 μg for Marquis reagent. To put this in perspective, one standard aspirin tablet contains 325 mg (milligrams) of aspirin, which is the same as 325,000 μg.

Table VII-1-2: *Color changes for selected compounds with Mandelin reagent, Marquis reagent, and Scott reagent (data from NIJ 183258)*

Compound	Mandelin reagent	Marquis reagent	Scott reagent
Acetaminophen	Moderate olive (10Y 5/8)	n/a	n/a
Aspirin (powder)	Grayish olive green (2.5GY 4/2)	Deep red (5R 3/10)	n/a
Benzphetamine HCl	**Brilliant yellow green (2.5GY 8/10)**	**Deep reddish brown (7.5R 2/6)**	Brilliant greenish blue (5B 7/8)
Brompheniramine maleate	Strong orange (7.5YR 7/14)	n/a	Brilliant greenish blue (5B 6/10)
Chlordiazepoxide HCl	n/a	n/a	Brilliant greenish blue (2.5B 6/8)
Chlorpromazine HCl	Dark olive (10Y 3/4)	**Deep purplish red (2.5RP 3/8)**	Brilliant greenish blue (5B 6/10)
Cocaine HCl	**Deep orange yellow (10YR 7/14)**	n/a	Strong greenish blue (5B 5/10)
Codeine	**Dark olive (10Y 3/4)**	**Very dark purple (7.5P 2/4)**	n/a
Contac (powder)	Strong yellow (2.5Y 6/10)	n/a	n/a
d-Amphetamine HCl	**Moderate bluish green (5BG 5/6)**	**Strong reddish orange (10R 6/12) to dark reddish brown (7.5R 2/4)**	n/a
d-Methamphetamine HCl	**Dark yellowish green (10GY 4/6)**	**Deep reddish orange (10R 4/12) to dark reddish brown (7.5R 2/4)**	n/a
Diacetylmorphine HCl (heroin)	**Moderate reddish brown (10R 3/6)**	**Deep purplish red (7.5RP 3/10)**	Strong greenish blue (7.5B 6/10)

Compound	Mandelin reagent	Marquis reagent	Scott reagent
Dimethoxy-meth HCl	Dark olive brown (5Y 2/2)	Moderate olive (7.5Y 5/8)	n/a
Doxepin HCl	Dark reddish brown (10R 2/4)	Blackish red (7.5R 2/2)	Brilliant greenish blue (5B 6/10)
Dristan (powder)	Grayish olive (7.5Y 4/4)	Dark grayish red (5R 3/2)	n/a
Ephedrine HCl	n/a	n/a	Strong greenish blue (5B 5/10)
Excedrin (powder)	Dark olive (7.5Y 3/4)	Dark red (5R 3/8)	n/a
Hydrocodone tartrate	n/a	n/a	Brilliant greenish blue (5B 6/8)
LSD	n/a	Olive black (10Y 2/2)	n/a
MDA HCl	Bluish black (10B 2/2)	**Black (Black)**	n/a
Meperidine HCl	n/a	Deep brown (5YR 3/6)	Strong greenish blue (5B 5/10)
Mescaline HCl	**Dark yellowish brown (10YR 3/4)**	**Strong orange (5YR 6/12)**	n/a
Methadone HCl	Dark grayish blue (5B 3/2)	Light yellowish pink (2.5YR 8/4)	**Brilliant greenish blue (5B 6/10)**
Methaqualone	Very orange yellow (10YR 8/14)	n/a	n/a
Methylphenidate HCl	Brilliant orange yellow (2.5Y 8/10)	Moderate orange yellow (10YR 8/8)	Brilliant greenish blue (10BG 6/8)
Morphine monohydrate	**Dark grayish reddish brown (10R 3/2)**	**Very deep reddish purple (10P 3/6)**	n/a
Opium (powder)	**Dark brown (7.5YR 2/4)**	**Dark grayish reddish brown (10R 3/2)**	n/a
Oxycodone HCL	Dark greenish yellow (10Y 6/6)	**Pale violet (2.5P 6/4)**	n/a
Phencyclidine HCL	n/a	n/a	Strong greenish blue (5B 5/10)
Procaine HCl	Deep orange (5YR 5/12)	n/a	**Strong greenish blue (5B 5/10)**
Propoxyphene HCl	Dark reddish brown (10R 2/4)	Blackish purple (2.5RP 2/2)	**Strong greenish blue (5B 5/10)**
Pseudoephedrine HCl	n/a	n/a	Strong greenish blue (5B 5/10)
Quinine HCl	Deep greenish yellow (10Y 9/6)	n/a	Strong blue (2.5 PB 5/12)
Salt (crystals)	Strong orange (5YR 7/12)	n/a	n/a
Sugar (crystals)	n/a	Dark brown (5YR 2/4)	n/a

PROCEDURE VII-1-1: TESTING SPECIMENS AGAINST PRESUMPTIVE REAGENTS

In this procedure, we'll test our known and questioned specimens against our three presumptive test reagents, recording the color change(s), if any, that occur. Although some protocols call for dissolving solid specimens in chloroform or methanol, that's not necessary for any of the three reagents we're using.

Prepare tablet specimens by carefully crushing them or by using a sharp knife to scrape off a small amount. If the tablet has a colored coating, scrape off and discard that coating and test only the interior material from the tablet. Prepare capsule specimens by carefully opening the capsule and emptying the powder it contains onto a clean sheet of paper. For poppy seeds and similar specimens, use the tip of the stirring rod to crush the seeds to a powder.

Use about the same amount of specimen for each test. For solid specimens, something in the range of about 10 to 50 milligrams—about as much as you can pick up on the end of a flat toothpick—is sufficient. If you are also testing liquid specimens of your own, use a sample of one or two drops.

1. If you have not already done so, put on your splash goggles, gloves, and protective clothing.

WARNING

Mandelin reagent and Marquis reagent contain concentrated sulfuric acid. They are toxic and extremely corrosive. Wear gloves and goggles, and do not allow these reagents to come into contact with anything other than their containers, pipettes, and the spot plate. If either of these reagents contacts your skin or eyes, immediately flush them off with cold running water for several minutes and seek medical assistance.

2. Using a clean toothpick for each specimen, transfer a small amount of each of the known and questioned specimens to a separate well of the spot plate. Use the two questioned specimens and the six known specimens supplied with the kit as well as table salt you supply as the seventh known. Use the outer wells for the known specimens and inner wells for the questioned specimens. Record the well numbers and their contents in your lab notebook.

Questioned specimens supplied with the kit may be one of the known materials, but they may instead contain a mixture of two of the known materials.

3. Place the spot plate under a strong light, note the start time, and then transfer two drops of Marquis reagent to the first known well (Figure VII-1-1). Record the initial color in the well in your lab notebook and observe and record any color transitions that occur over the next 60 seconds. After one minute has elapsed, record the "final" color present in the well in your lab notebook.

The color present in a well after 60 seconds is conventionally called the "final" color. In fact, color transitions may continue to occur over several minutes or more, but for many reagent/specimen combinations, these longer term color changes are not predictable or reproducible and so are of no value for identifying a questioned specimen.

4. Repeat step 3 for each other well that contains a known or questioned specimen. At this point, you may be able to tentatively identify one or both of the questioned specimens as being consistent with one or more of the known specimens.

5. Carefully place the populated spot plate in the sink and flood it gently with tap water from a beaker or similar container to rinse away the contents of the populated wells. Once the contents of all wells have been diluted and rinsed away, wash the spot plate thoroughly under running tap water and dry it completely. Make sure the sink is also rinsed thoroughly.

6. Repeat steps 1 through 5 using Mandelin reagent.

7. Repeat steps 1 through 5 using Scott reagent.

Figure VII-1-1: *Running presumptive drug tests in a reaction plate (using larger than normal volumes for visibility)*

PROCEDURE VII-1-2: VERIFYING TEST RESULTS

In the first procedure, we used a scattershot approach to test each of our known and questioned specimens against each of our three presumptive test reagents, observing and recording the color transitions that occurred with each combination. That should be sufficient to allow us to tentatively identify which, if any, of the known specimens is generally consistent with each of the questioned specimens.

> If the questioned specimen is a mixture of two known specimens, you will need to use multiple reagents to sort things out. For example, one known specimen may have reacted only with Marquis reagent to yield a particular final color, whereas another known specimen may have reacted only with Scott reagent. If one of your questioned specimens reacts with Marquis reagent to yield the same final color as one of your knowns, and also reacts with Scott reagent to yield the same final color as another of your knowns, the working assumption is that the questioned specimen contains a mixture of both knowns.

In this procedure, we'll test our questioned specimens against the known specimen or specimens that we tentatively identified as consistent by reacting each questioned specimen simultaneously, side by side with the candidate known specimen(s). By running the tests simultaneously in adjacent wells, we can observe the reactions of the two specimens with each presumptive test reagent. If the specimens are in fact consistent, the color transitions that occur should be identical or very closely similar.

1. If you have not already done so, put on your splash goggles, gloves, and protective clothing.

2. If your tentative identification suggests that the first questioned specimen is consistent with a mixture of two or more of your knowns, proceed to step 8.

3. If your tentative identification suggests that the first questioned specimen is consistent with only one of your knowns, and if those specimens reacted with only one of the presumptive test reagents, transfer a small amount of the known specimen to one well of the spot plate and a similar amount of the questioned specimen to an adjacent well. If both specimens reacted with two of the presumptive

test reagents, populate another pair of adjacent wells with the specimens. If both specimens reacted with all three reagents, populate a third pair of adjacent wells.

4. Place the spot plate under a strong light. Use a plastic pipette to draw up a small amount of a reagent that reacted with both specimens. Add two drops of this reagent to the first well that contains the known specimen. As quickly as possible, add two more drops of the reagent to the well that contains the questioned specimen.

5. Observe both wells. Note the color transitions that occur. If the two specimens are consistent, both the color transitions and the timing should be similar or identical. Record your observations in your lab notebook.

6. If your two specimens reacted with a second presumptive test reagent, repeat steps 3 and 4 with that second reagent. If your specimens reacted with all three test reagents, repeat steps 3 and 4 with the third reagent, as well.

7. Carefully place the populated spot plate in the sink and flood it gently with tap water from a beaker or similar container to rinse away the contents of the populated wells. Once the contents of all wells have been diluted and rinsed away, wash the spot plate thoroughly under running tap water and dry it completely. Make sure the sink is also rinsed thoroughly.

   If you have tentatively identified one of your questioned specimens as a mixture of two of your known specimens, verify the reactions of each of those two candidate knowns and the questioned specimen against each of the reagents that reacted with the questioned specimen and with either or both of the candidate known specimens.

8. If only two of the three reagents reacted with the questioned specimen, use clean toothpicks to populate the spot plate with two rows of three specimen samples, with candidate known #1 on the left, the questioned specimen in the center, and candidate known #2 on the right. If all three of the reagents reacted with the questioned specimen, populate three such rows.

9. Place the spot plate under a strong light. Use a plastic pipette to draw up a small amount of one of the reagents that reacted with the questioned specimen. Add two drops of this reagent to the first well that contains the candidate known #1 specimen. As quickly as possible, add two more drops of the reagent to the well that contains the questioned specimen, and two more drops of the reagent to the well that contains the candidate known #2 specimen.

10. Observe all three wells, noting the color transitions that occur. If one of the two known specimens is consistent with the questioned specimen, both the color transitions and the timing should be similar or identical. No reaction should occur in the well that contains the second known specimen. Record your observations in your lab notebook.

11. Use a plastic pipette to draw up a small amount of the second reagent that reacted with the questioned specimen. Add two drops of this reagent to the first unused well that contains the candidate known #1 specimen. As quickly as possible, add two more drops of the reagent to the adjacent unused well that contains the questioned specimen, and then two more drops of the reagent to the adjacent unused well that contains the candidate known #2 specimen.

12. Observe all three wells, noting the color transitions that occur. If one of the two known specimens is consistent with the questioned specimen, both the color transitions and the timing should be similar or identical. No reaction should occur in the well that contains the other known specimen. Record your observations in your lab notebook.

13. If your questioned specimen reacted with a third presumptive test reagent, repeat steps 11 and 12 with that third reagent.

14. Carefully place the populated spot plate in the sink and flood it gently with tap water from a beaker or similar container to rinse away the contents of the populated wells. Once the contents of all wells have been diluted and rinsed away, wash the spot plate thoroughly under running tap water and dry it completely. Make sure the sink is also rinsed thoroughly.

REVIEW QUESTIONS

Q1: A questioned specimen tests positive for codeine with Marquis reagent. Which additional reagent or reagents from among those we used in this lab session would you use to confirm the presence of codeine before submitting the specimen for instrumental analysis, and why?

Q2: The questioned specimen that tested positive for codeine with Marquis reagent tests negative with the other presumptive reagent or reagents. Would you assume that the Marquis reagent test yielded a false positive? Why or why not?

Q3: A questioned specimen tests positive for cocaine with Scott reagent. Which additional reagent or reagents from among those we used in this lab session would you use to confirm the presence of cocaine before submitting the specimen for instrumental analysis, and why?

Q4: What were the identities of questioned specimen #1 and questioned specimen #2? How do you know?

Detect Cocaine and Methamphetamine on Paper Currency

Lab VII-2

EQUIPMENT AND MATERIALS

You'll need the following items to complete this lab session. (The FK01 Forensic Science Kit for this book, available from *http://www.thehomescientist.com*, includes the items listed in the first group.)

MATERIALS FROM KIT

- Goggles
- Forceps
- Marquis reagent
- Pipettes
- Scott reagent
- Specimen: diphenhydramine
- Spot plate, 12-well

MATERIALS YOU PROVIDE

- Gloves
- Bag, small plastic zip-lock
- Cotton balls, swabs, or paper towel
- Currency specimens to be tested (see text)
- Hair dryer (optional)
- Water, distilled or deionized

WARNING

Use extraordinary care when using Marquis reagent. It contains a concentrated solution of sulfuric acid, which is extremely corrosive.

BACKGROUND

Presumptive drug test reagents are most often used to test questioned solids and liquids, but they may also be used to detect traces of illegal drugs that adhere to people, clothing, and other objects. In the United States, among the common objects that are frequently contaminated by illegal drugs are US currency notes, which frequently contain traces of cocaine or methamphetamine. The type of paper and the intaglio printing used for US currency notes is ideally suited to retain drug traces for long periods. As a result, many US currency notes have detectable levels of cocaine, methamphetamine, or both.

The likelihood that a particular note will be contaminated and the degree of contamination depends on several factors, including the age and denomination of the note and where it has circulated. For example, an elderly $100 note that has spent most of its life in Los Angeles or New York is much more likely to be heavily contaminated with cocaine than is a newish $1 note that has circulated primarily in Smalltown, USA. The degree of contamination varies significantly. An August 2008 article in the journal *Trends in Analytical Chemistry* reported that US currency notes contained an average of 2.9 to 28.8 micrograms of cocaine, depending on denomination, age, and location, with maximum levels of 1,300 micrograms on some 1996 notes.

DENNIS HILLIARD COMMENTS

US currency is considered forensically as a fabric rather than paper. It may better explain how the drug is retained within the fabric material.

In this lab session, we'll test paper currency for the presence of cocaine using Scott reagent and for the presence of methamphetamine using Marquis reagent. Because your currency specimens may or may not be contaminated at a level sufficient to yield positive tests with these reagents, we'll also do a trial run in which we intentionally contaminate a note with diphenhydramine (a substance that yields a false positive with Scott reagent) and then test that note by using Scott reagent.

You'll need currency specimens to test, none of which will be damaged by the testing. For the first procedure, you'll need one note of any denomination. For the second and third procedures, obtain as many notes as you can, as old as possible and of as high a denomination as possible. We used five well-worn $100 notes, but $50 or even $20 notes are also likely to be contaminated with drug traces.

PROCEDURE VII-2-1: TESTING A CONTROL SPECIMEN

Before we test a note for the presence of cocaine, we need to know what a positive test looks like. That's easily done by intentionally contaminating a note with diphenhydramine, which yields a false positive with Scott reagent, and then testing the note. The threshold sensitivity of Scott reagent for cocaine is 60 to 65 µg. A typical diphenhydramine tablet or capsule contains 25 mg (25,000 µg). We don't know exactly what the threshold sensitivity of Scott reagent for diphenhydramine is, but it must be on the same order as that for cocaine, so using an entire diphenhydramine tablet is certain to yield a positive test.

1. If you have not already done so, put on your splash goggles, gloves, and protective clothing.
2. Place a note inside the plastic zip-lock bag. (Any note can be used, including a fresh $1 note; save your high-denomination notes for later in the lab session.)

3. If your source of diphenhydramine is a capsule, separate the capsule carefully and add the powder it contains to the zip-lock bag. If your diphenhydramine is in tablet form, crush the tablet to a fine powder and transfer that powder to the zip-lock bag.

4. Seal the bag and shake it vigorously to ensure that the entire surface of the bill is exposed to the powder.

5. Remove the note from the bag, shaking off any excess powder from the note. Wipe the note with a paper towel to remove any loosely adhering powder. If necessary, wipe down your work surface with a wet paper towel to remove any trace of the powder. Dispose of the bag, powder, and paper towel in the household waste container.

6. Wash your gloved hands thoroughly in hot running water and dry them with a clean towel. (We want to test only the note for the presence of diphenhydramine, so it's important not to contaminate the working area with any diphenhydramine powder that adheres to your gloves.)

7. Diphenhydramine is readily soluble in water, so that's what we'll use to extract the diphenhydramine from the note. To begin, moisten a cotton swab or a small part of a cotton ball or paper towel with distilled water and squeeze out any excess water. You want the absorbent material to be moist, but not dripping wet.

8. Place the treated note on a flat, clean work surface. Wipe down both sides of the note with the moist absorbent material. (If the absorbent material dries out, moisten it as needed with another drop or two of distilled water.) Use the same part of the material for both sides of the note. The goal is to get as much as possible of the diphenhydramine concentrated in one area of the material. When you finish wiping down the note, you can return it to your wallet.

9. Place the absorbent material, contaminated side up, on a clean work surface. Add a few drops of Scott reagent to the contaminated area and observe any color change that occurs. (If the test is positive, as shown in Figure VII-2-1, the pink Scott reagent immediately turns bright blue.) Note your observations in your lab notebook.

Figure VII-2-1: *A cotton ball showing a positive test for diphenhydramine*

PROCEDURE VII-2-2: TESTING CURRENCY FOR COCAINE

In this procedure, we'll use the same method we used in the first procedure, but this time we'll test notes for the presence of actual cocaine.

1. If you have not already done so, put on your splash goggles, gloves, and protective clothing.

2. Make sure your work surface is completely clean. Any diphenhydramine contamination present from the previous exercise may produce a false positive.

3. Cocaine hydrochloride is readily soluble in water, so we'll use the same procedure we used in the previous exercise to extract any cocaine present on the notes. To begin, moisten a cotton swab or a small part of a cotton ball or paper towel with distilled water and squeeze out any excess water.

4. Place your first high-denomination note on a flat, clean work surface. Wipe down both sides of the note with the moist absorbent material. Use the same part of the material for both sides of the note to transfer as much as possible of the cocaine to one area of the material.

5. To maximize the probability of a positive test, repeat step 4 with several more high-denomination notes, using the same area of the same material. If the absorbent material begins to dry out, add a couple drops of distilled water to the contaminated area of the material. Dampen it very slightly. As you wipe the notes, you are not only transferring cocaine from the notes to the absorbent material. You are also transferring some of any cocaine that was extracted from one note onto the following notes. Working with absorbent material that is only slightly dampened favors the transfer from the notes to the material. Wetting the material too much may reverse that, transferring more cocaine from the material to subsequent notes. When you finish wiping down the notes, you can return them to your wallet.

6. Place the absorbent material, contaminated side up, on a clean work surface. Add a few drops of Scott reagent to the contaminated area and observe any color change that occurs. Note your observations in your lab notebook.

REAL WORLD RESULTS

We tested currency for cocaine many times over a period of months on many different specimens, with denominations ranging from $20 to $100. Perhaps because we live in Winston-Salem, North Carolina, which is not a hotbed of cocaine use, we got few positive results. We did get a striking result with one $100 bill. The cotton turned bright blue, leading us to believe that that bill had been used by someone to snort cocaine (or, perhaps, to snort Benadryl...). We also got a few semipositive results, in which the contaminated surface of the cotton remained mostly pink, but with a few bluish speckles. If we lived in New York City or Los Angeles or San Francisco or Las Vegas, we might well have seen more positive results. The lesson here is that your mileage may vary.

PROCEDURE VII-2-3: TESTING CURRENCY FOR METHAMPHETAMINE

Methamphetamine is sometimes called "poor man's cocaine" because it is used primarily by people in the lower socioeconomic classes. Methamphetamine is actually more readily available and widely used in rural areas than it is in towns and cities. Because methamphetamine is much less expensive than cocaine, meth residues are about as likely to be found on $1, $5, and $10 notes as on notes of higher denomination. Marquis reagent is an extremely sensitive test for methamphetamine, reliably yielding positive results—a deep reddish-orange color—if as little as 5 μg is present. Under ideal conditions, Marquis reagent may yield positive results for methamphetamine at levels as low as 1 μg.

In this procedure, we'll test notes with Marquis reagent for the presence of methamphetamine, but we'll have to modify our methods slightly. Marquis reagent is primarily concentrated sulfuric acid, so we can't apply the test reagent directly to the absorbent material, which would discolor and char from the action of the sulfuric acid. Instead, we'll use absorbent material to remove any methamphetamine residue present on the notes, and then extract that residue into solution, which we'll test by using a spot plate.

1. If you have not already done so, put on your splash goggles, gloves, and protective clothing.

2. Make sure your work surface is completely clean. If you have any doubts, work on a clean sheet of paper.

3. Methamphetamine hydrochloride is readily soluble in water, so that's what we'll use. To begin, moisten a cotton swap or a small tuft of cotton ball or paper towel with a drop or two of distilled water.

4. Place a note on the clean work surface. Wipe down both sides of the note with the moist absorbent material. Use the same part of the material for both sides of the note to transfer as much as possible of any methamphetamine present to one area of the material. If the material dries as you work, add a drop or two of water to the contaminated surface. Use just enough to dampen it. When you finish wiping down the note, allow it to dry and then return it to your wallet. Repeat this step with any other notes you want to test, using the same area of the same absorbent material, moistened with a minimum amount of distilled water.

5. Transfer the absorbent material to a well in the spot plate. If necessary, add a drop or two of distilled water to moisten the absorbent material. The goal is to extract as much of any methamphetamine present in the absorbent material to the well, while using a minimum amount of water.

6. Use the forceps to squeeze as much water as possible from the absorbent material into the well. Ideally, you'd like to have only a drop or two of water in the well, with that drop or two containing as much as possible of the methamphetamine originally picked up by the absorbent material. If there is excess water present in the well, allow some of it to evaporate naturally or speed up evaporation by using a hair dryer. Discard the absorbent material.

7. Add a few drops of Marquis reagent to the well and observe any color change that occurs. Note your observations in your lab notebook.

REAL WORLD RESULTS

We tested about 40 notes for methamphetamine with Marquis reagent. One note displayed a strong presumptive positive result—a striking reddish-orange color in the well, shown in Figure VII-2-2. One note yielded what we considered a "probable positive"—a much less intense but still distinct reddish-orange coloration. A few notes displayed what we considered "possible positives"—a pale reddish-orange coloration. (For better visibility, when we shot this image, we actually used a clear plastic reaction plate instead of a spot plate, and we increased the volume of Marquis reagent accordingly.)

Figure VII-2-2: *A positive Marquis reagent test for methamphetamine*

REVIEW QUESTIONS

Q1: **You have obtained a very strong positive Marquis reagent test for methamphetamine from a currency note confiscated during a raid. Which other reagent would you use to test another note to confirm the presence of methamphetamine on the notes? Why is that reagent a secondary reagent for methamphetamine rather than a primary reagent?**

Q2: **You have obtained a positive Scott reagent test for cocaine from a currency note confiscated during a raid. Which other reagent would you use to test another note to confirm the presence of cocaine on the notes?**

Analysis of Drugs by Chromatography

Lab VII-3

EQUIPMENT AND MATERIALS

You'll need the following items to complete this lab session. (The FK01 Forensic Science Kit for this book, available from *http://www.thehomescientist.com*, includes the items listed in the first group.)

MATERIALS FROM KIT

- Goggles
- Centrifuge tubes, 50 mL
- Chromatography paper
- Iodine crystals
- Pipettes
- Ruler
- Spot plate, 12-well
- Specimen: acetaminophen
- Specimen: aspirin
- Specimen: caffeine
- Specimen: questioned (chromatography)

MATERIALS YOU PROVIDE

- Gloves
- Ethanol (see text)
- Hair dryer (optional)
- Iodine fuming chamber (see text)
- Pencil
- Questioned OTC drug(s) (optional)
- Scissors
- Toothpicks
- UV light source (optional; see text)

WARNING

Iodine crystals and iodine vapor are toxic, corrosive, irritating, and stain skin, clothing, and other materials. You can remove iodine stains by dissolving a vitamin C (or multivitamin) tablet in a small amount of water. The vitamin C solution converts iodine to soluble, colorless, harmless iodide ions, which can be flushed from the material with water. Ethanol is flammable. Ultraviolet light may be hazardous to the eyes and skin. Follow the directions and heed the warnings supplied with your UV light source. Wear splash goggles, gloves, and protective clothing.

BACKGROUND

Chromatography (from the Greek for color-writing) is a procedure used to separate mixtures and to identify the components present in those mixtures. The Russian botanist Mikhail Semyonovich Tsvet invented chromatography in 1900 and used it to separate chlorophyll and other plant pigments. Tsvet named his new procedure chromatography because his chromatograms were literally colorful. In forensic labs, chromatography is often used to separate colorless compounds, so the original rationale for the name no longer applies.

In chromatography, a mixture (called the *analyte*) dissolved in a *mobile phase* (sometimes called the solvent or *carrier*) passes through a *stationary phase* (sometimes called the *substrate*). During the passage, various components of the analyte carried by the mobile phase selectively adhere to the stationary phase with greater or less affinity, separating the components physically across the stationary phase.

Professional forensics labs use various chromatography methods, including *thin-layer chromatography* (TLC), *gaseous chromatography* (GC), and *high-performance liquid chromatography* (HPLC). These methods have many advantages—including fast throughput, sharp separations, high sensitivity, and the ability to work with very small samples—but all of them require expensive equipment or materials that make them impractical for a home lab.

The simplest chromatography method—the one we use in this lab session—is called *paper chromatography*. Unlike other methods, paper chromatography requires no expensive equipment or special materials. In paper chromatography, the mobile phase is water or another solvent, and the stationary phase is a strip of paper. The solvent is absorbed by the paper, and dissolves a spot of the analyte. As the solvent is drawn up the paper by capillary action, the various dissolved components of the mixture are deposited on the paper at different distances from the original spot, a process called *developing a chromatogram*.

The key metric for paper chromatography is called the *retardation factor* (R_f), which is the ratio of the distance the analyte moves from the initial point to the distance the solvent front moves from the initial point. For example, if one component of the analyte migrates 1 cm from the initial point during the time it takes the solvent front to migrate 5.0 cm from the initial point and a second component migrates 2.5 cm, the R_f for the first compound with that particular solvent and that particular substrate is 0.2 (1 cm / 5 cm) and the R_f for the second compound is 0.5 (2.5 cm / 5 cm). The R_f of a particular compound is a dimensionless number that is fixed for any particular combination of solvent and substrate, but may vary dramatically for other combinations of solvent and substrate.

But most drugs are colorless compounds. How can you determine R_f values if you can't see the compound on the paper? The answer is to process the developed chromatograms in some way to make the invisible compounds visible, a process called *visualizing a chromatogram*.

There are many ways to visualize a chromatogram. We'll use two popular and effective methods, *iodine fuming* and *visualization with UV light*. Iodine fumes bind to many organic compounds, including most drugs, to form colored compounds that are easily visible on the paper. Many organic compounds, including drugs, that are colorless under white light fluoresce in various colors under an ultraviolet lamp ("black light"). Used alone or together, these two methods allow us to see what is normally invisible.

Although many professional forensics labs have purpose-built iodine fuming chambers, they're just as likely to use ad hoc chambers. We used a one-quart Gladware kitchen container with a snap-on lid to visualize our chromatograms.

Dennis Hilliard comments: "As an alternative method, we've found that fuming with iodine can easily be accomplished in a zip-lock bag. A few iodine crystals are placed in the corner of a bag and the strips are placed in the bag and sealed. The crystals are then heated slightly with a reading lamp light to increase sublimation of the iodine."

You can use any UV light source in a darkened room to visualize chromatograms, including the BLB "black light" fluorescent tubes available at Lowes or Home Depot. You can purchase BLB tubes in various sizes to fit ordinary fluorescent fixtures or portable fluorescent lights. A UV LED flashlight, available online for a few dollars, is an excellent (and portable) source of UV light. "Black light" sources emit only long-wave UV radiation, and are safe to use without protective goggles. Other UV sources may require protective goggles and skin protection. Follow the directions and observe the precautions provided with your UV light unit.

In this lab session, we'll produce chromatograms using known specimens of aspirin, acetaminophen, and caffeine, as well as one or more questioned specimens that may contain all, any, or none of the known compounds. We'll visualize the chromatograms using iodine vapor to make the otherwise invisible compounds visible, with white light and/or UV light, and we'll compare the chromatograms of the known compounds against those of the questioned specimens to determine which, if any, of the known compounds are present in each of the questioned specimens as well as if compounds are present in the questioned specimen(s) that do not correspond to one of our knowns.

If you want to test known and/or questioned specimens other than those provided in the kit, visit the drugstore or check your medicine cabinet. Good candidates for analysis by chromatography include other OTC pain relievers (ibuprofen, naproxen, and so on), allergy drugs, prescription drugs, veterinary drugs, and so on. It's also interesting to run a chromatogram on a multivitamin to see how many distinct compounds can be visualized.

PROCEDURE VII-3-1: PREPARE CHROMATOGRAPHY JARS AND STRIPS

Developing a chromatography strip takes a few minutes to half an hour, depending on temperature and other factors. To save time, we'll prepare several chromatography jars, in which we'll develop all of our chromatography strips at the same time, one strip per jar. We'll use 50 mL self-standing centrifuge tubes as chromatography jars. We want these tubes to be saturated with solvent vapor before we use them, so prepare them at least a few minutes before you intend to develop the chromatograms.

1. Remove the cap on one of the 50 mL centrifuge tubes, stand the tube on a flat surface, and then use a pipette to fill the tube with 70% ethanol to just slightly above the top of the tapered portion. Recap the tube and set it aside, standing vertically.

If you don't have 70% ethanol, you can substitute 70% isopropanol for this and the following procedures. Drugstores sell both of these alcohols as "rubbing alcohol." Buy the pure alcohol rather than a product that includes green dye or other contaminants.

2. Repeat step 1 to make additional chromatography jars. You'll need at least three jars to complete the following procedures. Each of them will contain a strip that has been spotted with two materials: the questioned specimen and one of the known specimens. If you intend to test additional known(s) and/or questioned specimen(s), prepare sufficient additional chromatography jars to accommodate them.

   While you wait for your chromatography jars to become saturated with solvent vapor, prepare as many chromatography strips as you'll need to test all of your specimens. Wear gloves when touching the chromatography paper to prevent your skin oils from being transferred to the paper.

3. Cut as many strips of chromatography paper to about 2.5 cm × 10 cm (1" × 4") as you need to run all of your specimens, with one known/questioned pair per strip. The width of each strip should be sufficient to allow the strip to stand freely upright in the tube without flopping into contact with the sides of the tube. The length of the strip should be such that it comes as close as possible to the rim of the tube without protruding above it.

4. Use the ruler and pencil to draw two light lines across the width (short dimension) of each strip, one line about 0.5 cm from one end and the second about 1.5 cm from the same end. Place two hash marks on the 1.5 cm line, each about 1/3 of the width of the strip in from the edge.

5. Use the scissors to cut the corners off the strip, from the 0.5 cm line to the end of the strip. This will allow the bottom of the strip to protrude down into the tapered portion of the tube, where it can draw up the alcohol.

PROCEDURE VII-3-2: PREPARE SOLUTIONS OF KNOWN AND QUESTIONED SPECIMENS

The next step is to prepare solutions of your known and questioned specimens. Chromatography is extremely sensitive, providing useful results with specimens of one milligram or less. Accordingly, we need make up only a few drops of each solution.

Aspirin and acetaminophen are only sparingly soluble in water. At room temperature, one liter of water dissolves only about 3 g of aspirin or about 5 g of acetaminophen. (That's why the instructions on the bottle tell you to take these analgesic drugs with a full glass of water.) Both compounds are considerably more soluble in ethanol. At room temperature, one liter of pure ethanol dissolves about 200 g of aspirin or about 275 g of acetaminophen. (And, no, that doesn't mean it's a good idea to chase an aspirin or acetaminophen tablet with a straight shot of vodka.) Conversely, caffeine is actually less soluble in ethanol, about 15 g/L versus about 22 g/L in water. These relative solubilities mean that ordinary drugstore 70% ethanol is a good choice of solvent for all of these compounds. Ethanol also evaporates quickly, which makes it easier to spot the chromatograms properly.

1. If you have not already pulverized the acetaminophen tablet, do so now. Transfer a tiny amount of the acetaminophen powder—about as much as half a grain of rice—to well 1 of the spot plate. Add five drops of alcohol to the well. Record the contents of the well in your lab notebook.

2. Repeat step 1 to make up solutions of aspirin in well 2, caffeine in well 3, and the questioned specimen in well 4. If you have additional specimens, make up solutions of them in other wells.

Move on to the next procedure without delay, before your solutions can evaporate from the wells.

PROCEDURE VII-3-3: SPOT AND DEVELOP THE CHROMATOGRAMS

The next step is to prepare the chromatography strips by transferring tiny amounts of the known and questioned specimens, a process called *spotting*. Once we have finished spotting the strips and allowing them to dry, we'll develop the chromatograms in the chromatography jars we made up earlier.

1. Wearing gloves to prevent skin oil from contaminating the chromatography paper, place the first strip on a clean sheet of paper or other clean work surface.

2. At the top (nonpointed) end of the strip, use the pencil to label the strip, "P" on the left side for the acetaminophen (paracetamol) specimen and "Q" on the right side for the questioned specimen.

3. Dip a clean toothpick into the solution in well 1 of the spot plate and touch the tip gently to the left hashmark on the strip, transferring a tiny amount of solution to the strip. Your goal is to make a very small spot—ideally about 2 to 3 mm or less—centered on that hashmark, as shown in Figure VII-3-1.

Figure VII-3-1: *Spotting a chromatography strip*

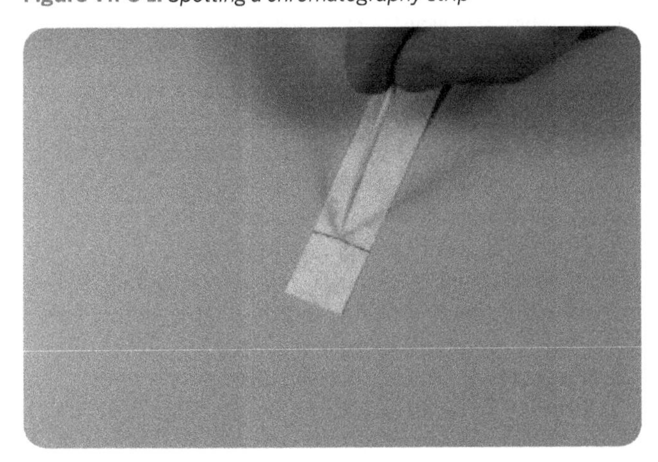

4. Repeat step 3 using a clean toothpick to spot the questioned specimen solution on the right hashmark.

5. Allow both spots to dry. You can (carefully) use a hair dryer to speed things up.

6. After both spots have dried, repeat steps 3 through 5 to re-spot the same solutions on the same spots on the strip. Your goal is to produce a concentrated spot of each specimen that is as small as possible.

7. Repeat step 6 until you have spotted each solution at least five times. Place the strip aside.

8. Repeat steps 1 through 7 with the second strip, producing spots of aspirin ("A") on the left hashmark and the questioned specimen ("Q") on the right hashmark.

9. Repeat step 8 with the third strip, producing spots of caffeine ("C") on the left hashmark and the questioned specimen ("Q") on the right hashmark.

With all of the strips spotted and dried, it's time to develop the chromatograms.

10. Working as quickly as possible (to avoid losing the alcohol vapor), remove the cap from the first chromatography jar and slide the first strip into the jar, pointed end down. Make sure the pointed end protrudes well into the alcohol in the tapered bottom of the tube. Recap the jar immediately.

11. Repeat step 10 to develop each of your other strips, each in its own chromatography jar.

12. As each strip develops, you'll see the solvent front climbing up the strip, drawn by capillary action. Observe development progress. When the solvent front nears the top of the strip, immediately take the strip out of the chromatography jar and place it flat on a clean sheet of paper. Use the pencil to make two hashmarks, one on each side of the strip, to mark the maximum progress of the solvent front. (Once the strips dry, the solvent front will no longer be visible.) Allow the strips to dry.

PROCEDURE VII-3-4: VISUALIZE THE CHROMATOGRAMS

At this point, you have three or more developed chromatograms, each of which looks like a plain strip of paper with a few pencil marks on it. The analytes present on the strips are colorless, and therefore invisible. To visualize them, we'll use ultraviolet light and iodine fuming, both techniques that are widely used in real forensic labs.

1. Working in a darkened room, illuminate the first strip with a UV light source. One or more of the analytes may fluoresce under UV light. If one or more of the analytes does fluoresce, use the pencil to mark the minimum and maximum extents on the strip.

2. Repeat step 1 for your other chromatograms.

   The next step is to measure and calculate R_f values for each of the detected analytes.

3. On the first strip, measure the distance from the center of the original spot to the line that marks the maximum progress of the solvent front and record that value in your lab notebook.

4. If the aspirin and/or the questioned specimen produced a fluorescent spot or spots, measure the distances from the center of the original spot to the lines that indicate the minimum and maximum extents of each fluorescent spot. Record those values in your lab notebook. If the minimum and maximum distances differed significantly, also calculate the average distance and record that value in your lab notebook.

5. Calculate R_f values for each of the spots by dividing the average distance the analyte moved by the total distance from the original spot to the line that indicates the maximum progress of the solvent front. For example, if the minimum and maximum distances of the analyte spot were 2.4 and 2.6 cm, respectively, and the maximum progress of the solvent front was 7.5 cm, calculate the R_f value for that analyte as (2.5 cm / 7.5 cm) = 0.33.

   You may find that only some (or none) of the analytes fluoresced under UV light. This often happens in chromatographic drug analyses. Fortunately, most analytes that do not fluoresce do bind tightly to iodine vapor, allowing them to be visualized with this alternative method. To use iodine vapor fuming, we'll need a fuming chamber.

6. Prepare your iodine fuming chamber by transferring a few crystals of iodine to a container such as a plastic bag or plastic kitchen container with a lid. (Really, just a *few* crystals; half the size of a grain of rice is plenty.) Place the developed chromatograms in the chamber, seal the chamber, and allow the strips to be exposed to the iodine vapor (*not* in direct contact with the iodine crystals) for several minutes to an hour or so. (You can speed things up by warming the iodine crystals with a hair dryer, in a tray of warm water, or simply by using your hands to heat them.)

7. Once iodine stains are showing on the chromatograms, open the container and quickly remove the chromatograms. Reseal the container immediately. You can use it again later without adding more iodine crystals.

8. Repeat steps 4 and 5 to measure and calculate the R_f values for the analytes using iodine fuming visualization.

REVIEW QUESTIONS

Q1: **Define retardation factor.**

Q2: **You have run a paper chromatogram. The solvent front moved 10 cm from the initial point. One of the analytes moved 3.7 cm from the initial point. What is the retardation factor for that analyte?**

Q3: **Using paper chromatography with an ethanol mobile phase to separate a mixture of drugs known to contain at least two analytes, you find that no separation occurs. What change would you make to obtain better separation?**

Observation of Drug Microcrystalline Structures and Precipitation Reactions

Lab VII-4

EQUIPMENT AND MATERIALS

You'll need the following items to complete this lab session. (The FK01 Forensic Science Kit for this book, available from *http://www.thehomescientist.com*, includes the items listed in the first group.)

MATERIALS FROM KIT

- Goggles
- Centrifuge tubes, 15 mL
- Copper(II) sulfate solution
- Pipettes
- Polarizing filters
- Slides, deep well
- Slides, flat
- Sodium carbonate solution
- Test tube rack
- Specimen: acetaminophen
- Specimen: aspirin
- Specimen: caffeine
- Specimen: questioned (chromatography)

MATERIALS YOU PROVIDE

- Gloves
- Camera with microscope adapter (optional)
- Ethanol (see text)
- Hair dryer (optional)
- Microscope
- OTC drug specimen(s) (optional)

WARNING

Although none of the materials used in this lab session present severe hazards, it is good practice to wear splash goggles, gloves, and protective clothing at all times when you are working in the lab.

BACKGROUND

Nowadays, quantitative forensic drug analysis is done instrumentally, using such techniques as high-performance liquid chromatography (HPLC), gas chromatography (GC), mass spectrometry (MS), Fourier transform infrared spectrometry (FTIR), and others. These instrumental techniques require expensive equipment that may be difficult on which to schedule time, so older presumptive techniques remain important and are frequently used even in the most up-to-date forensic lab.

Among those older techniques, *microcrystal tests* provide a fast, easy, cheap, sensitive way to test for specific chemical compounds, including drugs. To run a microcrystal test, the forensic scientist first dissolves a small amount of the questioned substance in water or another solvent, places a drop of the solution on a well slide, allows the solution to evaporate, and then examines the crystals microscopically with normal and polarized light. The appearance of those crystals (e.g., flat plates, needles, cubic, hexagonal, etc.) and their effect on polarized light can then be compared against standard reference exemplars for the compound in question. If the characteristics of the questioned specimen differ significantly from those of the reference exemplar, a match can be ruled out. If the characteristics are consistent in all respects, there is a reasonable presumption that the substances are identical.

If the questioned specimen matches the reference exemplar in the first phase, the forensic scientist can obtain additional confirming data by reacting the questioned substance with a test reagent that is known to form a characteristic precipitate with the drug suspected to be present in the specimen. The microscopic appearance of the precipitate crystals and their effect on polarized light is again compared against standard reference exemplars. If the two are consistent in all respects, there is a very strong presumption that the two substances are identical.

In this lab session, we'll examine the microcrystalline appearance of aspirin, acetaminophen, and caffeine under normal and polarized light, and compare our questioned specimen against each of those known specimens. From our observations, we'll determine which, if any, of our known compounds is consistent with the questioned specimen.

PROCEDURE VII-4-1: PREPARING SOLUTIONS OF KNOWN AND QUESTIONED SPECIMENS

To begin, we'll prepare solutions of our known and questioned specimens by using ethanol as the solvent (as in the preceding lab session). This time, we'll make up larger volumes of each specimen.

1. Label four 15 mL centrifuge tubes as follows: "acetaminophen," "aspirin," "caffeine," and "questioned."
2. Crush an acetaminophen tablet and transfer the powder to the corresponding tube. Fill the tube to the 2 mL line with ethanol and swirl the contents to dissolve as much as possible of the solid. Recap the tube and place it in the rack.

Don't be concerned if solid material remains in the tube; the tablet contains insoluble binders, and we want excess acetaminophen present to ensure a saturated solution.

3. Repeat step 2 to produce saturated ethanolic solutions of aspirin, caffeine, and the unknown in the corresponding tubes.

PROCEDURE VII-4-2: OBSERVING MICROCRYSTALLINE STRUCTURES

1. Label four flat microscope slides as follows: "acetaminophen," "aspirin," "caffeine," and "questioned."

2. Use a clean pipette to transfer one or two drops of the acetaminophen solution to the center of the corresponding slide. Allow the ethanol to evaporate completely, leaving the acetaminophen crystals visible as a cloudy area in the center of the slide. You can use a hair dryer to speed evaporation. If you use a hair dryer on one slide, use it on all slides to keep conditions consistent.

3. Repeat step 2 with the corresponding slides with the aspirin, caffeine, and questioned solutions.

4. Place the questioned slide on the microscope stage and turn on the illuminator. Scan the slide at 40X to locate an area with a good deposit of crystals. (Figure VII-4-1 shows aspirin crystals at 40X magnification.) Adjust the brightness and diaphragm to show maximum detail in the crystals. Record the appearance of the crystals in your lab book. Sketch the crystals or shoot an image for your records.

Figure VII-4-1: *Aspirin crystals at 40X*

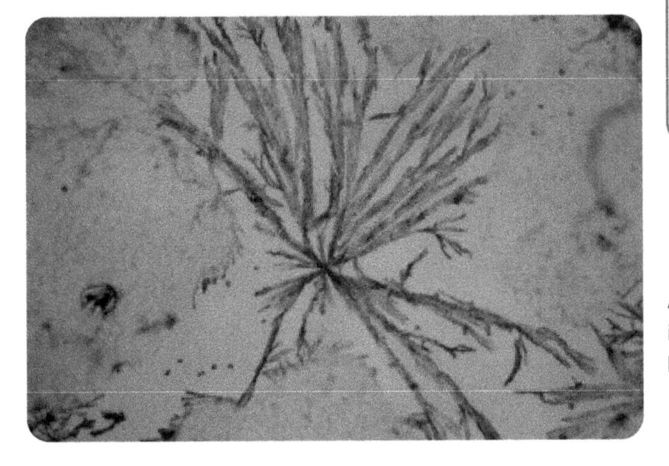

5. Place a polarizing filter flat on top of the illuminator and view the crystals by plane-polarized light. Note any visible differences in color, contrast, or other visual characteristics with polarized light. Record your observations in your lab notebook. Sketch the appearance of the crystals under polarized light or shoot an image.

6. While observing the crystals, slowly rotate the polarizing filter through 90°. Note the effect, if any, of rotating the plane of the polarized light on the appearance of the crystals, and record your observations in your lab notebook.

7. With the first polarizing filter still in place on top of the illuminator, hold a second polarizing filter between your eye and the eyepiece lens. Slowly rotate the second polarizing filter through 90°. Note the effect, if any, of rotating the second filter, and record your observations in your lab notebook.

As you rotate the second filter, the intensity of the illumination will change, from a maximum when the polarizing planes of the two filters are oriented at 0° to each other to a minimum when the planes are oriented at 90° to each other. This is expected behavior with polarizing filters, and is of no diagnostic value. What we're interested in is any change in the appearance of the crystals themselves as the plane of polarization is rotated.

8. Repeat steps 4 through 7 to observe the acetaminophen, aspirin, and caffeine slides.

At this point, you should have sufficient data to tentatively identify the questioned specimen as consistent with one of the known specimens. Verify your observations by performing a direct comparison between the questioned specimen and the known specimen you believe to be consistent with it. If your tentative identification is correct, the two specimens should be consistent in every respect, under both white and polarized light.

Note that the absence of any characteristic visual evidence under polarized light is itself evidence. For example, if one of your known specimens displays a rainbow of colors under polarized light and your questioned specimen does not, that is sufficient to rule out that known specimen as consistent with the questioned specimen.

PROCEDURE VII-4-3: ANALYSIS OF DRUGS BY PRECIPITATION

Having tentatively identified the questioned specimen as consistent with one of the known specimens by comparing their crystals under white and polarized light, the next step is to confirm that identification by using a second, entirely different test. If the results of this second test show that the two specimens are consistent, there is a high probability that the two specimens are in fact the same.

For our second test, we'll use a precipitation reaction. Until the last third of the 20th century, when instrumental methods came into common use, professional forensics labs routinely used precipitation tests to confirm tentative identifications made by microcrystalline tests. There were literally scores of precipitation reagents and protocols in regular use—many of which used salts of mercury, osmium, and other very toxic materials—and every forensic lab had volumes of comparison images showing the appearance of the precipitated crystals that resulted from different combinations of drug and precipitation reagent.

Before the advent of GC/MS (gas chromatography/mass spectrometry) and other instrumental methods, the combination of microcrystalline tests and precipitation reactions was legally considered to be conclusive, even though each of those test methods on its own was considered to provide only presumptive evidence. In particular, if a questioned specimen and a known specimen showed identical reactions with multiple precipitation reagents, courts considered the identification positive beyond all reasonable doubt.

1. Use a clean pipette to transfer sufficient questioned solution to a deep-cavity slide, filling the cavity about a third full.
2. Use a clean pipette to transfer five drops of 1 M sodium carbonate solution to the well.
3. Position the slide on the microscope stage with the well centered under the objective and observe the slide at 40X. Note the appearance of the specimen and record it in your lab notebook if any significant change has occurred.
4. While observing the slide through the microscope, add a drop of 1 M copper(II) sulfate solution. If a significant change occurs, continue adding copper(II) sulfate solution dropwise until no further change is evident. Observe the specimen with both white and polarized light. If an obvious change occurs, make a sketch or shoot an image. Record your observations in your lab notebook.
5. Repeat steps 1 through 4 using the solution of the known specimen that you tentatively identified as being consistent with the questioned specimen.
6. Repeat step 5 using the other known specimen solutions.

REVIEW QUESTIONS

Q1: Were you able to discriminate between the known specimens based on the appearance of their crystals?

Q2: Based on your observation of crystalline structures, what is the questioned specimen?

Q3: Did the precipitation test refute or confirm your tentative identification of the questioned specimen?

Q4: Using Internet resources, determine why we added sodium carbonate solution to our specimens.

Q5: If you dissolved and then crystallized a specimen that contained both aspirin and caffeine, what would you expect to see? How would this phenomenon bear on using micro-crystalline tests to identify an actual questioned specimen?

Assay Vitamin C in Urine by Iodometric Titration

Lab VII-5

EQUIPMENT AND MATERIALS

You'll need the following items to complete this lab session. (The FK01 Forensic Science Kit for this book, available from *http://www.thehomescientist.com*, includes the items listed in the first group.)

MATERIALS FROM KIT

- Goggles
- Ascorbic acid (vitamin C) tablet, 500 mg
- Graduated cylinder, 10 mL
- Graduated cylinder, 100 mL
- Iodine solution
- Pipettes
- Starch indicator solution
- Test tubes
- Test tube rack

MATERIALS YOU PROVIDE

- Gloves
- Water, distilled or deionized
- Soft drink bottle, 500 mL, clean and empty
- Specimen(s): urine (see text)

WARNING

We'll assume that you're using your own urine, which presents no biohazard to you. (In fact, urine obtained from any healthy person is sterile unless it is contaminated during collection.) Iodine solutions are irritating, toxic if ingested, and can stain skin and clothing.

BACKGROUND

Forensic labs often test urine, blood, or other body fluids to determine the presence and concentration of alcohol or illicit drugs such as cocaine, amphetamines, and opiates. Instrumental analysis is essential for confirmatory tests that will be used in court testimony, but initial "pass/fail" screening tests are often done using spot tests, titrations, and other wet chemistry techniques.

Although blood tests generally provide the most accurate results, blood specimens must be drawn by qualified medical professionals under sterile conditions and are therefore poorly suited for use in the field or for random no-notice drug tests. Conversely, a urine specimen can easily be obtained via a non-invasive procedure by someone without medical training, and provides reasonably reliable data about the drug intake of the person who provides the specimen. For that reason, and because urine specimens present a much smaller potential biohazard than blood specimens, urine tests are used whenever possible.

Depending on the drug, a urine specimen may contain the drug in its original form, or it may contain metabolites (breakdown products) of that drug. Field testing kits are available for most illicit drugs, in the form of spot-tests, test strips, or small reaction vials with premeasured volumes of test reagents. The technician applies a few drops of urine to the reagent vial or test strip. A specific color change indicates the presence of the drug in question (or its metabolites). These tests are usually qualitative or, at best, semiquantitative. Conversely, wet-chemistry drug tests done in forensics labs generally use fully quantitative techniques.

DENNIS HILLIARD COMMENTS

A positive urine drug screen is at best semiquantitative and is not sufficient to determine that the individual being tested was under the influence of the drug. Drugs of abuse and their metabolites show up in urine well after they have been eliminated from the blood. THC, the active ingredient in marijuana, can be detected for up to three weeks after last use in individuals who have frequently abused this drug.

In this lab session, we'll use a quantitative technique called *titration* to determine the concentration of a drug in urine specimens. The problem, of course, is that none of us use illegal drugs, so we'll have to find something else to test for. Fortunately, everyone's urine contains large amounts of at least one legal drug called ascorbic acid, better known as vitamin C. Very little vitamin C is metabolized in the human body; most is excreted unchanged. About 3% of excreted vitamin C is found in the feces, with the remainder excreted in the urine. The concentration of vitamin C in human urine can vary dramatically, from less than 10 milligrams (mg) per liter (mg/L) to several thousand mg/L. The concentration varies from person to person and from hour to hour for the same person, depending on the amount of vitamin C consumed, frequency and volume of urination, time of day, state of health, and so on. For healthy people, the normal concentration of vitamin C in fresh urine ranges from about 100 mg/L to about 1,000 mg/L.

Vitamin C is a strong reducing agent. We'll use this fact to do a quantitative assay of vitamin C in urine specimens by using a procedure called *iodometric titration*. An aqueous or alcoholic solution of iodine is brown. Vitamin C reacts quickly and quantitatively with iodine, reducing the brown elemental iodine to colorless iodide ions and oxidizing the vitamin C to dehydroascorbic acid, which is also colorless. We'll start with a solution that contains a known amount of iodine and slowly add urine until all of the iodine has been decolorized. By measuring how much urine is required to reach that point, we can determine how much vitamin C is present in a known volume of that urine specimen.

One molecule of vitamin C reacts with one molecule of iodine, producing one molecule of dehydroascorbic acid and two iodide ions. The gram molecular weight of vitamin C, $C_6H_8O_6$, is 176.126 g/mol, and that of iodine, I_2, is 253.809 g/mol. Because we know that one molecule of vitamin C reacts with one molecule of iodine, we also know that 176.126 grams of vitamin C reacts with 253.809 grams of iodine. Simplifying this ratio, 176.126:253.809, tells us that this reaction consumes about 0.69 milligrams (mg) of vitamin C per mg of iodine. Or, another way of looking at it, 1.44 mg of iodine reacts with 1 mg of vitamin C. We'll titrate a known volume of iodine solution of known concentration with a solution of urine of unknown vitamin C concentration, and use this ratio to calculate the concentration of vitamin C in the urine specimen.

FORMULARY

If you do not have the kit, you can make up the specialty solutions you'll need as follows:

Iodine solution The iodine solution supplied with the kit is an aqueous solution that is 0.1 M with respect to iodine and 0.1 M with respect to potassium iodide, which means the solution is about 1.26% w/v iodine and 1.66% w/v potassium iodide. We'll standardize our iodine solution against a known concentration of vitamin C before titrating our questioned specimen, so the exact concentration of iodine is not important. What is important is that we know the actual concentration.

Iodine tincture, available in nearly any drugstore, is a solution of iodine and sodium iodide in ethanol. Iodine tincture was formerly available in various concentrations, from a nominal 2% (w/v) iodine to about 8%. Because iodine is sometimes used by illegal methamphetamine labs, the DEA clamped down on iodine sales in 2007, limiting individual purchases to one 1 oz. or 30 mL bottle that contains 2.2% or less iodine. Iodine tincture now found on drugstore shelves typically contains 1.8 g to 2.2 g of iodine per 100 mL (1.8% to 2.2%) in 45% to 50% ethanol, with 2.1 g to 2.6 g of sodium iodide per 100 mL to increase the solubility of the iodine. Such drugstore iodine solutions are perfectly suitable for use in this lab session, as long as you standardize the solution before doing the titration.

Starch indicator solution The starch indicator solution supplied with the kit is an aqueous solution of soluble starch with thymol as a preservative. You can substitute any starch solution, including water in which cornstarch, potatoes, macaroni, or rice has been boiled. Such solutions do not keep for more than few days even if refrigerated, so make up fresh solution for each laboratory session.

PROCEDURE VII-5-1: PREPARE A STANDARD VITAMIN C SOLUTION

Our first task is to prepare a standard solution of vitamin C, which means the solution has a known concentration of vitamin C. Because the vitamin C concentration in urine from a healthy person may be 1,000 mg/L or more, we'll make the concentration of our standard solution 1,000 mg/L. We can do that simply by dissolving a 500 mg vitamin C tablet in 500 mL of water.

1. Use the 100 mL graduated cylinder to transfer as accurately as possible 300 mL of water to a labeled clean and empty 500 mL soft drink bottle.
2. Add one 500 mg vitamin C tablet to the bottle and swirl the contents until the tablet has dissolved. Don't be concerned if some undissolved solids remain. Most tablets contain insoluble binders. You can crush the tablet before adding it to the bottle to speed dissolution. If you do that, make sure to transfer all of the solid to the bottle.
3. Once the tablet has dissolved, add as accurately as possible 200 mL more water to the bottle, swirling frequently to make sure the solution is uniform.

PROCEDURE VII-5-2: TITRATE THE STANDARD VITAMIN C SOLUTION

The next thing we need to determine is how much iodine solution is needed to react with all of the vitamin C present in a known volume (called an *aliquot*) of our standard vitamin C solution. We'll find out by doing a microscale titration, adding iodine solution dropwise to the aliquot of vitamin C solution, counting the number of drops of iodine solution required to react completely with all of the vitamin C present.

Before we proceed, we need to determine an appropriate size for the vitamin C solution aliquot. We already know that our iodine solution is 1.26% w/v iodine, which can also be stated as 12.6 grams (12,600 mg) of iodine per liter or 12.6 mg of iodine per milliliter. We also know that 1 mg of iodine reacts with about 0.69 mg of vitamin C. That means that 1 mL of our iodine solution will react with about (12.6 * 0.69) = 8.7 mg of vitamin C. The standard vitamin C solution contains 1 mg per milliliter, so we can calculate that 1 mL of our iodine solution should react with about 8.7 mL of our standard vitamin C solution. Those are both convenient volumes for a microscale titration, so we'll use a 9 mL aliquot of our standard vitamin C solution and expect to use roughly 1 mL of our iodine solution to complete the titration.

One problem remains. Although the iodine solution is a distinct yellow-brown color, we'll be adding it dropwise to the vitamin C solution. All of the iodine solution prior to the last drop will react with vitamin C, producing colorless iodide ions, so the titration solution will remain colorless throughout the titration. Once all of the vitamin C is consumed, the *endpoint* of the titration is reached. At the endpoint, adding one more drop of iodine puts iodine in excess. But we'll have no way of knowing that the endpoint has been reached, because that one excess drop of iodine is insufficient to yield a visible yellow-brown color when diluted by several milliliters of colorless liquid.

Fortunately, there's an easy solution to that problem. Even the tiniest amount of free iodine reacts with starch to produce an intensely blue-black complex that serves as an indicator for the endpoint of the titration. Because that blue-black complex is distinctly visible at a concentration of only micrograms per liter, adding the final drop of iodine will cause the solution to turn suddenly from colorless to a distinct blue hue.

1. Use the 10 mL graduated cylinder and a clean pipette to transfer as accurately as possible 9 mL of the standard vitamin C solution to a test tube. Make sure to transfer all of the liquid from the cylinder to the tube. If necessary, you can use a pipette filled with distilled water to rinse any remaining solution from the cylinder to the tube.

2. Use a clean pipette to transfer about 0.25 mL (the bottom line on the pipette barrel) of the starch indicator solution to the test tube. Swirl to mix the solutions.

3. Fill a clean pipette with the iodine solution. Add the iodine solution dropwise to the test tube, counting drops and swirling the contents of the tube as you add the iodine. As each drop contacts the surface of the liquid in the tube, you'll notice a small blue area where the drop entered the solution in the tube, caused by a local excess of iodine reacting with the starch. This color will dissipate on swirling as the iodine solution mixes with the liquid in the tube.

4. As you continue adding drops of iodine, the blue color will continue to dissipate, but more slowly the more iodine you've added. Continue adding iodine solution dropwise until the blue color persists with swirling for at least 15 seconds. At that point, the endpoint of the titration has been reached. Record the number of drops of iodine solution required in your lab notebook.

PROCEDURE VII-5-3: TITRATE THE QUESTIONED URINE SPECIMEN

We now know the number of drops of iodine solution required to react with a 9 mL aliquot of standard vitamin C solution that contains 1,000 mg of vitamin C per liter. Our next task is to determine the vitamin C concentration in a questioned urine specimen. To do that, we'll repeat the same procedure we used with the standard specimen and compare the number of drops required to react with all of the vitamin C present in the urine specimen. With that information, it requires only a simple calculation to determine the concentration of vitamin C in the urine specimen.

Titrating the urine specimen requires more patience than titrating the standard vitamin C solution. With that solution, we had a pretty good idea of about how much iodine solution would be needed, so we could add iodine quickly at first, slowing down only as we reached the endpoint. With the urine specimen, the concentration of vitamin C might easily range from 100 mg/L to 1,000 mg/L or more. If, for example, titrating the aliquot of the standard vitamin C solution (1,000 mg/L) required (say) 35 drops of iodine, titrating the urine specimen aliquot might require anything from perhaps 3 or 4 drops to 35 or more drops of iodine. That means we have to be careful and go slowly when titrating the urine specimen.

> Titrate urine specimens as soon as possible after obtaining them. The concentration of vitamin C in urine rapidly decreases over time, particularly if the specimen is exposed to air and light. Or does it? We'll verify this later in this procedure.

1. Use the 10 mL graduated cylinder and a clean pipette to transfer as accurately as possible 9 mL of the questioned urine specimen to a test tube. Again, make sure to transfer all of the specimen, rinsing the cylinder if necessary with distilled water and transferring that rinse to the tube.

2. Use a clean pipette to transfer about 0.25 mL of the starch indicator solution to the test tube. Swirl to mix the solutions.

> Urine ranges in color from almost colorless to a relatively deep yellowish. Colorless urine yields the same blue color with the starch indicator as the aqueous vitamin C solution. Yellow urine, depending on the intensity of its color, may yield a color change ranging from yellowish green to greenish blue, but the color change is still obvious when the endpoint is reached. If you are working with colored urine, it's useful to have a second tube of urine as a control for color comparison.

3. Fill a clean pipette with the iodine solution. Add the iodine solution dropwise to the test tube, counting drops and swirling the contents of the tube as you add the iodine. As each drop contacts the urine, you'll notice a local color change at the point of contact. That color dissipates on swirling as the iodine solution mixes with the urine.

4. As you continue adding drops of iodine, the color will continue to disappear with swirling, but more slowly the more iodine you've added. Continue adding iodine solution dropwise until the color change persists with swirling for at least 15 seconds. At that point, the endpoint of the titration has been reached. Record the number of drops of iodine solution required in your lab notebook.

5. Calculate the concentration of vitamin C in the urine specimen using the ratio of drops required for the known and questioned specimens. For example, if a 9 mL aliquot of the 1,000 mg/L known specimen required 35 drops

of iodine to reach the endpoint, and a 9 mL aliquot of the questioned specimen required 21 drops of iodine, you can calculate the vitamin C concentration in the questioned specimen as [(21 drops / 35 drops) * 1,000 mg/L] = (0.6 * 1,000 mg/L) = 600 mg/L.

6. Repeat the titration using the same urine specimen after it has been exposed to air and light for 1, 2, 4, and 8 hours. If you find that the vitamin C concentration in the specimen changes over time, graph your results to determine if the concentration change is linear with time.

REVIEW QUESTIONS

Q1: **Analyzing urine specimens provides less accurate results than analyzing blood samples. Why, then, is urinalysis used so frequently for drug screening tests?**

Q2: **What did you determine the vitamin C concentration to be in your urine specimen?**

Q3: **What major assumption did we make in determining the concentration of vitamin C in the urine specimen based on values we obtained for the aqueous vitamin C in the known specimen? If that assumption is faulty, how might it affect the values we obtained for vitamin C concentration in the urine specimen?**

Q4: Based on your titration results with aged urine specimens, what conclusions, if any, can you draw about the importance of using fresh specimens for urinalysis? Other than time, what factors might effect the outcome? How might you modify your tests to eliminate various factors? Would the results you observed with vitamin C pertain to other drugs as well?

Forensic Toxicology Group VIII

Forensic toxicology is the analysis of *poisons* and *toxins*, both of which are substances that harm or kill living organisms. By convention, the word poison is used for toxic materials that are elements (such as arsenic, lead, mercury, and other *heavy metals*) or simple chemical compounds, such as potassium cyanide. The word toxin is used for a poisonous material that is produced by a living organism, such as aconitine, coniine, and other alkaloids. Toxins are generally much more complex molecules than are simple poisons, and many of them are lethal at much smaller dosages than simple poisons.

Forensic toxicology debuted in 1836, when the Scottish chemist James Marsh announced the eponymous Marsh Test, the first reliable forensic test for arsenic and antimony. In the early 19th century, arsenic compounds were widely used for many purposes, and were readily available at low cost. Would-be murderers then considered arsenic an ideal poison because arsenic compounds were lethal in small doses, readily soluble, odorless, tasteless, and impossible to detect. Because the symptoms of arsenic poisoning are very similar to the symptoms of gastroenteritis, an untold number of murder victims were instead believed to have died of natural causes.

But even if murder by arsenic was suspected, there was no easy way to prove it. Marsh developed his test in frustration after watching just such a murderer go free. In 1832, Marsh testified as an expert witness at the trial of John Bodle, who was accused of murdering his grandfather by putting arsenic in his coffee. Marsh used the then-standard test for arsenic, mixing a specimen of the suspect material with hydrochloric acid and hydrogen sulfide to precipitate the arsenic as insoluble arsenic trisulfide.

Unfortunately, by the time Marsh testified, the specimen had degraded and the jury was unconvinced by Marsh's scientific testimony. (This problem persisted for many years; until the late 19th century, many juries and even judges gave little weight to the testimony of forensics experts.) Bodle was acquitted, and Marsh watched a guilty man walk free. Determined not to let that happen again, Marsh devoted much of his free time over the next several years to developing a reliable test for arsenic. The test Marsh finally devised was revolutionary then, and is still used today by forensics labs in jurisdictions that cannot afford mass spectrometers and all of the other expensive instruments typically found in modern forensics labs.

Within a few decades, forensic toxicology was well developed. Hundreds of specific forensic tests had been devised. A forensic chemist still had to know what he was looking for, but if that substance was present, the tests would identify it. Reliable chemical tests had been devised for most common poisons, and would-be poisoners could no longer assume that poisonings would go undetected.

In the 1860s, the world-renowned forensic toxicologist, Robert Christison, was testifying for the prosecution at the trial of an accused murderer. He was asked by the prosecutor if common poisons could be detected reliably in the lab. Christison testified that all common vegetable and mineral poisons could easily be detected, with only one exception. At that point, the judge ordered Christison to stop speaking, concerned that he might reveal the name of this "undetectable" poison. (That poison was aconitine, extracted from the plant monkshood, which is lethal in doses too small to be detected by any means available at the time, and remains difficult to detect today.)

From the mid-20th century onward, chromatography, spectroscopy, fluorimetry, and other instrumental analysis techniques began to supersede wet-chemistry tests. Nowadays, at least in first-world forensics labs, instrumental analysis techniques dominate forensic toxicology, with wet-chemistry techniques largely relegated to use as presumptive "screening" tests. In many cases, that's because instrumental analysis is much more sensitive and accurate than the corresponding wet-chemistry procedures. In other cases, the wet-chemistry tests were and are as accurate and sensitive as the instrumental analysis methods, but were more time-consuming or required hazardous or expensive chemicals. Despite those disadvantages, wet-chemistry toxicology tests are still widely used, both for presumptive testing in the field and in forensics labs, and for confirmatory tests in less well-equipped forensics labs in poorer jurisdictions.

The laboratory sessions in this group use wet-chemistry techniques to explore various aspects of forensic toxicology.

Salicylate Determination by Visual Colorimetry

Lab VIII-1

EQUIPMENT AND MATERIALS

You'll need the following items to complete this lab session. (The FK01 Forensic Science Kit for this book, available from *http://www.thehomescientist.com*, includes the items listed in the first group.)

MATERIALS FROM KIT

- Goggles
- Graduated cylinder, 10 mL
- Pipettes
- Salicylate standard solution
- Salicylate reagent
- Spot plate

MATERIALS YOU PROVIDE

- Gloves
- Desk lamp or other light source
- Paper towel
- Soft drink bottle or other collection vessel
- Water, distilled or deionized
- Specimens: urine (see text)

WARNING

This lab session requires a urine specimen obtained from someone who has recently taken aspirin or another salicylate drug. Consumption of aspirin and other salicylate drugs by young people is associated with an increased incidence of Reye's Syndrome, a potentially fatal disease. In the United States, the Centers for Disease Control and Prevention, the US Surgeon General, the American Academy of Pediatrics, and the Food and Drug Administration recommend that aspirin and other products that contain salicylates not be administered to anyone aged 18 years or younger. Accordingly, if you are not at least 19 years of age, we recommend that you obtain the urine specimen from an adult who has recently used aspirin or another salicylate product such as a methyl salicylate muscle rub cream.

Always handle urine specimens using aseptic procedures, including wearing gloves and washing thoroughly with soap and water after handling specimens.

The salicylate test reagent, which contains iron(III) nitrate, is toxic, an oxidizer, and is hazardous by skin contact or ingestion. Wear gloves and goggles.

BACKGROUND

Colorimetry is the study of color. As it applies to forensics, colorimetry is used to determine the percentage of light that is transmitted by a solution that contains a colored solute, and from that to determine the concentration of the solute. In forensic toxicology, that solute may be a poison or (more likely) an intensely colored compound that is formed when a particular poison reacts with a particular reagent. Nowadays, most colorimetry is done instrumentally, using one of the following instruments.

Colorimeter

A relatively inexpensive instrument—entry-level colorimeters are available for less than $1,000—that allows the light transmission of a solution to be tested with white light or with a limited number of discrete wavelengths. Older colorimeters use an incandescent white light source, which may be used as-is or may be filtered to allow testing transmission with colored light. Typically, such colorimeters are supplied with three filters, one violet blue, one green yellow, and one orange red. Newer colorimeters often substitute colored LEDs for the incandescent light source. Before use, a colorimeter is calibrated by filling the specimen tube, called a *cuvette*, with distilled water and setting the colorimeter to read 100% transmission with that cuvette in place. The cuvette is then filled with the questioned solution and used to determine the percentage of light transmitted at one or several wavelengths. Colorimeters are widely used in industrial processes, such as wine making, where high accuracy and selectivity is not required.

Spectrophotometer

A *spectrophotometer* is a more sophisticated (and much more expensive) version of a colorimeter. Instead of using only white light or light at three or four discrete wavelengths, a spectrophotometer simultaneously tests a specimen at many wavelengths. Inexpensive spectrophotometers may sample every 10 or 20 nm over the visible spectrum from 400 to 700 nm. Spectrophotometers used in university, industrial, and forensics labs typically sample every nanometer (or better) over a range from 200 nm in the ultraviolet to 800 to 1,100 nm in the infrared. If a colorimeter is a meat cleaver, a spectrophotometer is a scalpel. The many data points determined by a spectrophotometer can be plotted on a graph to yield a "fingerprint" of the transmission characteristics of a particular solution over a wide range of wavelengths. By comparing this fingerprint to the fingerprints of known compounds, spectrophotometry can be used to identify questioned compounds in solution, rather than just the concentration of a known compound.

Despite the ubiquity of colorimeters and spectrophotometers in modern forensics labs, such instruments are not required to obtain useful data by colorimetry. Colorimetry was used long before such instruments became available. How? By using one of the most sensitive instruments available for determining color: the Mark I human eyeball.

In this lab session, we'll use colorimetry to do a quantitative determination of a poison that most people think of as a harmless drug: aspirin, also called acetylsalicylic acid. Accidental aspirin overdoses occur too often, with children the usual victims. Suicide by aspirin overdose is also common, despite the fact that death by aspirin overdose is extremely unpleasant. And, although most deaths from aspirin overdoses are accidents or suicides, murder by aspirin poisoning is by no means unheard of.

Aspirin is rapidly metabolized in the body to form salicylate ions. The therapeutic range for pain relief is 10 milligrams (mg) to 20 mg of salicylate per deciliter (dL, 100 mL) of blood serum. Symptoms of salicylate toxicity may appear at levels below 30 mg/dL, and are usually evident at levels higher than 40 mg/dL. Extreme salicylate toxicity—serum levels around 60 mg/dL for chronic overdoses and 100 mg/dL for acute overdoses—is an emergent condition that requires extreme measures such as dialysis. The adult human body typically contains five to six liters (50 to 60 dL) of blood, so acute salicylate toxicity sometimes occurs after consumption of only half a dozen 500 mg aspirin tablets over a period of a few hours. Small children, with their lower body mass and blood volume, may exhibit acute salicylate toxicity after swallowing only one or two 500 mg aspirin tablets.

Although aspirin and most salicylate compounds are colorless, salicylate ions form an intensely colored violet complex with iron(III) ions. In this lab session, we'll use that fact by making up solutions of various known concentrations of salicylate ions, reacting those solutions with iron(III) ions, and visually comparing the intensity of the color produced against a similarly prepared specimen with an unknown concentration of salicylate ions. By comparing the intensity of the violet color of the questioned specimen to the knowns, we can estimate the salicylate concentration of the questioned specimen quite closely.

FORMULARY

If you do not have the kit, you can make up the specialty solutions you'll need as follows:

Salicylate standard solution The standardized salicylate stock solution is simply a solution of sodium salicylate of known concentration. The kit includes a 500 mg/dL salicylate standard stock solution, but any concentration in that approximate range suffices, as long as you know the actual concentration. You can make a 500 mg/dL solution by dissolving 0.584 grams of sodium salicylate (14.36% of the gram molecular mass of sodium salicylate is sodium, so you need 584 mg to get 500 mg of salicylate ion) and making up the solution to 100 mL in a volumetric flask or graduated cylinder. If you have a centigram balance rather than a milligram model, 0.58 g is acceptable accuracy for this lab session. Sodium salicylate is available from laboratory chemical suppliers. Drugstores may carry sodium salicylate under that name or under various brand names. It is sometimes used as an analgesic by people who cannot tolerate aspirin. Pharmaceutical sodium salicylate is usually supplied in tablet form. If you use such tablets, remember that only 85.64% of the nominal mass is salicylate ion. For example, if you dissolve a 500 mg sodium salicylate tablet in 100 mL of water, the salicylate ion concentration is only (500 * .8564) = 428.2 mg/dL. Adjust your calculations accordingly.

Salicylate reagent The salicylate reagent is a solution of iron(III) ions, the exact identity and concentration of which is not critical. The kit includes a 10% w/v solution of iron(III) nitrate (ferric nitrate), made by dissolving 100 grams of the anhydrous salt (or 167 g of the hexahydrate salt) in water and making up the solution to one liter. You can substitute a roughly equivalent mass of any other soluble iron(III) salt you happen to have on hand, such as ferric chloride or ferric ammonium sulfate.

PROCEDURE VIII-1-1: PREPARE AN ARRAY OF SALICYLATE CONCENTRATIONS

Visual colorimetry involves comparing the questioned specimen against known specimens of various concentrations. That's much easier to do accurately if the questioned and known specimens are in close proximity. The best way to do that is to create a 3 × 3 array in the wells of a spot plate. The center well contains the questioned specimen, and the surrounding eight wells contain the known specimens. That places the questioned specimen immediately adjacent to all of the known specimens.

To get started, we'll use a procedure called *serial dilution* to produce the eight concentrations of salicylate ions, recording those values in our lab notebook as we proceed with the dilutions.

1. Using a clean pipette, transfer as accurately as possible 1 mL of the stock standard salicylate solution (500 mg/dL) to the upper left well of the array. (The top line of the pipette, just below the bulb, is the 1 mL line.)

2. Use the same pipette to transfer as accurately as possible 1 mL of the 500 mg/dL salicylate solution to the 10 mL graduated cylinder, followed by 1 mL of distilled water. Use the tip of the pipette to mix the solution. Use the pipette to draw up and expel the solution several times to ensure thorough mixing. The graduated cylinder now contains 2 mL of a 250 mg/dL salicylate solution.

3. Using the same pipette, transfer as accurately as possible 1 mL of the 250 mg/dL salicylate solution from the graduated cylinder to the upper-center well of the array. That well and the cylinder now each contain 1 mL of 250 mg/dL salicylate solution.

4. Transfer 1 mL of distilled water to the graduated cylinder, again using the pipette to mix the solution thoroughly. Transfer 1 mL of that 125 mg/dL salicylate solution to the upper-right well of the array.

5. Repeat the dilutions and transfers to populate the right-center well with 1 mL of 62.5 mg/dL salicylate solution, the lower right well with 1 mL of 31.25 mg/dL salicylate solution, the lower-center well with 1.0 mL of 15.625 mg/dL salicylate solution, the lower-left well with 1 mL of 7.8125 mg/dL salicylate solution, and the left-center well with 1 mL of 3.90625 mg/dL salicylate solution.

> As always, pay attention to significant figures. Given the inherent inaccuracies of our method, we actually recorded our concentrations as ~500, 250, 125, 63, 32, 16, 8, and 4 mg/dL.

PROCEDURE VIII-1-2: TEST THE REAGENT

The next task is to observe the reaction of the salicylate reagent with specimens that contain salicylate ions and to verify that the reagent produces no color change in the absence of salicylate ions by testing the reagent against a *blank*. Iron(III) ions react with salicylate ions to form a strongly purple-colored complex. If the reagent is present in excess, the intensity of the purple color is directly proportional to the concentration of salicylate ions present. The color is stable for at least an hour, so you'll have time to test several questioned specimens. If you are running tests for more than an hour, redo the serial dilution each hour to prepare fresh standards.

1. Transfer 1 mL of distilled water to the center well of the array. This distilled water well serves as a blank for testing the reaction of the reagent to a solution that contains no salicylate ions.

2. With the plate under a strong light, add two drops of the salicylate test reagent to each of the nine populated wells. Record your observations in your lab notebook.

3. Use the pipette to draw up the solution in the center well. Discard that solution, rinse the pipette with distilled water, and use the corner of a paper towel to remove any remaining droplets from the center well.

At this point, we've verified that the salicylate reagent produces no color change with distilled water, but we still don't know if or how it will react with a urine specimen that contains no salicylate ions. We can determine that by testing a urine specimen obtained from someone who has not recently ingested any aspirin or other salicylates.

4. Transfer 1 mL of the urine specimen known not to contain salicylate ions to the center well of the plate.

5. Add two drops of the salicylate reagent to the center well. Record your observations in your lab notebook.

6. Use the pipette to draw up as much as possible of the solution from the center well. Use the corner of a paper towel to remove any remaining liquid from the center well.

PROCEDURE VIII-1-3: TEST THE QUESTIONED SPECIMEN(S)

Now that we've determined how the salicylate reagent reacts in the absence and presence of salicylate ions, it's time to test questioned urine specimens for the presence and concentration of salicylate ions.

1. Transfer 1 mL of the questioned urine specimen to the center well of the plate.

2. Add two drops of the salicylate reagent to the center well.

3. Compare the intensity of the purple color in the center well against the surrounding wells to determine which of the surrounding wells is the closest match. You can interpolate values. For example, if the color of the questioned specimen is between that of the 32 mg/dL well and the 16 mg/dL well, but closer to the latter, you might estimate the salicylate concentration of the questioned specimen as 20 mg/dL. Or, if the color of the questioned specimen appears to be close to halfway between the 16 mg/dL well and the 8 mg/dL well, you might record the salicylate concentration of the questioned well as (16 + 8)/2 = 12 mg/dL. Record your best estimate of the concentration in your lab notebook, with notes of how you estimated the concentration.

REVIEW QUESTIONS

Q1: How accurately were you able to estimate the salicylate ion concentration in your questioned specimen?

Q2: Using only the materials and equipment listed for this lab session, describe an extension to the experimental procedure that would allow you to determine salicylate concentration more accurately.

Q3: If you test a urine specimen and determine that the salicylate concentration in that specimen is 400 mg/dL, can you conclude that the person from whom the specimen was obtained is dead? If not, why not? What additional test(s) would you run to verify or refute your hypothesis?

Q4: Drugs, including aspirin, are broken down and/or excreted unchanged over a period of time. Biological half-time (BHT) is the period over which the concentration of the drug is halved. For example, if the BHT is one hour, after one hour the concentration has fallen to half the original value, after two hours to one quarter the original value, after three hours to one eighth, and so on. Using only the materials listed for this lab session, describe an experimental procedure to determine the biological half-time of aspirin with respect to concentration in urine.

Detect Alkaloid Poisons with Dragendorff's Reagent

Lab VIII-2

EQUIPMENT AND MATERIALS

You'll need the following items to complete this lab session. (The FK01 Forensic Science Kit for this book, available from *http://www.thehomescientist.com*, includes the items listed in the first group.)

MATERIALS FROM KIT

- Goggles
- Dragendorff's reagent
- Centrifuge tubes, 15 mL
- Centrifuge tubes, 50 mL
- Chromatography paper
- Hydrochloric acid
- Pipettes
- Ruler
- Spot plate
- Test tubes
- Test tube rack
- Specimen: caffeine
- Specimen: poppy seeds

MATERIALS YOU PROVIDE

- Gloves
- Cotton swabs
- Pencil
- Scissors
- Toothpicks, plastic
- UV light source (optional)
- Water, distilled or deionized
- Specimen(s): alkaloids (optional; see text)

WARNING

Dragendorff's reagent is corrosive and toxic. Hydrochloric acid is corrosive and emits strong fumes. Most alkaloids are toxic, some of them extremely so. Always wear splash goggles, gloves, and protective clothing when working in the lab.

BACKGROUND

Before the use of spectroscopy and other instrumental analysis techniques became common in forensic toxicology, analysis was done strictly by wet-chemistry techniques. Because many different plant alkaloids—such as nicotine, strychnine, aconitine, and coniine—are fatal in small doses, forensic chemists during the 19th and early 20th centuries devoted considerable effort to developing reagents that could detect alkaloids at very low concentrations. Many of these reagents are general alkaloid reagents, providing positive results with many or most alkaloids. Others were used to differentiate specific alkaloids.

Most of these alkaloid reagents were named for the chemists who invented them, and most are now of only historical interest. A few of those reagents remain in use, however, and one of those is the wonderfully named Dragendorff's reagent, an acidic solution of bismuth and potassium iodide that yields a distinctive orange precipitate with most alkaloids.

There are many different formulations for Dragendorff's reagent. Even Dragendorff himself used different formulations of his own reagent. What all have in common is the presence of a bismuth salt in acidic solution with potassium iodide. The active species is probably some form of iodobismuthate complex, and it appears to make little difference which bismuth salt is used, or which acid. Various formularies list bismuth subnitrate, bismuth subcarbonate, bismuth chloride, and other bismuth salts, in combination with acetic, hydrochloric, sulfuric, or nitric acid of various concentrations. No doubt there is an optimum formulation that offers maximum sensitivity, but for our purposes, just about any potassium-bismuth-iodide solution suffices.

Dragendorff's reagent is still used for presumptive spot tests for alkaloids, but its primary use in modern forensics labs is as a color developer for chromatograms that have been used to separate mixtures that contain alkaloids. Because Dragendorff's reacts with almost every alkaloid to yield the orange positive, Dragendorff's is an excellent general alkaloid reagent. Dragendorff's reagent is sprayed on the paper or TLC chromatogram, where it develops any alkaloid spots present to a distinct orange hue. By calculating the R_f for each of those spots, a forensic scientist can unambiguously identify the specific alkaloids present on the chromatogram.

You can obtain alkaloid specimens from many sources. We used quinine (tonic water), nicotine (tobacco or cigarette butt), codeine (a few drops of prescription cough syrup), pseudoephedrine (allergy tablets), serotonin (health food store), emetine (an ancient bottle of syrup of ipecac, which we should already have discarded; ipecac is no longer recommended for any purpose), and various plants.

Many flowers, decorative plants, and cacti contain various alkaloids, which may be present throughout the plant or in only some parts, such as leaves or roots. Alkaloids in general are very poorly soluble in water. When present in drugs and food products, they are normally in the form of hydrochlorides, which are extremely soluble and can be used as is. If you obtain alkaloids from plant sources—be very careful and research the plants and alkaloids in question; some are toxic in extremely small doses—the best way to extract them is to soak the plant material in dilute hydrochloric acid.

FORMULARY

If you do not have the FK01 kit, you can make up Dragendorff's reagent as follows:

1. If you have not already done so, put on your splash goggles, gloves, and protective clothing.
2. Tare a small beaker and transfer 0.4 g of bismuth subnitrate to the beaker.
3. Transfer 10 mL of concentrated hydrochloric acid to the beaker and swirl until the solid salt dissolves.
4. Tare a second small beaker, and transfer 5 g of potassium iodide to the beaker. Add about 50 mL of distilled or deionized water to the beaker and swirl until the solid salt dissolves.
5. Pour the contents of the first beaker into the second beaker. Rinse the first beaker with a few milliliters of water and transfer the rinse to the second beaker.
6. Bring up the volume in the beaker to 100 mL with distilled water and transfer the resulting golden yellow solution to a storage bottle labeled Dragendorff's reagent.

If you're unable to obtain bismuth subnitrate, you can use the following alternative procedure to make up a usable solution of Dragendorff's reagent from Pepto-Bismol tablets or the generic equivalent. These tablets contain 262 mg of bismuth subsalicylate per tablet, which is equivalent in bismuth mass to about 212 mg of bismuth subnitrate. We can therefore substitute two Pepto-Bismol tablets for the 0.4 g of bismuth subnitrate. This alternative method isn't pretty, but it does use readily available materials.

1. If you have not already done so, put on your splash goggles, gloves, and protective clothing.
2. Transfer two Pepto-Bismol tablets and about 20 mL of water to a small beaker.
3. Swirl the beaker until the tablets have broken up into powder. Bismuth subsalicylate and the binders used in the tablets are both very insoluble in water, so don't be concerned if it looks as though none of the powder has dissolved.
4. Add about 10 mL of concentrated hydrochloric acid to the beaker. Swirl the beaker occasionally until foaming ceases. At this point, the liquid appears chalky white.
5. Allow the contents of the beaker to settle. Most of the solid matter precipitates, but enough remains suspended to give the liquid a cloudy white appearance.
6. Filter or carefully decant the liquid into the second beaker to remove as much as possible of the undissolved solids.
7. Dissolve 7 g of potassium iodide in a few milliliters of water and transfer that solution to the beaker that contains the bismuth solution. The solution immediately assumes a yellow-brown, cloudy appearance.
8. Bring up the volume in the beaker to about 100 mL with distilled water, allow the solid material to settle, and then pour off the clear yellow-brown solution into a storage bottle labeled Dragendorff's reagent.

The alternative formulation yields a yellow-brown solution rather than the golden yellow of the normal formulation, but the alternative formulation appears to work properly. Although we used Dragendorff's reagent prepared by the first method for this lab session, we did test the reagent produced by the second method by reacting it with quinine water. Both reagents produced the fine, bright orange precipitate characteristic of a positive Dragendorff's reagent test for alkaloids.

PROCEDURE VIII-2-1: PREPARE QUESTIONED ALKALOID SPECIMENS

In this procedure, we'll prepare specimens of various substances that may contain alkaloids and test them using Dragendorff's reagent for the presence of alkaloids. Dragendorff's reagent is non-selective in the sense that nearly any alkaloid yields a positive test (a bright orange suspension that settles as a bright red-orange precipitate). Conversely, the reagent is selective in the sense that few common compounds that are not alkaloids yield a positive test. Dragendorff's reagent is quite sensitive. It produces a positive test with some alkaloids at concentrations smaller than 1:100,000, and with many alkaloids at concentrations smaller than 1:10,000.

The degree of preparation necessary varies according to the source of the alkaloid. Some specimens, such as tonic water or cough syrup, can be tested as is, without further preparation, or if necessary, diluted with distilled water. To test for alkaloids present in pharmaceuticals, in which the alkaloid is normally present in water-soluble hydrochloride form, you can simply dissolve a small amount of the powder in a few milliliters of water. For alkaloids in free-base form, such as those present in plant material and in unprocessed foods such as poppy seeds, crush a small amount of the plant material and soak it in dilute hydrochloric acid to convert the free-base form of the alkaloid to the soluble hydrochloride form.

1. If you have not already done so, put on your splash goggles, gloves, and protective clothing.

2. Label 15 mL centrifuge tubes for each of your known and questioned specimens. (If you have more specimens than tubes, you can substitute small bottles or similar containers.)

3. For tablets, capsules, and other pharmaceutical specimens, crush the specimen to powder and transfer it to the tube. Add about 3 mL of distilled or deionized water to the tube and swirl to mix the contents. Stand the tubes in the test tube rack or a beaker and allow the specimens to dissolve. Don't worry if not all of the powder goes into solution. Record the contents of each tube in your lab notebook.

4. For seeds, leaves, and other vegetable specimens, use the clean stirring rod or a spoon to grind or crush each specimen thoroughly and transfer it to a tube. Add about 3 mL of distilled or deionized water and about 0.25 mL (the line nearest the tip of the pipette) of 6 M hydrochloric acid to the tube and swirl to mix the contents. Stand the tubes in the test tube rack or a beaker, swirl them occasionally, and then allow the dilute acid to extract any alkaloids present in the specimens. Record the contents of each tube in your lab notebook.

PROCEDURE VIII-2-2: TEST SPECIMENS FOR THE PRESENCE OF ALKALOIDS

In this procedure, we'll test our prepared specimens for the presence of alkaloids.

1. If you have not already done so, put on your splash goggles, gloves, and protective clothing.

2. Label six test tubes A through F and place them in the test tube rack.

3. Transfer about 2.5 mL of each alkaloid solution to the corresponding test tube. (Leave at least 0.5 mL of each specimen in its centrifuge tube. We'll use that in the next procedure.)

4. Transfer about 0.5 mL of Dragendorff's reagent to each test tube and observe the results. If an alkaloid is present in relatively high concentration, the liquid immediately assumes a bright yellow-orange, opaque appearance, as shown in Figure VIII-2-1. Within a few minutes, a bright orange precipitate settles to the bottom of the tube, leaving the supernatant liquid clear. If an alkaloid is present in low concentration, the liquid assumes a cloudy appearance, and it may take the precipitate some time to settle. At very low concentration, it may take a few minutes for slight cloudiness to develop. Record your observations in your lab notebook.

> The left test tube in Figure VIII-2-1 appears to have a slight precipitate in the bottom of the tube, but that's merely a photographic artifact. In fact, the contents of that tube were a clear golden color with no solids at the bottom of the tube.

Figure VIII-2-1: *Negative (left) and positive Dragendorff's reagent tests for an alkaloid*

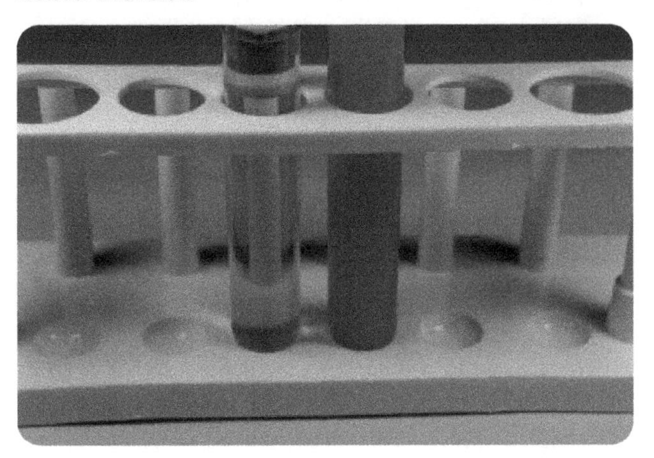

PROCEDURE VIII-2-3: ANALYZE ALKALOIDS USING PAPER CHROMATOGRAPHY

In the preceding procedure, we established that Dragendorff's reagent is a good precipitation reagent for alkaloid testing, but precipitation tests tell us nothing about the identity of the alkaloid. In this procedure, we'll use paper chromatography to produce chromatograms of known alkaloids, visualize those chromatograms with Dragendorff's reagent, and determine the R_f values for each alkaloid. We'll then produce spot a chromatography strip with a mixture of alkaloids, develop and visualize the chromatogram, and determine the R_f values for each alkaloid present in the mixture.

1. If you have not already done so, put on your splash goggles, gloves, and protective clothing.

2. Cut enough chromatography paper strips for each of your specimens and one spare. Each strip should be about 2.5 x 10 cm, just long and wide enough to fit inside a 50 mL centrifuge tube. (Wear gloves while you prepare the strips; skin oils interfere with developing the chromatograms.) Cut two corners from the same end of each strip to allow the bottom of the strip to protrude into the conical part at the bottom of the tube. Draw a pencil line across the width of each strip, about 2 cm from the pointed end of the strip. Put a small dot near the center of each line. At the top end of each strip, label it for the corresponding alkaloid.

3. We'll use the 50 mL centrifuge tubes as chromatography jars. Transfer enough hot water to each tube to fill it just to the top of the conical section at the bottom of the tube. Replace the caps on the tubes and set them aside for now.

> We use hot water and immediately cap the tubes because we want the air inside the tubes to be saturated with water vapor. Otherwise, the dry air in the tubes can evaporate water from the chromatograms as they develop, interfering with the development process.

4. Using a clean toothpick for each solution, spot each of the chromatography strips with the corresponding alkaloid solution from each centrifuge tube. Transfer the tiniest possible drop from the tip of the toothpick to the center mark on the line of each strip. Allow each spot to dry as you spot the other strips.

5. When you've spotted all of the strips with your known alkaloid solutions, repeat the spotting for all of the strips. Try to place each new spot directly on top of the dried spot. The goal is to get the smallest and most concentrated spot possible on each strip. Respot each strip until you have spotted it at least five times and then allow the strips to dry completely.

6. Choose two or three of your known alkaloid solutions. Transfer three drops of each solution to the same well on the spot plate. Use the tip of a clean toothpick to mix the solutions.

7. Spot the extra chromatography strip with the solution of mixed alkaloids, allow it to dry, and then respot it a total of at least five times.

8. Holding the first strip outside one of the 50 mL centrifuge tubes, verify that the water level in the tube is not high enough to reach the spot on the strip. As quickly as possible, remove the cap from the tube, slide the strip into the tube (pointed end down), and recap the tube.

9. Repeat step 8 for the other strips.

10. Keep a close eye on the tubes to watch development progress. The paper strip draws up water from the bottom of the tube by capillary action. As development progresses, the solvent front line climbs higher and higher on the strip. When the solvent front line approaches the top of a strip, uncap the tube and remove the strip.

> Do not allow the solvent front to reach the top of the strip. If it does, that chromatogram is ruined and must be redone.

11. Place the strip aside to dry on a clean sheet of paper or a paper towel. Before it dries, use the pencil to make a hash mark on either side of the strip to indicate the maximum progress of the solvent front. (Once the strip dries, the solvent front will no longer be visible.)

> When the strips are dry, there's no obvious difference from the original appearance because the alkaloids are colorless. To make the alkaloid spots visible, we'll visualize the chromatograms by treating them with Dragendorff's reagent.

12. If you have an ultraviolet (UV) light source available, you should always use it first to visualize a chromatogram, before you attempt chemical means of visualizing the chromatogram, such as Dragendorff's reagent or iodine fuming. Many colorless analytes, including some alkaloids, fluoresce under UV light. Take your developed chromatograms into a closet or other darkened area, and view them under your UV light source. If any analyte traces are visible under UV, use the pencil to mark their positions and extents. Figure VIII-2-2 shows the striking blue-green fluorescence of the alkaloid quinine, which is present at only 83 parts per million in a bottle of tonic water.

13. Place the first chromatogram on a clean sheet of paper or a paper towel. Transfer a couple drops of Dragendorff's reagent to a cotton swap. Put the tip of the swab at the top center of the strip and wipe downward, dampening the center of the strip along its length until you reach the spot. If necessary, add more reagent to the cotton swab. The goal is not to soak the entire strip, but to dampen the area of the strip where the alkaloid traces are present. Depending on the particular alkaloid and the amount present in the trace, a yellow-orange patch may appear immediately, or it may take a few minutes to become visible.

14. Using a fresh swab each time, repeat step 13 for each of the other chromatograms.

Figure VIII-2-2: *Quinine in tonic water fluorescing under UV light*

15. After the strips have visualized, measure the distance on the first strip from the center of the original spot to the line marking the maximum progress of the solvent front. Record that value in your lab notebook. Then measure the distance from the original spot to the center of the visualized alkaloid trace, and record that value in your lab notebook.

16. Calculate the R_f value for that alkaloid by dividing the second value by the first. For example, if the distance from the original spot to the maximum progress of the solvent front is 7.5 cm and the distance from the original spot to the alkaloid trace is 2.5 cm, calculate the R_f value for that alkaloid as (2.5 / 7.5) = 0.33. Record the R_f value for that alkaloid in your lab notebook.

17. Repeat steps 15 and 16 for each of the other chromatograms of your known alkaloids.

18. Repeat steps 15 and 16 for the chromatogram of your mixed alkaloid solution. In this case, you should have multiple alkaloid traces present. Calculate separate R_f values for each.

19. Compare each of the R_f values from your mixed alkaloids chromatogram with those of the known alkaloids, and attempt to identify each of the mixed alkaloids based on the similarity of their R_f values to those of your knowns.

REVIEW QUESTIONS

Q1: Which of your known alkaloid specimens did not precipitate Dragendorff's reagent?

Q2: If the R_f value of a questioned specimen is the same as that of one of your known specimens, can you identify the questioned alkaloid specimen as matching the known specimen?

Gunshot and Explosive Residues Analysis

Group IX

Analysis of gunshot residues (GSR) and explosive residues is an important part of the work of most forensic labs. In professional forensic labs, definitive analyses of such residues are done instrumentally, using expensive technologies such as neutron activation analysis and scanning electron microscopy. But, while these instrumental analyses are essential, the demand persists for inexpensive, fast presumptive tests that can be used to screen potential evidence before incurring the inevitable delays and expense of instrumental analysis.

In this group of lab sessions, we'll use a variety of presumptive color tests to detect the presence of gunshot residues and explosive residues.

Presumptive Color Tests for Gunshot Residue

Lab IX-1

EQUIPMENT AND MATERIALS

You'll need the following items to complete this lab session. (The standard kit for this book, available from *http://www.thehomescientist.com*, includes the items listed in the first group.)

MATERIALS FROM KIT

- Goggles
- Bottle, sprayer
- Lead nitrate, 0.1% w/v
- Modified Griess reagent, part A
- Modified Griess reagent, part B
- Sodium nitrite, 1% w/v

MATERIALS YOU PROVIDE

- Gloves
- Acetone
- Camera (optional)
- Clothes iron
- Cotton swabs
- Firearm and ammunition (optional; see text)
- Measuring tape or yard/meter stick
- Paper (see IX-1-2)
- Paper towels
- Pencil
- Permanent marker or grease pencil
- Plastic wrap
- Scanner and software (optional)
- Scissors
- Sodium rhodizonate and buffer (optional)
- Tape, masking
- Vinegar, distilled white
- Specimens, cardboard (see IX-1-1)
- Specimens, cloth (see IX-1-1)

WARNING

Wear splash goggles, gloves, and protective clothing. Read the MSDS for each chemical you use and follow the recommended safety precautions. Observe all safety precautions, laws, and regulations when handling and firing firearms.

BACKGROUND

Pulling the trigger of a loaded firearm sets a chain of events in motion. The hammer or striker strikes the firing pin, which in turn strikes the primer in the base of the cartridge. The primer contains a shock-sensitive explosive, which is detonated by the firing pin strike. The flame produced by the primer detonation ignites the propellant—black or smokeless gunpowder—which burns in a tiny fraction of a second, producing a large amount of very hot gas under very high pressure. That gas expands, pushing the bullet out of the cartridge case, down the barrel, and out of the muzzle.

The gas—which contains a mixture of burned, unburned, and partially burned propellant, as well as reaction products from the primer and possibly material from the bullet or its jacket material—exits the muzzle at high velocity. In autoloading pistols, rifles, and shotguns, some propellant gas may also exit the breech as the spent cartridge is ejected. In revolvers, some propellant gas is vented from the flash gap between the front of the cylinder and the forcing cone at the rear of the barrel. As the gas expands and cools, some of it is deposited as solid residue on everything in the vicinity—the skin and clothing of the shooter, the target, the firearm itself, and anything else nearby. This condensed solid material is called *gunshot residue*, usually abbreviated GSR.

Gunshot residue commonly contains the following materials:

Nitrates
: All common propellants contain nitrates (NO_3 groups). For example, potassium nitrate is one of the major components of black powder and black powder substitutes such as Pyrodex. Single-base smokeless powders contain guncotton (nitrocellulose or cellulose nitrate), and double-base smokeless powders contain guncotton and nitroglycerin (glyceryl trinitrate). When a cartridge is fired, some propellant is always unburned and can be detected in gunshot residue by testing for the presence of nitrates.

Nitrites
: When any firearm propellant burns, the nitrates present are converted to nitrites (NO_2 groups). These nitrites make up the majority of solid gunshot residue, and are easily detected by color tests.

Primer residues
: Primer residues make up a small fraction of GSR, and may or may not be detectable by color tests, depending on the distance the propellant gas travels before being deposited as solid GSR, the surface upon which it is deposited, and the type of primer.

Corrosive primers, based on mercury fulminate, were used in almost all ammunition until WWII and are still sometimes encountered today, particularly in surplus military ammunition. (WWII-era ammunition is about as reliable as it was the day it was made; corrosive primers remained in use so long because they are much more reliable after decades-long storage than are non-corrosive primers.) The presence of mercury in GSR is easily detected, and is a strong indication that the GSR was produced by ammunition with a corrosive primer. Ammunition with corrosive primers is sufficiently uncommon nowadays that detection of corrosive primer residues is significant forensically because of their relative rarity.

Noncorrosive primers have been in wide use since the 1950s. Most noncorrosive primers are based on lead styphnate or lead azide, so the presence of lead in GSR suggests that the cartridge used a standard noncorrosive primer. There are other potential sources of lead in GSR, including the bullet itself (unless it is fully jacketed), lead splash at a revolver's forcing cone, and bore deposits that are vaporized by subsequent shots (even a fully jacketed bullet fired through a badly leaded barrel can produce more lead residue than is produced by the primer). The ubiquity of lead in GSR is the reason that lead tests are always done on surfaces that are suspected of containing GSR.

Lead-free primers have become increasingly common with the growth of concerns about environmental lead. Initially, lead-free primers were used primarily in .22 rimfire ammunition and in centerfire ammunition intended for use in indoor shooting ranges, but many ammunition manufacturers now use lead-free primers exclusively or nearly so. Detecting primer residue from a .22 rimfire cartridge or a centerfire cartridge with a lead-free primer may be difficult or impossible with color tests, particularly if the primer residue was deposited on a surface distant from the muzzle.

Miscellaneous residues

Other chemicals sometimes present in GSR are sometimes detectable by color tests. Among the most common of these other chemicals are barium, antimony, copper, chlorates, perchlorates, and halides. Because these chemicals may be present in anything from relatively large quantities to only trace amounts, they are not normally analyzed by presumptive color tests but are instead analyzed by atomic absorption (AA) spectroscopy, scanning electron microscope (SEM), and other instrumental methods. The combination and relative amounts of these miscellaneous compounds present in GSR may be useful in characterizing GSR to confirm that it is consistent with the primer and propellant present in a particular type of cartridge. Some primer compounds and propellants are intentionally contaminated by the manufacturer with tiny amounts of specific compounds, called *tracers* or *taggants*, whose presence in GSR allows the forensic scientist to characterize that GSR specimen as having been produced by one of these tagged primer compounds or propellants.

It's important to understand that the presence of any or all of these chemicals is not definitive evidence that a firearm has been discharged. Nitrates, nitrites, chlorates, perchlorates, and halides, for example, are common in food, household products, fertilizers, and so on. Small amounts of lead are also easy to pick up from non-firearms sources, for example, from solder, wheel weights, or an automobile battery. You can pick up enough lead to give a positive GSR test for lead just by raising the hood of your car to check the oil. If you happened to move a bag of fertilizer out of the way while checking your oil, your hands may also show the presence of nitrates. And, if you had bacon for breakfast, there may be detectable levels of nitrites on your hands.

This simple fact seems to have escaped most forensic scientists until only a couple of decades ago. Before that time, many people were convicted of serious crimes based on the presence of "GSR" on their hands that may not have been GSR at all. The *dermal nitrate test*—testing for nitrates on the hands of suspects—was entirely discredited on this basis. Two high-profile bombing cases in Britain in the mid-70s—those of Judith Ward and the Birmingham Six—called the validity of the Griess test for nitrites (which is also used for GSR tests) into question. The prosecution obtained convictions in both cases, but those convictions were overturned on appeal when it was found that forensic scientist Frank Scuse had misinterpreted Griess test results, claiming that positive results in both cases proved that the defendants had handled explosives. In fact, as the defense demonstrated, positive Griess test results could occur if the defendants had handled shoe polish or other common materials. The presiding judge concluded that the results of such tests were useful only as presumptive evidence.

Despite these uncertainties, GSR color tests continue to be used today because they are fast, inexpensive, and provide useful (if not definitive) information. Although the mere presence of nitrates, nitrites, lead, and other compounds present in GSR does not prove that GSR is present, the distribution pattern of those materials on clothing or other items may be very significant.

For example, if a person is found dead of a gunshot wound in the chest, it may be unclear whether that victim died by accident, suicide, or murder. Because the GSR pattern varies in size and shape with the distance of the muzzle from the victim's shirt, forensic scientists may be able to rule out one or more of these scenarios. For example, if little or no GSR is present on the victim's shirt and hands, it's very likely that the firearm was discharged at some distance from the victim, making accident or suicide very unlikely.

Conversely, if the GSR pattern suggests that the firearm was discharged in close proximity to the victim, any of the three scenarios is possible. If further tests establish the presence of GSR on one of the victim's hands, murder becomes less likely and accident or suicide more so. If no GSR is detected on the victim's hands, murder becomes more likely, although accident or suicide cannot be ruled out. Why? Because firearms vary in how much GSR they deposit on the shooter's hands, if any. For example, our .45 ACP Colt Combat Commander deposits relatively heavy GSR residues on our hands when we fire it, as does our Ruger .357 Magnum revolver. Conversely, our Ruger .22 target pistol deposits very little GSR on our hands, even after repeated firing.

For this reason, forensic technicians always test-fire the weapon, if it is available. Their goal is not just to obtain fired bullets for rifling mark comparisons, but also to note the GSR characteristics of the weapon.

Although many different spot-test color tests are used for GSR analysis, the two color tests most frequently used by real forensics labs are:

Modified Griess test

The *modified Griess test* detects the presence of nitrites by using a reagent made from sulfanilic acid and alpha napthol. A solution of these chemicals is coated on a sheet of paper and allowed to dry. The treated surface of the paper is placed in contact with the clothing or other material suspected to contain GSR. That sandwich is then covered by a sheet of cloth that is saturated with dilute acetic acid and pressed with a hot iron. Any nitrites present in the specimen react with the sulfanilic acid, alpha naphthol, and acetic acid vapor to produce an azo dye with a characteristic orange color.

Rhodizonate test

The *rhodizonate test* detects the presence of lead by its reaction with the sodium or potassium salt of rhodozinic acid. At pH 2.8, provided by a tartrate/tartaric acid buffer solution, rhodizonate ions react with lead to form a bright pink stain. Spraying the treated area with a dilute hydrochloric acid solution causes the color to change to blue-purple, confirming the presence of lead. Although lead is the most common metal tested for with the rhodizonate test, other metals also provide characteristic colors at different pH levels.

We considered including the materials required for the rhodizonate test in the FK01 Forensic Science Kit, but doing so would have significantly increased the cost of the kit. The modified Griess test and the rhodizonate test are quite similar in concept and method. The only real difference is that Griess detects nitrites and rhodizonate lead, so the extra cost would have provided minimal educational benefit. If you do want to do the rhodizonate test, you can purchase the necessary materials from any law enforcement forensic supply vendor. If you don't, simply read through the rhodizonate test procedure and imagine that you're actually doing it.

FORMULARY

If you don't have the FK01 Forensic Science Kit, you can make up the required solutions as follows: (Wear gloves and safety goggles while making up these solutions.)

Modified Griess reagent part A Dissolve 0.12 g of sulfanilic acid in 25 mL of distilled or deionized water. This solution is stable until it is mixed with the part B solution.

Modified Griess reagent part B Dissolve 0.07 g of alpha naphthol in 25 mL of methanol, ethanol, or isopropanol. This solution is stable until it is mixed with the part A solution.

Lead nitrate, 0.1% w/v Dissolve 0.1 g of lead nitrate in 100 mL of distilled or deionized water. This solution is stable. You do not need this solution unless you are using the alternate method in Procedure IX-1-1 for making up simulated GSR specimens and you intend to complete the optional rhodizonate test procedure.

FORMULARY (CONTINUED)

Sodium nitrite, 1% w/v Dissolve 1 g of sodium nitrite in 100 mL of distilled or deionized water. This solution is stable. You do not need this solution unless you are using the alternate method in Procedure IX-1-1 for making up simulated GSR specimens.

The following solutions are needed only if you intend to complete the optional rhodizonate test procedure. Although it is expensive, you can purchase potassium rhodizonate or sodium rhodizonate from any law enforcement forensics supply vendor. The other chemicals needed can be purchased from any lab supply vendor.

Rhodizonate reagent Dissolve a few crystals of potassium or sodium rhodizonate in about 10 mL of ice-cold distilled or deionized water to make a solution about the color of iced tea. This solution is unstable. At room temperature, this solution degrades within a few minutes to an hour or so. Even kept in an ice bath, it remains usable for only a couple of hours. If you intend to do the rhodizonate test, test all of your specimens in one batch, working as quickly as possible.

Rhodizonate reagent buffer Dissolve 1.9 g of sodium bitartrate and 1.5 g of tartaric acid in 100 mL of distilled or deionized water. The solids take some time to dissolve. You can speed the process by heating the solution gently and swirling it frequently. This solution has a limited shelf life, which can be extended by keeping it chilled. We make up a fresh buffer solution for each session.

Hydrochloric acid wash Add 5 mL of concentrated (37%) hydrochloric acid to 95 mL of distilled or deionized water.

PROCEDURE IX-1-1: PRODUCE GUNSHOT RESIDUE (GSR) SPECIMENS

Our first task is to produce specimens that contain GSR residue. If you do not have access to a firearm and a safe place to shoot it, you can use the alternate method described at the end of this procedure to produce simulated GSR specimens.

WARNING

Always wear eye and ear protection, and always obey safety rules when discharging a firearm. Make sure the backstop is adequate to prevent the bullet from penetrating. Note that the cloth may smolder or catch fire, particularly if the muzzle of the firearm is in close contact with the specimen.

DENNIS HILLIARD COMMENTS

Only someone who is fully trained in the safe and legal use of a firearm should be allowed to create the GSR samples. In addition to requirements for safely discharging a firearm, there are requirements for the safe handling of the firearm, the need for certification in some jurisdictions to possess a firearm, and laws in many jurisdictions that forbid discharging a firearm in an urban area.

Unless you meet all of these requirements, you **MUST** use an alternative method, such as the one described in "Producing simulated GSR specimens." Another alternative method is to visit a gun club and ask for their help in collecting GSR on shooters' hands and maybe having them shoot into cloth at varying distances and with different calibers.

1. If you have not already done so, put on your shooting glasses and ear protection.

2. Tape a white cloth swatch to a clean sheet of cardboard. Use tape at all four corners, and cover only enough of the cloth to secure it to the cardboard. Use the marker to label the cloth "contact" near one edge. Suspend the cloth/cardboard target from a target frame with a safe backstop.

3. Place the muzzle of the firearm nearly in contact with the surface of the cloth, and fire a shot, as shown in Figure IX-1-1. Figure IX-1-2 shows a close-up image of the GSR deposits on the cloth.

Figure IX-1-1: *Firing a test shot with the muzzle in close contact with the cloth specimen*

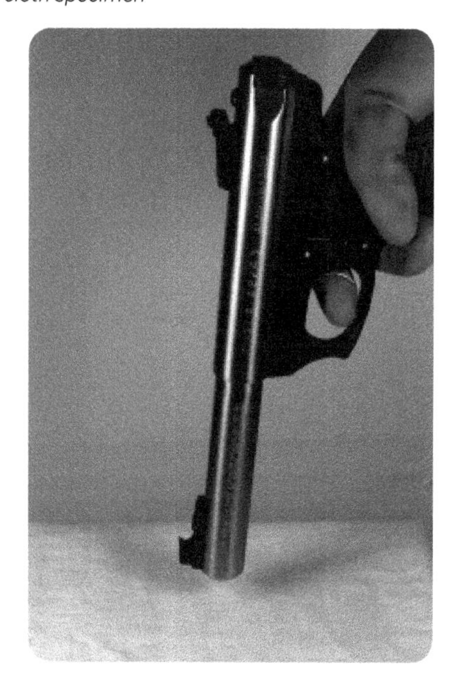

Figure IX-1-2: *GSR deposits produced with the muzzle in close contact with the cloth specimen*

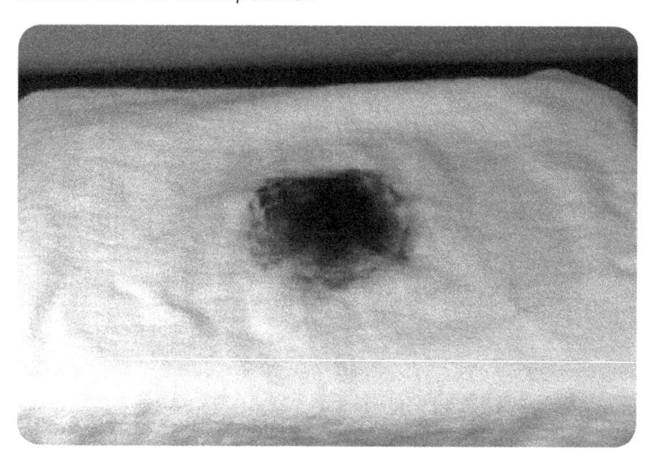

4. Leaving the cloth taped to the cardboard, cover the cloth with plastic wrap and tape the plastic wrap in place to protect the cloth from GSR produced subsequently.

5. Repeat steps 2 through 4 with a fresh piece of white cloth labeled "30 cm" and a fresh piece of cardboard, this time with the muzzle 30 cm (~1 foot) from the cloth.

6. Repeat steps 2 through 4 with a fresh piece of white cloth labeled "60 cm" and a fresh piece of cardboard, this time with the muzzle 60 cm (~2 feet) from the cloth.

7. Repeat steps 2 through 4 with a fresh piece of white cloth labeled "120 cm" and a fresh piece of cardboard, this time with the muzzle 120 cm (~4 feet) from the cloth.

8. Repeat steps 2 through 7, substituting your colored/patterned/striped cloth for white cloth. (If possible, have someone else produce these GSR specimens for you, labeling them randomly 1 through 4 or A through D, and keeping track of which label corresponds to which distance. This will allow you to examine true "questioned specimens" to try to determine the distance at which the shots were fired.)

For those of you who have watched the old television series *Police Squad*, we should note that test firing a weapon into a stack of old Elvis records is *not* a procedure used by real forensics labs. None that we know of, anyway.

If you have sufficient time and materials and access to more than one firearm and/or to different types of ammunition for the same firearm, you can produce GSR specimens with each of them. We used a Ruger .22 target pistol with recent commercial ammunition, a .45 ACP Colt Combat Commander with recent commercial ammunition, old commercial ammunition, and old military ammunition, and a .357 Magnum Ruger revolver with recent commercial .38 Special and .357 Magnum ammunition and old hand-loaded .357 Magnum ammunition.

PRODUCING SIMULATED GSR SPECIMENS

The GSR reagents we're using actually test the cloth swatches for the nitrites and lead that are present in GSR, whatever the source of those materials. If you don't have access to a firearm or a safe place to fire it, you can still observe the color reactions of these reagents for positive tests, although you won't be able observe the characteristic patterns of gunshot residue. To make up these substitute GSR specimens, take the following steps:

If you don't intend to do the optional rhodizonate test, you needn't produce simulated GSR specimens using the lead nitrate solution.

1. If you have not already done so, put on your splash goggles, gloves, and protective clothing.

2. Tape a white cloth swatch to a clean sheet of cardboard. Use tape at all four corners, and cover only enough of the cloth to secure it to the cardboard. Use the marker to label the cloth "nitrite known" near one edge.

3. Place the specimen flat on your work surface. Aim a sprayer that contains 1% w/v sodium nitrite solution near the center of the cloth. Hold the sprayer 20 to 25 cm above it, pointed downward at the cloth, and press the sprayer pump several times. Your goal is not to drench the cloth, but only to moisten it slightly with the sodium nitrite solution.

4. Remove the treated cloth/cardboard sandwich from your work area to a location where it will not be contaminated by subsequent sprayings, and allow the cloth to dry completely.

5. Repeat steps 2 through 4 with a cloth labeled "lead known" and spray it with a 0.1% w/v solution of lead nitrate or lead acetate.

PROCEDURE IX-1-2: MAKE UP MODIFIED GRIESS REAGENT TEST PAPER

Modified Griess reagent test paper is used to test for the presence of nitrites in GSR. The modified Griess test traditionally uses standard black-and-white photographic paper that has been fixed and thoroughly washed. We used 8 × 10 Oriental Seagull double-weight (non-RC) photographic paper that had been in our freezer for about 20 years, but you can substitute ordinary white bond or copy paper.

Griess reagent is also needed for the following lab session, so don't use it all up in this lab session.

1. Combine equal volumes of a few milliliters each of modified Griess reagent parts A and B in a small spray bottle and swirl to mix the solutions. (The two stock solutions can be stored indefinitely, but once they are mixed, they should be used within a day or so. Don't make up more than you need for immediate use.)
2. Place as many sheets of paper as you want to prepare on a flat surface that is protected with old newspapers or paper towels.
3. Spray the surface of the paper sheets until they are evenly coated with the mixed modified Griess reagent solution. Don't soak the paper, but make sure the entire surface is lightly dampened.
4. Allow the paper to absorb the solution and dry completely. The treated sheets remain usable indefinitely as long as they are kept dry.

PROCEDURE IX-1-3: TEST FOR NITRITE RESIDUE IN GSR SPECIMENS

The Modified Griess test for nitrites is always done first, because it does not interfere with later tests.

1. If you have not already done so, put on your splash goggles, gloves, and protective clothing.
2. To verify the sensitivity of a sheet of your modified Griess reagent test paper, clip one small corner from it. Apply a cotton swab saturated with the 1% sodium nitrate solution to the surface of the paper, followed by a swab saturated with vinegar. An orange stain develops quickly if the paper is sensitive to nitrites.
3. Place two thicknesses of paper towel on a flat, heat-resistant surface.
4. Place a sheet of test paper face up on top of the paper towels. Use a pencil (not a marker) to label the test sheet with the GSR specimen number.
5. Place the GSR specimen face down on top of the test paper.
6. Saturate a paper towel with vinegar, squeeze out the excess liquid, and then place it on top of the GSR specimen. (You want the paper towel wet with vinegar, but not dripping.)
7. Set the clothes iron to medium heat. Once it has heated, press the entire surface of the sandwich for a minute or so, as shown in Figure IX-1-3. Your goal is to vaporize the acetic acid in the vinegar, forcing acetic acid vapor through the GSR specimen and into contact with the test paper. In the presence of nitrites, the acetic acid vapor reacts with the sulfanilic acid and alpha naphthol present in the test paper to form an orange azo dye.

Figure IX-1-3: *Pressing the Griess reagent test paper in contact with the cloth specimen*

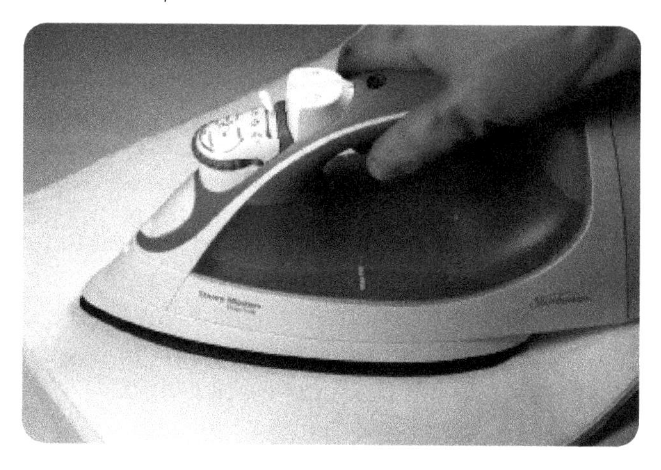

8. Separate the layers. Place the GSR specimen aside for further testing. Discard the paper towels and examine the test sheet. Orange stains on the test sheet, as shown in Figure IX-1-4, indicate the presence of nitrites in the GSR specimen.

Figure IX-1-4: *Orange stains indicate the presence of nitrites in a simulated GSR specimen (sodium nitrite on paper)*

9. Record your observations in you lab notebook. Shoot images or make scans of the test results for your lab notebook.

The orange reaction resulting from nitrite residues in GSR is ordinarily blotchy and distinct. An even light-orange appearance where the specimen contacts the test paper indicates that the test paper sheets were contaminated with nitrites that were not produced by GSR. Nitrites are relatively common. Some blue jeans and other clothing contain detectable levels of nitrites, as do some laundry detergents, cleaning solutions, disinfectants, and other common household materials.

PROCEDURE IX-1-4: TEST WHITE GSR SPECIMENS FOR LEAD RESIDUE

The sodium rhodizonate test is used to detect the presence of lead in GSR. At about pH 3 in the presence of lead and certain other metals, sodium rhodizonate produces a pink stain. Lead is confirmed by spraying the stain with dilute hydrochloric acid. The stain turns blue only if lead is present.

Lead detected on the GSR specimen may originate from several sources. If the cloth has been penetrated by an unjacketed lead bullet or a lead shotgun pellet, there will likely be a "wipe" mark surrounding the hole, where lead from the projectile has been physically abraded onto the cloth. A gas check (a thin metal plate crimped to the base of a lead bullet to protect it from hot propellant gases) or its fragments may produce similar wipe marks. Particulate lead from the primer compound or from bore deposits usually causes small, relatively sharp blotches randomly distributed across the GSR area. Gaseous lead produced from the primer compound or vaporized from bore deposits or the bullet itself may appear as relatively large, diffuse areas of light pink staining or as blotches resembling those produced by particulate lead.

The *limit of detection* (LOD), also called the *detection threshold*, for the sodium rhodizonate test for lead is less than 200 nanograms per square centimeter (ng/cm^2) for the pink stain and less than 350 ng/cm^2 for the blue stain. One nanogram is a billionth of a gram. To visualize this level of sensitivity, consider that a 500 mg extra-strength aspirin tablet contains 500 million ng of aspirin. If the rhodizinate test were sensitive to aspirin, the aspirin in that tablet would be detectable if spread evenly across an area of (500,000,000 / 200) = 2,500,000 square centimeters or 250 square meters, about the floor area of a typical suburban four bedroom home.

Despite this very high sensitivity, the amount of lead present in GSR is typically small enough that the sodium rhodizonate test can be used to estimate at least roughly the distance between the muzzle of the firearm and the surface upon which the GSR was deposited. At the close ranges typical of suicides and accidents, gaseous lead and particulate lead are likely to be detected unless lead-free ammunition was fired in a weapon that was free of lead residues. At more distant ranges, GSR from particulate lead may be present, but GSR from gaseous lead is typically at a level too low to be detected. Although the sodium rhodizonate test cannot provide exact muzzle-target distances, if the weapon is available, it can be test fired with identical ammunition to provide comparison specimens. This may allow the forensic scientist at least to quantify the likely minimum or maximum muzzle-target distance.

In solution, sodium rhodizonate is quickly reduced to tetrahydroxyquinone, so the sodium rhodizonate solution must be made up immediately before use. At room temperature, the solution is usable for an hour or so after it is made up. If the solution is kept in an ice bath between uses, it may remain usable for at most a few hours.

1. If you have not already done so, put on your splash goggles, gloves, and protective clothing.

2. Add a few crystals of sodium rhodizonate to a few milliliters of ice-cold distilled or deionized water in a small, clear, colorless sprayer bottle and swirl to dissolve the crystals. The solution should be about the color of strong tea.

Place the sprayer bottle in a beaker or other container filled with water and crushed ice, and return the sprayer bottle to the ice bath any time it is not actually being used. Make up only a few milliliters of sodium rhodizonate at a time. Although it's inconvenient to make up the solution several times over the course of a lab session, the fresher the solution, the better the results.

3. Place one of your actual or simulated GSR specimens on white cloth on a flat surface that is protected by old newspapers or paper towels. Use a piece of lead metal—a shotgun pellet, lead bullet, wheel weight, etc.—to mark the GSR specimen near the edge, outside the area of GSR residue. (You may use either a fresh GSR specimen or one that has been tested with modified Griess reagent; if possible, test both types of GSR specimens and compare the results.)

4. Spray the GSR specimen evenly from a distance of a few centimeters until the white cloth is stained yellow.

5. Immediately overspray the GSR specimen with the tartrate buffer solution. Any lead present, including the known mark you made, causes a pink stain, as shown in Figure IX-1-5.

Figure IX-1-5: *Pink stains indicate the presence of lead in a simulated GSR specimen (lead nitrate on paper)*

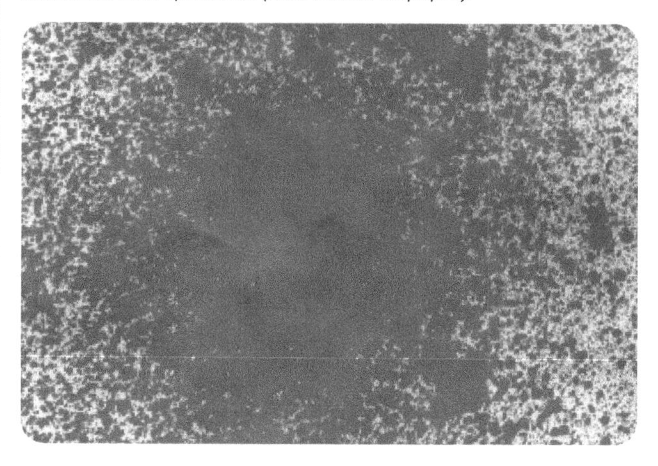

Some forensic scientists reverse the order of application, dampening the GSR specimen first with buffer solution and then spraying with sodium rhodizonate solution. Either method works. Also, although we did not try it, some forensic scientists report that the usable life of the sodium rhodizonate solution can be extended to several hours by dissolving the sodium rhodizonate crystals in the buffer solution rather than in water. In this case, a single spraying does the job; you need not apply buffer solution separately.

6. Shoot an image or make a scan of the treated GSR specimen for your lab notebook.

7. Overspray the GSR specimen with the dilute hydrochloric acid solution, or use a cotton swab to apply dilute hydrochloric acid directly to the pink stains. If the pink stains turn blue, the presence of lead is confirmed.

8. Shoot an image of the oversprayed GSR specimen for your lab notebook. Sodium rhodizonate stains may fade very quickly, so be prepared to shoot the image as soon as the test is complete.

PROCEDURE IX-1-5: TEST COLORED OR PATTERNED GSR SPECIMENS FOR LEAD RESIDUE

Positive rhodizonate test results are easily visible on white or light-colored cloth, Caucasian skin, and other light materials, but are difficult or impossible to see on dark or patterned materials. The workaround is to transfer the GSR residue from the dark material to white paper and run the rhodizonate test on the paper. Although we can't quantify it, our impression is that the transfer reduces the sensitivity of the test significantly, but it is still sensitive enough to be useful. Note that the results of a transfer test are a mirror image of the results in testing the GSR specimen directly. Usually, this is insignificant, but it should be noted when you write up the results of the test. To do a transfer test, take the following steps:

1. If you have not already done so, put on your splash goggles, gloves, and protective clothing.
2. Place one of your colored GSR specimens face up on a flat surface that is protected by old newspapers or paper towels.
3. Cover the area that contains GSR with a transfer sheet of ordinary white bond or copy paper.
4. Label the transfer sheet with pencil to indicate the specimen number or other identifier. Mark the transfer sheet lightly with pencil to indicate the position of the bullet hole.
5. Spray the transfer sheet with vinegar until it is saturated.
6. Cover the transfer sheet with several layers of paper towels.
7. Using a clean clothes iron at medium heat, iron the sandwich evenly until the transfer sheet is dry.
8. Use a piece of lead metal to mark the transfer sheet near the edge, outside the GSR area.
9. Spray the transfer sheet evenly from a distance of a few centimeters with sodium rhodizonate solution until the transfer sheet is stained yellow.
10. Immediately overspray the transfer sheet with the tartrate buffer solution. Any lead present, including the known mark you made, causes a pink stain.
11. Shoot an image of the treated transfer sheet for your lab notebook.
12. Overspray the transfer sheet with the dilute hydrochloric acid solution or use a cotton swab to apply dilute hydrochloric acid directly to the pink stains. If the pink stains turn blue, the presence of lead is confirmed.
13. Shoot an image or make a scan of the transfer sheet for your lab notebook. Again, rhodizonate stains may fade quickly, so be prepared to shoot the image as soon as the test is complete.

REVIEW QUESTIONS

Q1: **In addition to the substances we tested for in this lab session, what other substances are often found in GSR?**

Q2: **What does the presence of mercury in GSR suggest, and why is it significant?**

Q3: **Why are GSR color tests considered presumptive?**

Q4: **What useful information can be gained by observing the distribution pattern of GSR?**

Q5: **Why are firearms test fired whenever possible?**

Presumptive Color Tests for Explosives Residues

Lab IX-2

EQUIPMENT AND MATERIALS

You'll need the following items to complete this lab session. (The standard kit for this book, available from *http://www.thehomescientist.com*, includes the items listed in the first group.)

MATERIALS FROM KIT

- Goggles
- Diphenylamine reagent
- Lead nitrate, 0.1% (nitrate known)
- Methylene blue reagent, part A
- Methylene blue reagent, part B
- Modified Griess reagent, part A
- Modified Griess reagent, part B
- Pipettes
- Sodium nitrite, 1% w/v (nitrite known)
- Spot plate, 12-well
- Specimen: chlorate known
- Specimen: perchlorate known

MATERIALS YOU PROVIDE

- Gloves
- Acetone
- Cotton swabs
- Rocks, broken tile or flowerpot, etc. (see text)
- Vinegar, distilled white
- Water, distilled or deionized

WARNING

Diphenylamine reagent contains concentrated sulfuric acid, which is extremely corrosive. Acetone is flammable. Chlorate and perchlorate knowns are strong oxidizers; do not allow them to contact any other material. Wear splash goggles, gloves, and protective clothing. Read the MSDS for each chemical you use and follow the recommended safety precautions.

BACKGROUND

Explosives residue testing is similar in many respects to gunshot residue testing. The main difference between GSR analysis and explosives residue analysis is a matter of scale. GSR is produced in small amounts by controlled explosions, and is deposited only upon objects in the immediate vicinity of the gunshot. Explosives residues typically result from much larger, uncontrolled explosions and are accordingly deposited in much larger quantities over much larger areas.

Most explosions produce large amounts of rubble, only some of which contains explosive residues. Presumptive color tests for explosives residues provide a quick, inexpensive way to determine which pieces of rubble should be subjected to further instrumental testing.

An explosion may occur in a one-, two-, or three-stage process:

Three-stage detonation

The first stage of a three-stage detonation is called the *initiator*, a small amount of a very sensitive explosive that can be detonated by flame, electricity, or impact. The second stage, called the *booster*, is a larger amount of a less-sensitive explosive. Detonation of the initiator causes the booster charge to detonate. Detonation of the booster charge in turn detonates the *main charge*, which is a large amount of relatively insensitive explosive. ANFO (ammonium nitrate/fuel oil) is a typical example of a three-stage detonation. A blasting cap (the initiator) detonates a stick of dynamite or other high explosive (the booster), which in turn detonates the ANFO (main charge).

Two-stage detonation

If the explosive used for the main charge is sufficiently sensitive, no booster is required; the main charge can be detonated directly by the initiator. Commercial dynamite is a typical example of a two-stage detonation. A blasting cap (the initiator) detonates the dynamite main charge directly.

One-stage detonation

Some explosives are so sensitive that they require no initiator. In effect, the main charge is its own initiator. Such explosives have no commercial or military use (other than in small quantities packaged as initiators) because they are simply too dangerous to handle in the amounts required for a main charge. Such explosives are used only by terrorists. TATP (triacetone triperoxide), used in the 2005 London subway bombings and called Mother of Satan by Islamic terrorists, is a good example of a single-stage explosive.

COMBUSTION VERSUS DETONATION

The difference between low explosives (such as black powder or smokeless gunpowder) and high explosives (such as dynamite, ANFO, or TATP) is that low explosives combust, albeit very quickly, while high explosives detonate. In the rapid combustion of a low explosive, technically called deflagration, a subsonic flame front moves through the explosive initiating burning by heat transfer. In the detonation of a high explosive, a supersonic shock wave moves through the explosive at as much as 9,000 meters per second, initiating combustion by compression-induced heating. In effect, during a detonation, combustion occurs almost instantly throughout the entire body of the explosive rather than incrementally.

This uncontrolled and almost instantaneous release of all of the energy contained in the explosive in a tiny fraction of a second is why high explosive detonations are so destructive. Low explosives, conversely, release their energy in a more controlled fashion and over a much longer (although still very short) period of time, which is why they can be used as propellants. For example, in a pistol cartridge, the low-explosive propellant burns to produce hot gas that expels the bullet from the cartridge, down the barrel and out of the muzzle. If you substituted a high explosive for the propellant, the detonation of the charge would simply shatter the pistol before the bullet could be expelled from the cartridge case.

Regardless of the number of stages and the type of explosive, any explosion leaves residues that can be detected by color tests. The chemicals used in initiators are often similar to those used in firearms primers, and in theory, their residues should be detectable. In practice, the relatively tiny amount of initiator compound used with a large main charge often means that initiator residues are so widely scattered that they are undetectable, other than perhaps instrumentally.

Residues left by the booster and main charge are a different story. Most commercial and military explosives are nitrate-based. These explosives produce residues that include nitrates from the explosive itself and nitrites from the reaction products produced in the explosion. Some explosives, particularly pyrotechnics, include significant amounts of chlorates or perchlorates. All of these can be detected with color tests.

Color tests are normally done by swabbing suspect materials, particularly rubble in the immediate area of the center of the explosion, with cotton swabs or cotton balls dampened with acetone, which is a good solvent for most explosives residues. Many of the standard color-test reagents can be applied directly to the cotton, which is then observed for any color change. Other color-test reagents contain concentrated sulfuric acid or other corrosive materials that quickly char cotton, obscuring any color change. To make tests with these reagents, any residues present in the cotton swab are rinsed with acetone into a spot plate and then the acetone is evaporated to leave only the residues in the well of the spot plate. Those residues can then be tested directly or, in some cases, after extraction with water to remove organic compounds that char in the presence of sulfuric acid.

We had two problems in designing this lab session, one obvious and one less so. Obviously, explosives are tightly controlled and effectively impossible to ship even in the tiny amounts we might otherwise have included in the kits as knowns. Less obviously, some of the reagents used for color testing explosives residues are very expensive or extremely dangerous to handle, requiring fume hoods and other gear found only in professional labs. Fortunately, there are usable substitutes for actual explosives and several illustrative color tests we can do with more readily available materials. We'll use the following color-test reagents, whose reactions are summarized in Table IX-2-1.

Table IX-2-1: *Color change reactions of three standard explosives residue reagents*

	Modified Griess	Diphenylamine	Methylene blue
Bromides	n/a	Yellow (solid only)	Pale violet
Chlorates	n/a	Blue to blue-black	n/a
Iodides	n/a	Violet	Violet
Nitrates	Orange	Blue to blue-black	n/a
Nitrites	Orange	Blue-black	n/a
NC/NG/NS	Orange	Blue to blue-black	n/a
Perchlorates	n/a	n/a	Violet

Modified Griess reagent

Modified Griess reagent is always the first and often the only presumptive color test used to scan for explosives residues. First, because it doesn't interfere with subsequent tests using other reagents. Only, because the vast majority of forensic bombing cases involve explosives and residues that contain nitrates and nitrites, to which modified Griess reagent is very sensitive. (There are exceptions, such as acetone peroxide, which contains no nitrogen, but it's rare to encounter these exceptions in forensic bombings.)

Modified Griess reagent detects nitrates and nitrites in residues from MAN (methyl ammonium nitrate), NC/NG/NS (nitrocellulose/nitroglycerin/nitrostarch), PETN (pentaerythritol tetranitrate), RDX (1,3,5-trinitroperhydro-1,3,5-triazine), tetryl (2,4,6-trinitrophenyl-N-methylnitramine), TNT (trinitrotoluene), and nearly any other nitrated explosive. Modified Griess reagent provides an orange positive reaction with most common explosives and explosive residues, and is less subject to false positives than most other color-change reagents.

Diphenylamine reagent

Diphenylamine reagent is used to detect chlorates, nitrates, and nitrites, as well as explosives residues from MAN (deep blue), NC/NG/NS, PETN (blue), and tetryl (blue). Diphenylamine reagent provides a blue or blue-black positive reaction with most explosives and explosive residues. It also reacts with bromides and iodides to yield a color change (yellow and violet, respectively), but those color changes are easily discriminated from true positives.

Methylene blue reagent

Methylene blue reagent is used to test for tetryl, with which it yields an orange to red color change, and for perchlorates used in certain flash powders and initiators, with which it yields a violet color change or a violet precipitate, depending on concentration. Bromide and iodide ions yield false positives for perchlorates that are difficult or impossible to discriminate from true positive reactions.

These three reagents among them cover all of the chemical species that are of primary interest in forensic bombing cases: chlorates, nitrates, nitrites, and perchlorates, all four of which—particularly nitrates and nitrates—are commonly present in explosives residues.

With the exception of state-sponsored terrorist bombings, which may use military-grade explosives such as TNT, RDX, or PETN, most criminal bombings use one of five more readily available explosives: blasting explosives (dynamite and other types of explosives stolen from construction sites), ANFO (ammonium nitrate fertilizer and diesel fuel), smokeless powder (purchased from sporting goods stores), black powder (purchased from sporting goods stores or homemade), and flash powders (homemade). All of these explosives and their residues can be detected by the three color test reagents we use in this lab session. The first four explosives give positive tests for nitrates, and their residues test positive for nitrates and nitrites. Homemade flash powders vary greatly in composition, but nearly all of them use chlorate or perchlorate oxidizers and powdered aluminum or magnesium or sugar or another carbon compound as the fuel. These flash powders and their residues test positive for chlorates and/or perchlorates.

FORMULARY

If you don't have the FK01 Forensic Science Kit, you can purchase the necessary reagents from a law enforcement forensic supply vendor, or make them up yourself as follows:

Diphenylamine reagent Dissolve 0.1 g of diphenylamine in 10 mL of concentrated sulfuric acid.

Griess reagent Make up Griess reagent part A and part B according to the instructions in the preceding lab session.

Methylene blue reagent To make up part A, dissolve 5 g of zinc sulfate in 10 mL of distilled or deionized water and add two drops of 1% aqueous methylene blue solution. to make up part B, dissolve 4 g of potassium nitrate in 10 mL of distilled or deionized water.

The FK01 Forensic Science Kit includes specimens of the four common explosives residues: nitrate (lead nitrate solution

FORMULARY (CONTINUED)

and methylene blue reagent part B), nitrite (sodium nitrite solution), chlorate (solid specimen), and perchlorate (solid specimen). If you don't have the kit, you can obtain these four knowns as follows:

Chlorate You can use any chlorate salt, the most common of which is potassium chlorate. If you don't have potassium chlorate, you can substitute chlorate-based tablets sold in pet stores for oxygenating aquariums, some of which are nearly pure potassium chlorate.

Nitrate You can use any nitrate salt, such as potassium nitrate or ammonium nitrate (available in nitrate fertilizers and some drugstore cold-packs).

Nitrite You can use any nitrite salt, such as potassium nitrite or sodium nitrite.

Perchlorate You can use any perchlorate salt, such as potassium perchlorate. Pyrodex, used by many black-powder shooters as a substitute for black powder, is sold in sporting goods stores and contains potassium perchlorate. You may be able to obtain a small specimen of Pyrodex from a black-powder hobbyist. Most communities have reenactment groups that use black powder firearms.

For completeness, you may also want to obtain the following optional known specimens:

Bromide You can use any bromide salt, such as potassium bromide. If you don't have a bromide salt, don't worry about it. It's needed only to illustrate positive reactions with diphenylamine reagent and methylene blue reagent.

Iodide You can use any iodide salt, such as potassium iodide. Once again, iodide is needed only to illustrate false-positive reactions with diphenylamine reagent and methylene blue reagent, so don't worry too much about obtaining a specimen.

NC/NG/NS Nitrocellulose, nitroglycerin, and nitrostarch are chemically similar, so you can use a specimen of any one of them to represent the group. This specimen really is optional, because positive tests on all of these explosives result from their nitrate content, and positive tests on their residues result from their nitrite content.

Single-base smokeless gunpowders are made up primarily of nitrocellulose, with small amounts of stabilizers and lubricants added. Double-base smokeless gunpowders are made up of nitrocellulose and nitroglycerin in roughly equal proportions. You can obtain smokeless powders at sporting goods suppliers, or beg a tiny specimen from someone who reloads ammunition. (Although we recommend against doing so for safety reasons, we obtained one of our specimens by carefully pulling the bullet from an unfired .22 rimfire round.) Nitroglycerin is used in small amounts in tablets used by heart patients, so you may be able to obtain an outdated tablet from a friendly pharmacist. PETN is also used in small amounts in some heart medications, so you may be able to get a specimen of this, as well.

For the purposes of this lab session, you don't need any actual explosives. You can simply test the knowns and observe the reactions, which are the same reactions produced by actual explosives and explosives residues. If you want to test actual explosive residues, here are some options:

Smokeless powder Test a few grains of smokeless powder (single- or double-base) for the presence of NC/NG/NS. To test residues, place a small pile (about a quarter the size of a pea) on a rock, flowerpot fragment, or piece of unglazed ceramic tile and carefully ignite it. (Use a long match or some other means to keep well clear of the fire; the smokeless powder will flare strongly as soon as you ignite it.)

Black powder You can obtain black powder at sporting goods stores, or simply make up a small amount for testing. To do so, combine 0.75 g of potassium nitrate, 0.15 g of charcoal, and 0.1 g of sulfur, all finely powdered, in a small container. Moisten the mixture with water that contains a drop or two of water-soluble glue, honey, or corn syrup to the consistency of a very thick paste. Mix thoroughly and extrude the mixture through a piece of window screen or similar mesh. Allow the granules to dry thoroughly. Test a few granules of your black powder for the presence of nitrates. To test residues, carefully ignite a very small amount of black powder on a resistant surface. Again, use a long match or some other means to keep well clear of the fire; the black powder will flare strongly as soon as you ignite it. Burning black powder produces a characteristic "sulfur" odor, so do this outdoors.

Flash powder These are simply too dangerous to make up at home. However, "party poppers" (small crackers detonated by pulling a string) are not considered fireworks and so are legal in most states, are available in novelty and party supply stores, and provide a convenient source of flash powder residue.

PROCEDURE IX-2-1: TEST KNOWN SPECIMENS

1. If you have not already done so, put on your splash goggles, gloves, and protective clothing.

2. Place the white spot plate under strong illumination.

3. Place a crystal of your first known into each of the four wells in the first row of the spot reaction plate. An amount the size of a sand grain is sufficient.

4. To the first well, add two drops of water. This well is your control well, against which you'll compare results in the other wells with different reagents.

5. To the second well, add one drop of modified Griess reagent solution A, one drop of modified Griess reagent solution B, and one drop of distilled white vinegar. Observe the well for several seconds, and note the results in your lab notebook. An orange color shift is a positive test for nitrate and/or nitrite.

6. To the third well, add one drop of diphenylamine reagent. Observe the well for several seconds and note the results in your lab notebook. A blue to blue-black color shift is a positive test for chlorate, nitrate, and/or nitrite.

7. To the fourth well, add one drop of methylene blue reagent A and one drop of methylene blue reagent B. Observe the well for several seconds and note the results in your lab notebook. A violet color shift is a positive test for perchlorate. Figure IX-2-1 shows some resulting reactions.

Figure IX-2-1: *Testing known explosives residues in a reaction plate*

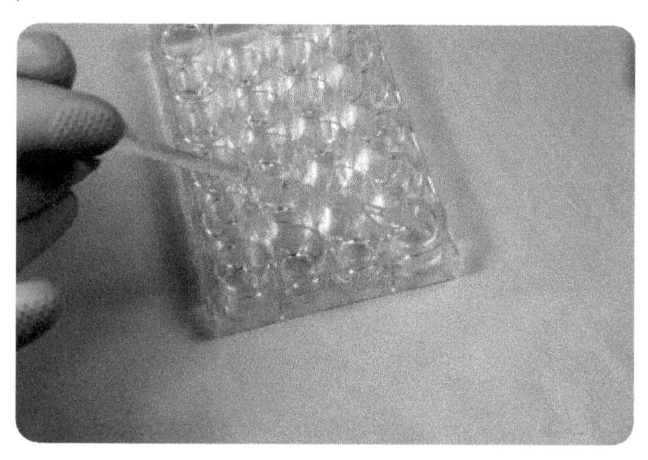

8. Using fresh rows of the reaction plate, repeat steps 3 through 7 with your second and third known.

9. Flood the reaction plate under the faucet to dilute the contents of all wells, which can be flushed down the drain. Wash and dry the reaction plate.

   Be careful with the wells that contain diphenylamine reagent, which contains concentrated sulfuric acid.

10. Repeat steps 3 through 7 for each of your remaining knowns.

PROCEDURE IX-2-2: EXTRACT EXPLOSIVES RESIDUES

Before you begin this procedure, ask a friend, family member, or lab partner to prepare questioned specimens for you on small rocks or pieces of a broken flowerpot or unglazed ceramic tile. Your volunteer should label each of the specimens, 1, 2, 3, 4, or 5. Prepare each specimen by dampening its entire surface with one of the following solutions or combinations of solutions.

- A few crystals of the chlorate known dissolved in a milliliters or so of water
- Sodium nitrite solution or methylene blue solution part B (either one is fine)

- A few crystals of the perchlorate known dissolved in a milliliters or so of water
- A mixture of one, two, or all three of the above solutions
- A different mixture of one, two, or all three of the above solutions

You can allow the questioned specimens to dry naturally or use a hair dryer to speed things up.

1. If you have not already done so, put on your splash goggles, gloves, and protective clothing.
2. Acetone is the solvent of choice for dissolving explosive residues because organic and inorganic residues are both usually soluble in acetone. Place a couple of drops of acetone on a clean cotton swab. You want the cotton wet, but not dripping.
3. Apply the tip of the swab to the surface of specimen 1. Scrub about one quarter of the surface gently to transfer the residues onto the cotton swab, as shown in Figure IX-2-2.
4. Label the swab. One easy way to do that is to use a marker to make one line on the shaft of the swab to indicate that the swab was taken from specimen 1, two lines to indicate that the swab was taken from specimen 2, and so on. Place the swab aside to allow the acetone to evaporate.

Figure IX-2-2: *Obtaining explosives residue specimens with a cotton swab*

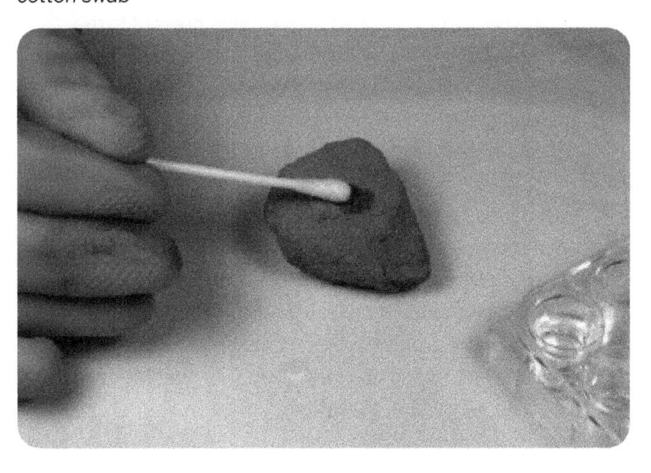

5. Repeat steps 2 through 4 with a second swab and a fresh area of the surface of specimen 1. Again, swab about one-quarter of the surface of the specimen.
6. Repeat step 5 with a third swab and a fresh area that covers about one quarter of the surface of specimen 1.
7. Repeat the preceding steps to produce three swabs from each of the remaining four specimens.

PROCEDURE IX-2-3: TEST SWABS FOR EXPLOSIVES RESIDUES

1. If you have not already done so, put on your splash goggles, gloves, and protective clothing.
2. To the tip of one of your specimen 1 swabs, apply one drop of modified Griess reagent A, one drop of modified Griess reagent B, and one drop of distilled white vinegar. A color change to orange is positive for the presence of nitrate and/or nitrite. Record your observations in your lab notebook.
3. Repeat step 2 with one swab from each of your remaining four questioned specimens.
4. To the tip of the second of your specimen 1 swab, apply one drop of methylene blue reagent A and one drop of methylene blue reagent B. A color change to violet is positive for the presence of perchlorate (or a false positive for bromide and/or iodide), nitrates, and/or nitrites. Record your observations in your lab notebook.
5. Repeat step 4 with a fresh swab from each of your remaining four questioned specimens.

We have one unused swab remaining from each of the five questioned specimens, and we want to test those swabs with diphenylamine reagent for the presence of chlorate. Unfortunately, we can't do that test directly on the swabs, because the concentrated sulfuric acid present in the diphenylamine reagent quickly chars cotton, obscuring any other color change. The way around this problem is to use acetone to extract any explosives residues present on the cotton swabs to wells in the spot plate.

6. Fill a pipette with acetone. While holding the contaminated tip of the final swab from specimen one over well 1 in the spot plate, dribble acetone onto the tip of the swab, allowing excess acetone to drain into well 1. Add the acetone a few drops at a time to the tip of the swab, and continue rinsing the tip of the swab with acetone until the well is about half full of acetone.

7. Repeat step 6 to rinse the residues from the specimen 2 swab into well 2, the residues from specimen 3 into well 3, the residues from specimen 4 into well 4, and the residues from specimen 5 into well 5.

8. Allow the acetone in the wells to evaporate. You can speed up this process by (carefully) using a hair dryer.

9. Add one or two drops of diphenylamine reagent to each of the five wells and observe those wells for any color change. A yellow color indicates the presences of bromide, a purple color indicates the presence of iodide, and a blue or blue-black color indicates the presence of chlorate, nitrite, and/or nitrate. Record your observations in your lab notebook.

REVIEW QUESTIONS

Q1: **What is the primary use of presumptive color tests for explosives residues?**

Q2: **You find that a questioned explosives residue specimen yields a blue-black color with diphenylamine reagent but no color change with Griess reagent or methylene blue reagent. Which explosives residue, if any, is likely to be present in the specimen?**

Q3: **You find that a questioned explosives residue specimen yields a violet color with methylene blue reagent. What explosive residue do you presume is present, and what substance may have produced a false positive? What other reagent would you use to determine whether the positive methylene blue test was a false positive, and what results would you expect from this reagent?**

Q4: **In Procedure IX-2-2, we made three swabs from each specimen. Why did we swab only one quarter of the specimen surface with each of the three swabs?**

Q5: **In Procedure IX-2-3, you observe a strong blue-black reaction with diphenylamine reagent. Does this color change mean chlorates are present?**

Q6: **In Procedure IX-2-3, you observe a violet reaction with methylene blue reagent. Does this color change mean perchlorates are present?**

Detecting Altered and Forged Documents

Group X

Forensics labs are frequently asked to examine documents to determine if they have been altered or forged. An *altered document* is an original, valid document that has been changed in some way. For example, one common type of altered document is a check in which the numbers have been changed to increase the amount drawn. A *forged document* is one that is created from scratch to masquerade as a valid document, or a valid document to which a page or pages has been added to alter the meaning or intent of the original document. For example, a later copy of a will or a codicil may be forged, or a page or pages added to a legitimate will to alter its terms.

Forensic document examiners use several methods to determine the validity of a questioned document. The first step is always to examine the document with the naked eye and under low magnification. A surprisingly large percentage of forgeries are so crudely done that the forgery is obvious even on cursory examination. If the document passes this first "sniff test," forensic document examiners have many other tools available.

Examination with alternate light sources

Papers and inks that are indistinguishable under white light may have very different appearances under ultraviolet or infrared wavelengths. Erasures, additions, and other alterations that appear normal under white light may stand out starkly under UV or IR illumination. Examination with an ALS is normally the first step in any document analysis. Often, no other steps are needed to reveal alterations.

Chemical analysis

Forensic document examiners use numerous chemical tests to compare inks and papers. For example, although two inks may appear to be consistent visually under white and ALS illumination, chemical tests of those inks may reveal that they are quite different. Similarly, although two sheets of paper may appear to be identical, chemical analysis may reveal differences in the types and relative amounts of the fibers that make them up.

Microscopic analysis

Even if two specimens appear to be consistent under ALS examination and chemical testing, microscopic analysis may reveal differences. For example, two pages of a questioned document may be on the same type of paper and use the same ink. But microscopic examination may reveal that the two ink specimens being compared were made with different pens, one with a tip narrower than the other.

And then there's handwriting analysis. Although many people conflate the two, forensic handwriting analysis and graphology are completely unrelated.

Forensic handwriting analysis is science-based, and depends as much as possible on objective data, such as the angle of handwritten letters; the relative lengths of ascending and descending strokes; the relative size, shape, and angle of loops; and so on. Although inevitably forensic handwriting analysis incorporates some subjective judgments made by the examiner, those are skilled and qualified opinions. If you submit known and questioned handwriting specimens to 10 qualified document examiners, there is a high probability that all 10 will make the same conclusion, that the specimens either were or were not probably produced by the same person.

Conversely, graphology is pseudoscientific woo. Graphologists claim to discern sex, age, personality characteristics and other information from handwriting specimens. If you submit a handwriting specimen to 10 graphologists, you're likely to get 10 different, conflicting opinions that are all over the map.

In this group, we'll examine several of the techniques used by forensic labs to determine if a document has been altered or forged.

Revealing Alterations in Documents

EQUIPMENT AND MATERIALS

You'll need the following items to complete this lab session. (The standard kit for this book, available from *http://www.thehomescientist.com*, includes the items listed in the first group.)

MATERIALS FROM KIT

- Goggles
- Bottle, sprayer
- Coverslips
- Iodine crystals
- Iodine/glycerol reagent
- Magnifier
- Pipettes
- Slides, flat

MATERIALS YOU PROVIDE

- Gloves
- Acetone
- Camera (optional)
- Chlorine laundry bleach
- Cotton swabs
- Ethanol
- Ink remover (optional)
- Iodine fuming chamber (from Lab IV-2)
- Microscope
- Pencil
- Scanner and software (optional)
- Scissors
- Ultraviolet (UV) light source (optional)
- Xylene (or toluene, paint thinner, etc.)
- Specimens, paper, white
- Specimen, pen, erasable ink
- Specimens, pens/markers, assorted black

WARNING

Perform this experiment only in a well-ventilated area. Acetone, alcohols, xylene, and other solvents are flammable. Iodine is toxic and corrosive, and iodine vapor is irritating. Chlorine bleach is toxic and corrosive. Read the MSDS for each chemical you use, and follow the instructions for safe handling and proper disposal. Wear splash goggles, gloves, and protective clothing.

BACKGROUND

Forensic labs are frequently tasked with determining if a document such as a check or a will has been altered. Such alterations may be handwritten, typewritten, or printed by computer, and may take either or both of two forms:

Adds

Handwritten material added to an existing document may be analyzed by a handwriting examiner to determine if the questioned text was written by the same person who wrote the accepted text. However, because handwriting analysis is perceived by many to be extremely subjective, the conclusions of a handwriting examiner are frequently challenged by opposing attorneys, often successfully. For that reason, a final determination of the validity of questioned handwritten text usually comes down to a chemical, chromatographic, or instrumental analysis of the ink or inks used for the accepted and questioned text.

Typewritten material added to an existing document is usually easy to detect, even if the same typewriter and ribbon was used for the alterations, for the simple reason that once a page has been removed from a typewriter, it is nearly impossible to reposition the page so that the added text aligns precisely with the original text. Careful measurements of the horizontal and vertical alignment of the text nearly always reveal such alterations. If all of the text aligns properly but the validity of the document remains questioned, the questioned text may be examined microscopically and tested chemically to determine if it is consistent with the accepted text.

Computer-printed material added to an existing document may or may not be easy to detect, depending on the precision of the paper feed of the printer in question and the skill of and the degree of care taken by the person who makes the alteration. In the worst case, it may be difficult or impossible to determine definitively that questioned text was added after the fact. In most cases, questioned text can be detected, either because of alignment problems or because the questioned text was added using a different model of printer or type of toner, which can be detected by microscopic or instrumental examination.

Deletes

Text may be removed from a document by physical means such as erasure or by chemical bleaching. Either method, no matter how carefully done, leaves traces that can be identified by careful examination by visible light or an ALS, or by chemical tests. In many cases, the original text can be made readable by these methods.

The proliferation of inexpensive PC scanners has added a new tool to the arsenal of forensic document examiners. By scanning a questioned document at high resolution and bit depth and then using graphics software such as Photoshop to tweak the contrast, alterations that are invisible to the naked eye or even under an ALS can sometimes be made clearly visible.

Some document alterations are obvious at first glance. Subtler alterations may be revealed by microscopic analysis, by examining the document under an ALS, by iodine fuming, or by chemical treatment. We'll use all of those methods in this lab session.

This lab has six procedures. In the first, we'll test various chemical means of removing various inks from paper. In the second, we'll produce questioned document specimens by removing ink using both physical means (erasure) and chemical means. When examining a questioned document for possible alterations, it's always important to use nondestructive methods first, so in the third procedure, we'll examine the questioned documents by visible and ultraviolet light and by scanning with contrast enhancement, and in the fourth, we'll examine the questioned documents microscopically. In the fifth procedure, we'll use iodine fuming, which is fully reversible, to reveal alterations to the questioned documents. Finally, in the last procedure, we'll use chemical treatment to reveal alterations to the questioned documents.

FORMULARY

If you don't have the FK01 Forensic Science Kit, you can make up the iodine/glycerol reagent by dissolving 0.1 g of potassium iodide in 3 mL of distilled or deionized water. Add about 5 mL of glycerol and swirl to mix the solution.

PROCEDURE X-1-1: TEST INK SOLVENTS

When a forger sets out to alter a document, the first problem is to determine how best to remove ink from the document. Physical erasure is a last resort, because it usually leaves obvious abrasions on the document as well as traces of the ink. Chemical means are preferable, because removing ink by chemical means may leave no traces that are obvious to the naked eye.

For documents printed on ordinary paper, the forger seeks a chemical method—solvent or bleach—that can remove the ink without making any obvious changes to the paper. For documents printed on security paper, such as checks, the problem is more difficult, because the forger needs a method to remove only the ink required to make the alteration, without removing any of the ink from the printed security pattern on the paper. (For this reason, security paper is printed with inks that are affected by all common chemical methods.)

In this part of the lab session, we'll test various chemical treatments against specimens of ink from our various pens. With the exception of chlorine laundry bleach and some commercial ink removers, all of the liquids we'll use work by dissolving the ink. Bleach and some commercial ink removers do not dissolve the ink, but oxidize some inks to colorless forms.

We'll use cotton swabs for our preliminary testing, although a real forger would never do so. A forger must determine the best chemical method to use by testing as many methods as necessary on the original document, but those tests must leave no obvious traces. The traditional method is by touching a short length of cotton thread or the frayed tip of a wooden toothpick dampened with liquid, which allows testing various chemical methods using only microliter (one one-thousandth of a milliliter) to picoliter (one one-millionth of a milliliter) quantities of liquid. If possible, the forger does these preliminary tests on a portion of the document distant from the intended alteration site.

1. If you have not already done so, put on your splash goggles, gloves, and protective clothing.

2. On one sheet of paper, draw a line about 10 cm long with each of your pens. Keep the lines separated by a few centimeters, and allow the inks to dry thoroughly before proceeding.

3. Dampen a cotton swab with acetone and touch it lightly to the first ink specimen, as shown in Figure X-1-1. (Do not rub it against the ink specimen.)

Figure X-1-1: *Using a moistened cotton swap to test ink solubility*

4. Examine the swab to determine if the ink specimen is insoluble, slightly soluble (faint stain), or very soluble (strong stain) in acetone. Note your observations in your lab notebook.

5. Repeat steps 3 and 4 for each of the other ink specimens. If an ink specimen stains the swab, either replace it with a new swab or use an unstained area of the swab to test other ink specimens.

6. Repeat steps 3 through 5 using xylene and any other solvents you have available and want to test.

7. Repeat steps 3 through 5 using the chlorine bleach and commercial ink remover. With these liquids, the ink may be bleached colorless rather than transferring to the swab, so what you're looking for is a lightening of the ink on the paper rather than staining of the swab.

PROCEDURE X-1-2: PRODUCE QUESTIONED DOCUMENT SPECIMENS

For subsequent testing, we'll need two or three copies each of documents altered by physical erasure and by chemical means. To produce questioned documents altered by physical erasure, take the following steps:

1. Label a sheet of paper as your erasable pen questioned specimen.

2. Use the erasable ink pen to write several figures on that sheet of paper.

3. After the ink has dried completely, use the pen's eraser to remove part of what you wrote, being careful to abrade the surface of the paper as little as possible.

4. Use the pen to write different text over the erasure.

   Figure X-1-2 shows our original text and Figure X-1-3 our altered text, with the contrast enhanced to make the alteration more visible.

Figure X-1-2: *The original text, before physical erasure (contrast enhanced)*

Figure X-1-3: *The text after physical erasure and alteration (contrast enhanced)*

Although pencils are not used for legal documents, alterations to documents written in pencil may sometimes be forensically significant.

5. Label a sheet of paper as your pencil questioned specimen.

6. Use a pencil to write several figures on that sheet of paper.

7. Working carefully, use the pencil's eraser to remove part of what you wrote, being careful to abrade the surface of the paper as little as possible.

8. Use the pencil to write different text over the erased area.

To produce questioned documents altered by chemical means, take the following steps:

1. If you have not already done so, put on your splash goggles, gloves, and protective clothing.

2. Based on your results in the first procedure, choose the combination of pen and solvent or bleach that worked best to remove the writing.

3. Use that pen to write several figures on a sheet of paper.

4. After the ink has dried completely, use the appropriate solvent or bleach to remove part of the ink from the specimen. Work slowly and carefully, as shown in Figure X-1-4, using as little liquid as possible to prevent the liquid from spreading beyond the area you are treating.

5. Allow the treated area to dry completely, and then use the same pen to write different text over the treated area.

Figure X-1-4: *Removing ink by chemical treatment*

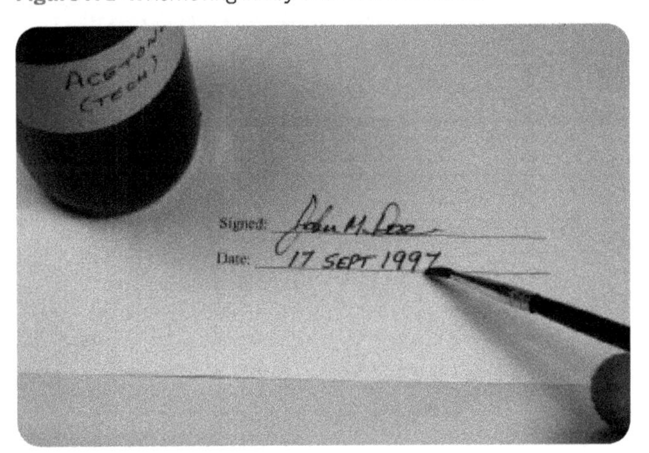

PROCEDURE X-1-3: EXAMINE QUESTIONED DOCUMENTS BY VISIBLE AND ULTRAVIOLET LIGHT

The first step in any questioned document examination is to observe the document by both white and ultraviolet light. Such examinations are nondestructive, and so are used before any other tests that may alter the specimen.

For the white-light examination, use a desk lamp or other strong, directional light source and examine the questioned specimens with the magnifier by both flat and oblique, grazing light. Try to discern any faint remnants of the original writing, and in particular any visible changes to the structure of the paper.

For the UV examination, again use flat and oblique lighting for each specimen, and look for any differences in fluorescence in the questioned area. Chemical treatment may cause an increase or decrease in fluorescence, or a change in the color of the emitted light. Remaining traces of the original ink may fluoresce. Also, most recent papers are treated with fluorescent brighteners to make them appear whiter, and any liquid coming into contact with the paper may alter the appearance of that fluorescence.

Record your observations in your lab notebook.

> Professional forensics labs also use other alternative light sources, particularly infrared, to examine and/or photograph questioned documents. Not all alterations are revealed by an ALS, but when this method succeeds, it often does so spectacularly, making the alterations clearly visible and often revealing the original text in readable form.

If no alterations are apparent with the magnifier under visible or UV light, the next step is to scan the document and use image-manipulation software to enhance the contrast, gamma, and other image properties. Make color scans of each of your questioned specimens at the highest resolution and bit depth supported by your scanner. Open the scanned images in your image-manipulation software and adjust the contrast, gamma, and other image properties to determine if any alterations are visible in the enhanced images.

PROCEDURE X-1-4: EXAMINE QUESTIONED DOCUMENTS MICROSCOPICALLY

> The preceding procedure may reveal the alterations in your questioned specimens. Even if the alterations in the questioned specimens were revealed by that or a later procedure, simply pretend that those procedures were unsuccessful and continue working your way through all of the procedures in the remainder of this lab session.

If alterations are not visible in the questioned document during the preliminary visual examination by white and UV light, the next step is to examine the questioned area of the document microscopically. To do so, place the questioned document on your microscope stage and examine it by both reflected and transmitted light at 40X and 100X. First, examine an area of the document that is known not to have been altered as a basis of comparison. Look for crushed, broken, or abraded fibers as an indication of physical erasure and for changes in the pattern of the fibers or differences in transparency as indications of chemical or solvent erasure. Look also for dark spots that may be remaining pigment from pigment-based inks and for stains that may be the residue from dye-based inks. Record your observations in your lab notebook.

PROCEDURE X-1-5: EXAMINE QUESTIONED DOCUMENTS BY IODINE FUMING

Iodine fuming, which we used in Lab Session IV-2 to reveal latent fingerprints, is also useful for revealing document alterations. Set up your iodine fuming chamber and follow the procedure described in Lab Session IV-2 to iodine-fume your questioned documents. Note your observations in your lab notebook.

PROCEDURE X-1-6: EXAMINE QUESTIONED DOCUMENTS BY CHEMICAL TREATMENT

The preceding methods are nondestructive and usually succeed in revealing most document alterations. If these methods fail, the next step is to use chemical means to reveal any alterations. Using chemical means may damage the specimen, so these means are always a last resort.

1. If you have not already done so, put on your splash goggles, gloves, and protective clothing.
2. Fill the small sprayer bottle with 70% to 95% ethanol. Place the questioned document specimen on a flat surface and spray the area in question with a fine mist to dampen it. If this method succeeds, some or all of the alterations may become faintly visible within a few seconds. Note your observations in your lab notebook.

Allow the questioned specimen to dry, or use a new questioned specimen for the following test.

3. Use a fresh cotton swab to moisten the questioned area of the document with the iodide/glycerol reagent, as shown in Figure X-1-5. Blot the solution gently onto the paper. Do not rub it in. Observe the treated area for at least one minute. Note your observations in your lab notebook.

Figure X-1-5: *Swabbing with iodide/glycerol solution to reveal alterations*

REVIEW QUESTIONS

Q1: **Why is chemical treatment preferred to physical erasure for altering documents?**

Q2: **How far into the sequence did you get before the alterations to your questioned documents were revealed?**

Analysis of Inks by Chromatography

Lab X-2

EQUIPMENT AND MATERIALS

You'll need the following items to complete this lab session. (The standard kit for this book, available from *http://www.thehomescientist.com*, includes the items listed in the first group.)

MATERIALS FROM KIT

- Goggles
- Centrifuge tubes, 50 mL
- Chromatography paper
- Forceps
- Pipettes
- Ruler
- Spot plate, 12-well

MATERIALS YOU PROVIDE

- Gloves
- Acetone
- Ethanol
- Pencil
- Scanner and software (optional)
- Scissors
- Toothpicks, plastic
- Xylene
- Specimens, black felt-tip pens (3)

WARNING

Wear splash goggles, gloves, and protective clothing. Acetone, ethanol, and xylene are extremely flammable. Read the MSDS for each chemical you use and follow the recommended safety precautions.

BACKGROUND

Inks can be broadly divided into pigment-based and dye-based varieties. The earliest inks were pigment-based, usually using lampblack or finely-divided minerals as pigments. Most modern writing inks, with the exception of some fine-arts inks, are dye-based. The key difference is that, while pigments are essentially insoluble, dyes are soluble and so can be separated and analyzed by chromatography.

Chromatography is a method used to separate mixtures of soluble compounds. Chromatography uses a *substrate*, sometimes called the *fixed phase* or *static phase*, and an *eluent*, sometimes called the *mobile phase*. A small amount of the *analyte* (the material to be separated) is spotted on the substrate at the *origin* and the eluent is run through the substrate by capillary action, physical pressure, or other means. As the eluent front passes through the analyte, it dissolves the analyte and draws it along with the eluent through the substrate. Different components of the analyte have different relative attractions to the two phases. If a component is strongly attracted to the eluent and less so to the substrate, that component passes quickly through the substrate. If a component of the analyte is more strongly attracted to the substrate than to the eluent, that component passes relatively slowly through the substrate.

Because different components of the analyte migrate at different speeds through the substrate, a developed chromatogram shows a series of lines or bands, each a different distance from the origin. The distance that the eluent migrates during a chromatography run provides a fixed point of comparison for the bands. For example, the maximum progress of the eluent front during a given chromatography run might have been 8 cm from the origin. Component A of the analyte may have formed a band at 2.4 cm from the origin, and component B may have formed a band at 6.4 cm from the origin.

Component A therefore migrated (2.4 cm / 8 cm) = 0.3 times the distance of the solvent front. This ratio, which is a dimensionless number, is referred to as the *retardation factor* (R_f) value for that material with that specific combination of substrate and eluent. By a similar calcuation, component B has an R_f value for that combination of substrate and eluent of (6.4 cm / 8 cm) = 0.8.

Although the R_f value for a material is characteristic of that material, it is not necessarily unique to that material. For example, one dye component of an ink may have an R_f value of 0.38 using a particular combination of substrate and eluent, but many other materials (including other dyes) may have the same or a very similar R_f value with that combination of substrate and eluent. That means that the most a forensic analyst can say on the basis of one chromatography run of a questioned material is that the results are consistent with a tested known material.

In chemistry, *polarity* refers to the degree of separation of electrical charges on a molecule, which is determined by the types of atoms present in the molecule and their physical arrangement. Some molecules are highly polar. For example, water is highly polar because there is a strong negative charge on the oxygen atom on one side of the molecule and two positive charges on the two hydrogen atoms on the opposite side of the molecule. Other molecules, such as xylene, have their electrical charges dispersed relatively evenly around the surface of the molecule.

In chemistry, the rule of thumb is that like-likes-like. For example, water and ethanol (both highly polar molecules) mix freely because their similar polarity causes the two types of molecules to attract each other. Conversely, highly polar water and nonpolar vegetable oil do not mix well because the polar and nonpolar molecules repel each other.

The application of polarity to chromatography is that polar analytes are much more strongly attracted to polar eluents than to nonpolar eluents, and vice versa. Similarly, polar analytes are much more strongly attracted to polar substrates than to non-polar substrates, and vice versa.

So, for example, if you develop a chromatogram of a polar analyte (e.g., a polar dye) with a polar eluent (e.g., ethanol) on a non-polar substrate (e.g., paper), the analyte will likely migrate very quickly through the substrate. The polar dye and the polar ethanol are strongly attracted to each other, and neither is attracted to the nonpolar paper substrate, so the R_f value for that combination will likely be very high, possibly even approaching 1 (the analyte progresses through the substrate at the same rate as the eluent front, or nearly so).

Conversely, if you run that polar dye with a nonpolar eluent (such as xylene), the polar dye is not attracted to the nonpolar eluent. The R_f value is likely to be extremely low, possibly even approaching 0 (the spot doesn't move from the origin), because the dye simply doesn't bond with the migrating eluent.

By using mixed solvents, you can adjust the polarity of the eluent across the range from extremely polar to extremely non-polar. Just keep in mind that a little bit of polar goes a long way. For example, water is extremely polar and acetone is moderately polar. Adding just a few drops of water to 10 mL of acetone produces a mixture that is closer in polarity to pure water than to pure acetone. Similarly, adding a few drops of moderately polar acetone to 10 mL of nonpolar xylene produces a mixture that is closer in polarity to pure acetone than to pure xylene.

The advantage of using a mixed eluent is that it allows separations of all components of a mixed analyte that contains both polar and non-polar components. If you develop a chromatogram of such an analyte with a pure polar solvent, the polar component(s) of the analyte migrate right along with the eluent front, while the nonpolar component(s) of the analyte just sit there at the origin, not budging. Conversely, if you use a nonpolar eluent, the nonpolar component(s) of the analyte migrate with the eluent, while the polar component(s) just sit there at the origin. A mixed eluent allows you to separate both types of analytes.

Because the same analyte compound has different R_f values with different eluents, making chromatograms with two or more eluents allows a forensic analyst to resolve ambiguities. For example, two different dyes may have the same or very similar R_f values with one eluent but very different R_f values with a second eluent. If questioned and known specimens have similar R_f values with one eluent, the forensic analyst will run chromatograms with a second eluent. If the R_f values are also similar with the second eluent, the analyst can conclude with some certainty that the two dyes are in fact the same. If any doubt remains, running chromatograms with a third eluent should resolve them.

Forensic analysts routinely analyze various types of questioned specimens by using several different types of chromatography, including *gas chromatography* (GC), *high-performance liquid chromatography* (HPLC), and *thin-layer chromotography* (TLC). Professional forensics labs can obtain accurate ink chromatography results using incredibly small specimens. An ink specimen the size of the period at the end of this sentence is sufficient. The standard sampling method is to use a hypodermic needle to cut tiny disks from the specimen. The ink present in those tiny discs is then extracted by a solvent and analyzed.

Some inks, particularly older ones, use only one dye; others, including many modern inks, use a mixture of two or more dyes. In either case, chromatography can be used to identify the ink by comparing a questioned specimen against various known specimens. In the case of a single-dye ink, the R_f value can be determined for several combinations of substrate and/or eluent. In the case of a multi-dye ink, the relative R_f values of the component dyes can identify the ink, ordinarily even if only one combination of substrate and solvent is tested.

In this lab session, we'll use paper chromatography to separate and analyze ink specimens from three black-ink felt-tip pens. We'll use specimens considerably larger than those that a professional lab requires to make a determination, but we'll be analyzing microgram quantities of dyes nonetheless. We'll use just one substrate, paper, but we'll test several solvents to determine which provides the best separation for our questioned ink specimen. By then producing chromatograms for each of our known ink specimens using that same solvent, we should be able to discriminate among ink specimens that appear visually similar.

PROCEDURE X-1-1: PREPARE CHROMATOGRAPHY JARS

We'll run our chromatograms using three different eluents, all of which are volatile. It's important to have the chromatography jars saturated with eluent vapor before we develop the chromatograms. Otherwise, as it climbs up the chromatography strip, the eluent may evaporate from the strip, ruining the chromatogram.

1. If you have not already done so, put on your goggles and gloves.
2. Label three 50 mL tubes for the eluents you'll be using.
3. To the first tube, transfer sufficient ethanol to fill the tube to the 5 mL line. Cap the tube and set it aside upright. This tube is our polar eluent tube. Record the contents of that tube in your lab notebook.
4. To the second tube, transfer sufficient xylene to fill the tube to the 5 mL line. Cap the tube and set it aside upright. This tube is our nonpolar eluent tube. Record the contents of that tube in your lab notebook.
5. To the third tube, transfer sufficient xylene to fill the tube to the 5 mL line and then add five drops of acetone. This tube is our mixed-polarity eluent tube. Cap the tube and set it aside upright. Record the contents of that tube in your lab notebook.

PROCEDURE X-1-2: PREPARE THE QUESTIONED INK SPECIMEN

Ask a family member, friend, or lab partner to prepare a questioned ink specimen for you by writing with one of the black-ink pen specimens on an ordinary sheet of paper.

1. If you have not already done so, put on your goggles and gloves.
2. Cut a piece of the questioned specimen that will fit easily into one well of the spot plate. The piece should contain as much of the questioned ink as possible.
3. Use a pipette to transfer a drop of xylene to the surface of the specimen in the well.
4. If the ink is freely soluble in xylene, it should immediately begin to spread and smear. In that case, add a few more drops of xylene and use the forceps to swirl the paper around to extract as much of the ink as possible from the specimen. When all of the ink appears to have dissolved in the xylene, use the forceps to remove the paper from the well, squeezing the paper to return as much xylene as possible to the well.
5. If the ink is unaffected by the xylene or only partially dissolves, add a couple of drops of acetone to the well and observe the effects. If the ink appears to be soluble in acetone, add a few more drops and swirl the specimen to remove as much of the ink as possible. If all of the ink appears to have dissolved, use the forceps to remove the specimen from the well, squeezing as much solvent as possible from the specimen back into the well.
6. If the ink is unaffected or only partially dissolves in the acetone, repeat the preceding step using ethanol.
7. When you have extracted as much ink as possible from the specimen, set the spot plate aside and allow the solvent to evaporate. Record your observations in your lab notebook.

Most pen inks contain dyes, but some may include pigments, which are insoluble. Professional forensics labs may use other solvents—including methanol, pyridine, THF, and others, many of which are hazardous, expensive, or both—to solubilize inks before they assume that a pigment is present in the ink. We'll simply assume that any ink that fails to dissolve in any of our solvents is either a pigment or simply a dye that's insoluble. In either case, that persistent component of the ink will simply not migrate when we run a chromatogram with it.

PROCEDURE X-1-3: PREPARE AND SPOT CHROMATOGRAMS

1. Wearing gloves to prevent your skin oils from being transferred to the paper, cut four strips of chromatography paper, each about 2.5 cm (1") wide by 10 cm (4") long. The strips should be just wide enough to slide without friction into the 50 mL centrifuge tubes.

2. Draw a pencil line across the width of each strip, about 2.5 cm (1") from one end of the strip. Use the pencil to put a short hash mark at the center of each line. This labels the origin.

3. Cut both corners from the same end of the strip to allow the bottom of the strip to protrude down into the tapered portion of the tube. Label each strip at the bottom (in the trimmed portion) with the specimen that strip is to be used for: questioned, K1, K2, and K3.

4. Use the K1 pen specimen to put a small spot of ink at the hash mark on the K1 strip. If the pen is an ultra fine-point, simply touch the tip to the hashmark and allow the paper to absorb ink until the spot is roughly 1 mm in diameter. If the pen has a broader point, touch the point very gently to the hash mark to avoid making too large a spot.

5. Repeat step 4 with the K2 pen specimen on the K2 strip and the K3 specimen on the K3 strip.

6. After the spots have dried, repeat steps 4 and 5 to re-spot each specimen, keeping the spot as small as possible.

7. Redissolve the questioned ink specimen by adding one drop of whatever solvent you determined in the preceding procedure to be the best solvent for the questioned ink. Roll the drop of solvent around the well to dissolve as much as possible of the questioned ink.

8. Use a plastic toothpick to pick up a tiny drop of the questioned ink solution, and use it to spot the questioned chromatography strip. Allow the spot to dry, and re-spot until you have a good accumulation of ink on the spot.

PROCEDURE X-1-4: DEVELOP CHROMATOGRAMS

With all four of the strips spotted and dried, the next task is to develop the chromatograms.

1. Working as quickly as possible to avoid losing vapor, slide the questioned strip, pointed side down, into the chromatography jar that contains the optimum eluent you determined earlier and recap the jar.

2. Observe the eluent front as it climbs up the strip. When the front has almost but not quite reached the top of the strip, remove the cap from the tube, extract the strip, and then immediately recap the tube. Place the developed strip on a flat surface and use the pencil to make hash marks on both edges of the strip to indicate the maximum progress of the eluent front. Draw a pencil line to connect the two hash marks.

3. Repeat steps 1 and 2 with the K1, K2, and K3 strips.

4. When all of the strips have dried, place them side by side and compare them. Unless two or all three of your pen specimens use identical inks, it should be immediately obvious which of the known pen specimens is consistent with the questioned specimen.

   Depending on the makeup of the inks in your pen specimens, the developed chromatograms may show separation of various colored dyes, or they may show only a single color. In the former case, it should be easy to identify which of the known pens is consistent with the questioned ink specimen. In the latter case, you may still be able to identify the pen that was used to produce the questioned specimen by comparing the distance that the single color dye moved up the strip.

   Figure X-2-1 shows three chromatograms developed in acetone, one from the questioned ink specimen and two from known ink specimens. All three black inks were visually indistinguishable, but the chromatograms clearly show that three different inks were used. The Sharpie chromatogram (top) shows that Sharpie ink contains two black component dyes, one of which has a blue-purple tinge and migrated sharply with the acetone solvent front, and the other of which appears pure black and barely migrated from the initial spot. The questioned ink contains only one black dye (or perhaps two dyes with very similar characteristics) that has a yellowish tinge and migrates with the solvent front, although not sharply. The Identi Pen ink contains one dye, also with a yellowish tinge, that lags the solvent front noticeably and spreads more than any of the other dye components shown.

Figure X-2-1: *Developed chromatograms for a questioned ink specimen and two known specimens*

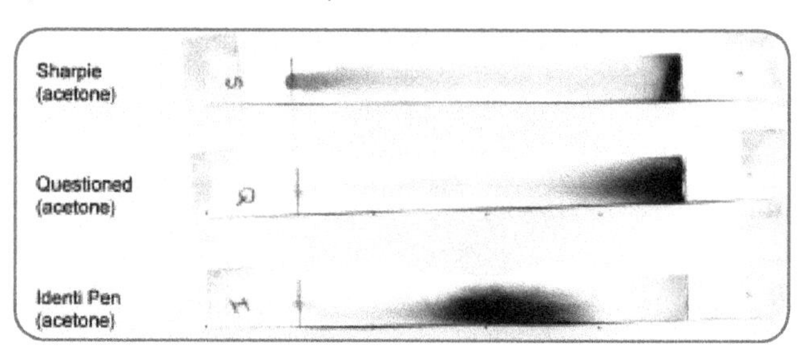

   For comparison/matching purposes, it's not necessary to calculate R_f values, because differences and similarities are obvious with the chromatograms positioned side by side. But for record-keeping purposes, it's still important to record the specifics of each chromatogram—specimen ink, substrate, and eluent—and to calculate R_f values for each dye present in each specimen.

5. Use the ruler to measure the distance from the origin to the eluent front line on the questioned specimen, and record that value in your lab notebook.

6. Make similar measurements from the origin to the center of each dye trace on the questioned specimen, and record those values in your lab notebook.

7. Calculate R_f values for each of the dye traces on your strips by dividing the distance from the origin to the dye trace by the distance from the origin to the eluent front line. Record these values in your lab notebook.

REVIEW QUESTIONS

Q1: **You are presented with a questioned writing specimen and a pen that is believed to have been used to produce the specimen. You will be permitted to take only a tiny sample of the questioned ink specimen, so it will not be possible to do preliminary testing on the questioned ink specimen. How might you modify the procedure we used in this lab session to determine if the ink in the pen provided is consistent with the questioned ink specimen?**

Forensic Analysis of Paper

Lab X-3

EQUIPMENT AND MATERIALS

You'll need the following items to complete this lab session. (The standard kit for this book, available from *http://www.thehomescientist.com*, includes the items listed in the first group.)

MATERIALS FROM KIT

- Goggles
- Coverslips
- Forceps
- Herzberg's stain
- Jenk's stain
- Magnifier
- Pipette
- Slides, flat

MATERIALS YOU PROVIDE

- Gloves
- Camera with microscope adapter (optional)
- Desk lamp or other incident light source
- Microscope
- Paper towels
- Scissors
- Ultraviolet light source (optional)
- Watch or clock with second hand
- Specimens, known and questioned paper

WARNING

Herzberg's stain and Jenk's stain are toxic and corrosive. Wear splash goggles, gloves, and protective clothing. Read the MSDS for each chemical you use and follow the recommended safety precautions.

BACKGROUND

Although it is made up primarily of cellulose fibers, paper is not the simple material that most people think it is. The basic material may be new or reprocessed fibers of cotton, linen, straw, or another natural fiber, or it may be chemically pulped or mechanically pulped wood fibers. Paper may be *laid* (produced on a patterned, directional screen; now used almost exclusively for handmade and art papers) or *wove* (produced on a fine, nondirectional screen; now used for 99% of all paper production). Various coatings, binders, and fillers are used in different papers, as are visible dyes (to change the tint) and ultraviolet brighteners. Better quality papers are often watermarked, which is a process that introduces a mark that is invisible by reflected light but visible by transmitted light.

Most people are aware that high-quality papers often contain high percentages of cotton or linen (so-called "rag") fibers. In fact, high-quality paper is often marketed using such terms as "100% cotton bond." But the original source of the fibers actually has little bearing on the quality of paper produced from them. How those fibers, regardless of their source, are extracted and treated determines the quality of the paper.

Most paper is produced using primarily or exclusively wood pulp, which is the least expensive source of cellulose fibers. Wood pulped mechanically, which essentially means grinding it up, gives high yields—one kilogram of wood yields about one kilogram of paper—but the quality of the paper is low because the lignin content is high. Such mechanical pulp papers are used primarily for newsprint, paperback books, and similar purposes, where paper longevity is not an issue. Wood pulped chemically—by treatment with concentrated sulfite or soda solutions—has most of the lignin removed. The downside of that is lower yields—as little as 50%—and accordingly higher costs, but the upside is that paper produced from chemical pulp is of excellent quality, as good as that produced from rag.

It is sometimes important forensically to determine if two paper specimens are closely similar or have distinguishable differences. For example, a question may arise as to whether a page was added to a contract or will. If the paper of the questioned page appears identical to the accepted pages under forensic examination, the questioned page may or may not be a part of the original document, because the person who added the questioned page may simply have used paper from the same stock as was used for the original document. But if the questioned page uses different paper from the rest of the document, it may reasonably be assumed that the questioned page was added later and was not a part of the original document.

A thorough forensic examination of a questioned paper specimen is typically done in four phases:

Visual examination

A preliminary visual examination of the gross physical characteristics of the paper specimen—color, weight, texture, transparency, watermarking, and so on—is occasionally sufficient to establish that the questioned specimen differs from the accepted specimen.

Microscopic examination

If the gross visual examination is inconclusive, the next step is to examine the questioned paper specimen microscopically by reflected and transmitted light. Two specimens that appear identical on gross visual examination may appear very different at 40X to 100X magnification. Differences in the type, length, or structure of the fibers, the size and shape of sizing or coating particles, or the dye absorption pattern of a colored paper are often sufficient to establish that two specimens differ.

Differential staining

It may be difficult or impossible to discriminate among the types of fibers present in the paper specimen using only visual and microscopic examination. Using *differential stains*—those that dye different types of fibers different colors—allows the examiner to determine which types of fibers are present in the specimen and in what relative numbers. We'll use two differential paper stains in this lab session, called *Herzberg's stain* and *Jenk's stain*.

Instrumental analysis

If all other methods are inconclusive—and if the matter is sufficiently important to justify it—instrumental analysis may be used to compare the questioned paper specimen against an accepted specimen. The most common techniques used for this purpose are electron microscopy and neutron-activation analysis, both used to compare the minor constituents of the paper specimen, such as sizing and coating particles.

In this lab session, we'll use the first three methods to compare various paper samples and discriminate one from another.

For your known and questioned paper specimens, obtain examples of as many similar types of white paper as possible. That is, try to obtain paper specimens that have no gross differences that would make it easy to discriminate between them at first glance. Specimens about 5 cm square are ideal, because they're large enough to allow visual comparison but not so large as to include gross differentiators such as watermarks. Try to obtain specimens ranging from inexpensive copy paper to high-quality bond paper with differing rag percentages. We got our specimens from sources ranging from junk mail to old letters to old hardback books that were destined for the discard pile. For your questioned specimen, ask a friend or relative to choose one of the sources of your known specimens and produce the questioned specimen from it.

FORMULARY

If you don't have the FK01 Forensic Science Kit, you can purchase Herzberg's stain and Jenk's stain from a law enforcement forensics supply vendor or make them up yourself. Wear gloves and safety goggles while making up and using these stains. Both are hazardous, particularly Herzberg's stain, which is extremely corrosive.

It's particularly important to use accurate weights and measures in making up Herzberg's and Jenk's stains, because small differences in the ratios between the components can cause major differences in staining behavior. For that reason, specimens stained with one batch of one of these stains cannot be compared to specimens stained with a different batch.

Herzberg's stain and Jenk's stain are reasonably stable in tightly capped containers stored in a cool, dark place. We've used samples of both that were five or more years old, and they worked properly. Note, however, that paper tests done years apart with the same batch of stain cannot be compared because the stains do gradually age and change properties.

Herzberg's stain Dissolve solid zinc chloride in 25 mL of distilled or deionized water to produce a saturated solution. (Caution: Corrosive!) Zinc chloride is hygroscopic and available in several hydration forms. It is also extremely soluble in water. Continue adding solid zinc chloride to the solution until undissolved crystals remain.

Separately dissolve 5.25 g of potassium iodide in about 5 mL of distilled or deionized water. Add 0.25 g of iodine crystals to the iodide solution and swirl until the iodine dissolves. Make up the iodide/iodine solution to 12.5 mL with distilled or deionized water.

Add 25 mL of the saturated zinc chloride solution to the 12.5 mL of the iodide/iodine solution with swirling to mix the solutions. Allow the mixed solutions to sit undisturbed overnight and then carefully decant off the red liquid, leaving any sediment present behind. Label the storage container Herzberg's stain and date it.

Jenk's stain Dissolve magnesium chloride in 50 mL of distilled or deionized water to produce a saturated solution. Continue adding solid magnesium chloride to the solution until undissolved crystals remain.

Separately dissolve 2 g of potassium iodide in about 4 mL of distilled or deionized water. Add 1.15 g of iodine crystals to the iodide solution and swirl until the iodine dissolves. Make up the iodide/iodine solution to 20 mL with distilled or deionized water.

While swirling to mix the solutions, add 2.5 mL of the iodide/iodine solution to 50 mL of the saturated magnesium chloride solution. Allow the mixed solutions to sit undisturbed overnight and then carefully decant off the liquid, leaving any sediment present behind. Label the storage container Jenk's stain and date it.

PROCEDURE X-3-1: EXAMINE PAPER SPECIMENS VISUALLY

1. Examine the questioned specimen visually by transmitted light and by reflected flat and oblique light, both with the naked eye and using your magnifier. Note all visible characteristics, including color, texture, pattern, or weave, finish, and so on. If you have an ultraviolet light source, repeat your observations by ultraviolet light. Record your observations in your lab notebook.
2. Repeat step 1 for each of your known specimens, and attempt to determine if the questioned specimen is consistent visually with one or more of the known specimens.

PROCEDURE X-3-2: EXAMINE PAPER SPECIMENS MICROSCOPICALLY

1. Cut a small sliver of the questioned specimen, place it on a slide and cover it with a coverslip. Do not use any water or other mountant; we'll examine the specimen first using a dry mount.
2. Examine the questioned specimen at 40X and 100X by incident (reflected) light. If solid particles are visible at 100X, they are probably coating or filler material. Increase magnification to 400X and try to determine the size, shape, color, and other characteristics of the solid particles. Record your observations in your lab notebook. If you are equipped to do so, shoot images of the specimen.
3. Turn off the incident light source and add one drop of water under the cover slip (or more, if necessary to wet the specimen thoroughly).
4. Examine the questioned specimen again at 40X and 100X by transmitted light. Record your observations in your lab notebook. If you are equipped to do so, shoot images of the specimen.
5. Repeat steps 1 through 4 for each of your known specimens, and attempt to determine whether the questioned specimen is consistent microscopically with one or more of the known specimens.

PROCEDURE X-3-3: EXAMINE PAPER SPECIMENS BY DIFFERENTIAL STAINING

1. If you have not already done so, put on your splash goggles, gloves, and protective clothing. (Jenk's stain and particularly Herzberg's stain are hazardous; although you will use them dropwise, you still need to protect your eyes, hands, and clothing.)
2. Label one microscope slide "Q-H" for your questioned specimen that is to be stained with Herzberg's stain, and a second slide "Q-J" for the questioned specimen to be stained with Jenk's stain.
3. Position a small sliver of your questioned specimen centered on each of the two slides.
4. Note the time, and put one drop of Herzberg's stain on the "Q-H" slide and one drop of Jenk's stain on the "Q-J" slide. (Use more stain if necessary to thoroughly wet each specimen.)
5. Allow the stains to work for one minute by the clock.
6. Use the corner of a paper towel to wick up the excess stain from each slide.

7. Using the disposable pipette, apply several drops of distilled water to each specimen to rinse out excess stain. Use a clean corner of a paper towel to wick off the rinse water. Repeat as necessary until all excess dye has been rinsed out.

8. Place a coverslip over the specimen and observe it at 40X magnification. Note your observations in your lab notebook. If you are equipped to do so, shoot images of the specimen.

9. Repeat steps 2 through 8 for each of your known specimens, and attempt to match the questioned specimen to one or more of the known specimens. Figure X-3-1 shows a typical differentially stained specimen at 40X. This specimen, stained with Jenk's stain, reveals a combination of chemical wood pulp and chemical straw pulp fibers.

The effect of these two stains depends on their exact composition as well as the particular types of fibers being stained.

Herzberg's stain dyes mechanical (ground) wood pulp—as well as other wood-like fibers that contain significant lignin, including jute, flax, and hemp—a yellow shade that may vary from lemon yellow to egg-yolk yellow. Chemical (sulfite or soda) straw or wood pulp and similar wood-like fibers from which most or all of the lignin has been removed are dyed a blue shade, from sky blue to navy blue. Cotton and linen fibers (from high-rag content bond papers) are dyed a wine-red color.

Jenk's stain dyes mechanical wood pulp fibers yellow, chemical wood pulp fibers anything from colorless to deep red, chemical straw pulp fibers blue to violet, and cotton or linen fibers brown.

Figure X-3-1: *A differentially stained paper specimen at 40X*

REVIEW QUESTIONS

Q1: Which, if any, of your tests allowed you to determine that a possible match existed between your questioned specimen and one or more of your known specimens? How?

Q2: **You have treated a paper specimen with Herzberg's stain and Jenk's stain and found that both stains dye all of the fibers in the specimen a yellow color. What do you conclude about the fiber type and general quality of the specimen?**

Q3: **You have a paper specimen thought to be a high-quality bond paper with 50% rag content. What results would you expect if you treated this paper with Herzberg's stain and Jenk's stain?**

Forensic Biology

In the past 20 years, forensic biology has grown from a relatively minor forensic sub-discipline to become a crucial part of the work of any forensics lab, and one that gets the lion's share of media attention. We're talking, of course, about DNA analysis, a forensic specialty that did not exist until 20 years ago.

DNA analysis is based on the fact that every living thing is composed of cells, all of which contain DNA in some form. Broadly speaking, there are two types of DNA.

Nuclear DNA (nDNA) is found in the nucleus of eukaryotic organisms, which include humans and other animals, plants, fungi, and protists. With the exception of protists, some of which reproduce asexually, eukaryotes transmit DNA to their descendants by sexual reproduction. Because an individual organism's nuclear DNA is produced by combining the DNA inherited from both parents, nuclear DNA is unique to the particular organism from which it originated, and is therefore considered individualized evidence.

DENNIS HILLIARD COMMENTS

An exception is the case of identical twins; they would have the same nuclear DNA profile. There have been documented cases where identical twins were both implicated in a crime since the victim was unable to discriminate between them on a visual ID and the DNA matched both.

Mitochondrial DNA (mtDNA) is passed unchanged (other than by mutation) from generation to generation matrilineally (through only the female line). In other words, your mtDNA (whether you are male or female) is identical to your mother's mtDNA, which is identical to her mother's mtDNA. Because mtDNA is not unique, it is considered class evidence rather than individual evidence. Despite this limitation, forensic labs often analyze mtDNA, for several reasons:

- mtDNA is often available in larger amounts than nDNA, and may in fact be the only usable form of DNA available for analysis. For example, a rootless hair is rich in mtDNA, but does not contain useful amounts of nDNA.

- The mtDNA in cells is hugely redundant, with many identical copies residing in each cell, and is accordingly robust. The nDNA in a specimen—particularly a specimen that is years or decades old—may be so degraded that it is useless for analysis purposes, but that specimen may still provide mtDNA in usable amounts.

- Because mtDNA is passed unchanged matrilineally, it is often the only means available to identify badly burned crash victims, decomposed bodies, and so on. By comparing the questioned mtDNA specimen against a known mtDNA specimen provided by an ancestor or descendant in the same matrilineal line, a forensic biologist can unambiguously establish the matrilineal origin of the questioned specimen, which in the proper circumstances also provides a strong presumption of identity.

- mtDNA analysis is faster and easier (and cheaper) than nDNA analysis. Accordingly, mtDNA analysis can be used for fast preliminary screening to rule out matches.

Despite its potential, DNA analysis did not become a mainstream forensic technique until the early 1990s, by which time two key developments had occurred, both of which were important enough to win Nobel Prizes for their inventors:

Restriction enzymes
: Daniel Nathans, Werner Arber, and Hamilton Smith won the 1978 Nobel Prize in Medicine for discovering *restriction enzymes*, which cut huge, monolithic DNA molecules at precisely known locations into relatively small base-pair fragments that can subsequently be separated and analyzed by *gel electrophoresis*, a method conceptually similar to standard chromatography. On a developed and visualized DNA gel, these base-pair fragments appear as sets of lines, the positions of which allow a questioned DNA specimen to be individualized against known specimens.

PCR replication
: DNA evidence is often recovered in only trace amounts that are insufficient for direct analysis. In 1984, Kary Banks Mullis invented *PCR polymerase chain reaction* (PCR) DNA replication, an achievement for which he shared the 1993 Nobel Prize in Chemistry. PCR allows even tiny DNA specimens to be replicated repeatedly to produce samples large enough for analysis.

In combination, these two technologies allow forensic scientists to process even small DNA-containing specimens—such as traces of saliva present on a cigarette butt or a tiny spot of blood—and produce DNA maps that are accepted by courts as individualized evidence. DNA evidence is often conclusive—establishing guilt or innocence beyond reasonable doubt—in both current cases and cold cases dating back years or even decades. Many serial killers, murderers, and rapists who thought their crimes could never be proven against them have been charged, convicted, and sentenced on the basis of later DNA tests. Just as important, many people who were wrongly convicted of crimes have been freed as a result of later DNA testing that established that their DNA did not match questioned specimens from the crime scenes.

Forensic biology is more than just DNA analysis, however. Experts in various forensic biology subspecialties gather and analyze numerous types of biological evidence. In this group, we'll examine and attempt to match pollen and diatom specimens. We'll also extract DNA from various sources, and use gel electrophoresis to analyze our DNA samples.

Pollen Analysis

EQUIPMENT AND MATERIALS

You'll need the following items to complete this lab session. (The standard kit for this book, available from *http://www.thehomescientist.com*, includes the items listed in the first group.)

MATERIALS FROM KIT

- Goggles
- Coverslips
- Glycerol
- Slides, flat

MATERIALS YOU PROVIDE

- Gloves
- Camera with microscope adapter (optional)
- Collection containers (small plastic bags, etc.)
- Immersion oil (with 1000X microscope only)
- Microscope (ideally 1000X)
- Toothpicks, plastic
- Specimens, pollen (see text)

WARNING

None of the materials used in this lab session present any serious hazard. However, it is good practice to wear splash goggles and gloves when collecting and handling specimens.

BACKGROUND

Pollen is a granular powder produced by seed plants for sexual reproduction. The biological payload in a pollen grain is surrounded and protected by a hard coating that protects the male gametophytes present in pollen grains during their journey from the stamen (male sex organ) of one flower to the pistil (female sex organ) of another flower. Pollen may be waterborne, airborne, or may be transferred between plants by insects (notably, bees), small mammals, or by other means.

The size of pollen grains differs according to the species of plant that produces them. The smallest pollen grains are about 3 μm in diameter. The largest are about 2,500 μm (2.5 mm) in length, and are readily visible to the naked eye. Typical pollen grains range from about 25 μm to 250 μm in length or diameter.

Zoogamous plants—those whose pollen is distributed by bees or other animals—typically produce relatively small numbers of relatively large pollen grains, because this pollen distribution method is very efficient. Conversely, anemophilous (wind-pollinated) plants typically produce huge numbers of relatively small pollen grains, because their pollen grains must be carried on the wind and only a minuscule percentage of the pollen grains will eventually reach another plant to fertilize it.

These differences profoundly affect pollen distribution patterns. Zoogamous pollen grains, because of their relative scarcity and tendency to be distributed only in the immediate vicinity of the plant that produced them, may, if found, be very significant forensically. Conversely, anemophilous pollen grains, because they are produced in such huge numbers and widely scattered on the wind, are often of limited forensic significance. Marijuana pollen, for example, is so common and so widely distributed that the presence of a few marijuana pollen grains indicates nothing at all. You probably have several marijuana pollen grains adhering to your clothing as you read this.

The morphology of pollen grains also varies by species, ranging from roughly spherical to tubular to various irregular shapes. Many types of pollen include spikes, knobs, or other protrusions, which help them adhere to insects and assist in their transfer from flower to flower.

This wide variation in morphology from species to species means it's possible to identify the species from which a particular pollen grain originated. Such identification, however, is a matter for a palynologist, a biologist who specializes in pollen. For example, Figure XI-1-1 is a scanning electron microscope (SEM) image from the Dartmouth Electron Microscope Facility, which identifies the pollen types as "sunflower [*Helianthus annuus*], morning glory [*Ipomea purpurea*], hollyhock [*Sildalcea malviflora*], lily [*Lilium auratum*], primrose [*Oenothera fruticosa*], and castor bean [*Ricinus communis*]." Looking at the image, we have no clue which is which, and neither would most biologists, including most forensic biologists. An expert, however, can identify the species from which an individual pollen grain originated.

Figure XI-1-1: *Scanning electron microscope image of several types of pollen*

Because they can be carried on the wind, individual anemophilous pollen grains can be found anywhere from urban high rises to arctic ice floes, so the presence of one or a few anemophilous pollen grains is ordinarily of no forensic significance. But finding numerous pollen grains of the same type is a different matter. Such heavy accumulations ordinarily occur only if the clothing or other specimen from which the pollen is obtained was in direct contact with or at least in close proximity to the plant that produced the pollen.

Individualizing pollen grains to the species level can provide important forensic evidence, particularly if pollen from several species is present, and even more particularly if one or more of those species is zoogamous and relatively uncommon or not widely distributed. For example, examining the clothing of someone suspected of a violent assault may identify half a dozen or more types of pollen grains present in relatively large numbers. If the plant species associated with those types of pollen are present at the crime scene—and, again, in particular if one or more of the species is not common—that evidence provides reasonable cause to believe that the suspect was at some time present at the crime scene.

As the newest of the forensic science subdisciplines, forensic palynology is still in an embryonic stage. With the exception of New Zealand, where forensic palynology testimony is commonly used in criminal cases, forensic palynology is not widely accepted or practiced, despite the potential usefulness of the evidence it can provide.

As of today, pollen is accepted—if at all—as class evidence, because one pollen grain from a particular species of plant is much like another pollen grain from another plant of the same species. However, we expect that forensic palynologists will eventually use DNA analysis to individualize pollen grains to the individual plants that produced them. Such testing presents real challenges, because it is currently painstaking, time-consuming, and expensive to extract and replicate sufficient DNA from individual pollen grains to complete the tests. We expect that the nanodroplet real-time PCR system announced in December 2008 in a paper by Kim, Dixit, Green, and Faris will soon be applied to DNA analysis of pollen grains.

In this lab session, we'll obtain various known types of pollen and compare them microscopically to a questioned specimen to determine which, if any, of the known specimens are consistent with the questioned specimen.

Depending on the season, you may be able to obtain pollen specimens from flowers, trees, and other plants in your own yard. (Different plants produce pollen at different times of year, from early spring to late autumn.) If you do this lab session in the dead of winter, as we did, one good source for pollen is a florist. You need only a few grains of each type of pollen, which you can obtain by touching the internal structure of a flower with the tip of a plastic toothpick or similar instrument or by tapping it gently over a collection surface such as a small envelope or plastic bag. Label each specimen as you collect it. Try to obtain several different pollen specimens from different kinds of plants (trees, bushes, ornamental flowers, garden vegetables, etc.) to show the diversity in pollen sizes and morphology.

PROCEDURE XI-1-1: EXAMINING KNOWN AND QUESTIONED POLLEN GRAINS

1. Label a clean specimen container "Questioned" and give it to a friend or lab partner, along with your known pollen specimens. Ask your lab partner to transfer a small amount of one of your known pollen specimens to the "Questioned" specimen container.

2. Label microscope slides for each of your known specimens and your questioned specimen.

3. Using a plastic toothpick, transfer a few grains of the questioned pollen to the center of the appropriate microscope slide. Add a drop of glycerol or another mounting fluid and place a coverslip over the drop of fluid.

4. Repeat step 3 to produce wet mounts for each of your known pollen specimens.

5. Observe the questioned specimen at 100X, 400X, and (if available) 1000X magnification. Note the visible characteristics of the specimen in your lab notebook, including the size, shape, color, surface texture, protuberances, and so on. If you have the necessary equipment, shoot an image of the specimen and include it in your lab notebook.

6. Repeat step 5 for each of the known specimens and attempt to determine which of the known specimens is consistent with the questioned specimen.

REVIEW QUESTIONS

Q1: **In examining a soil specimen from the boot of someone suspected of a violent assault, you find half a dozen pollen grains from a particular anemophilous plant, several specimens of which were found growing near the crime scene. What, if any, forensic significance do you attribute to the presence of these pollen grains?**

Q2: **In examining that same soil specimen, you find half a dozen pollen grains from a relatively uncommon zoogamous plant, one specimen of which was found growing near the crime scene. What, if any, forensic significance do you attribute to the presence of these pollen grains?**

Diatom Analysis

Lab XI-2

EQUIPMENT AND MATERIALS

You'll need the following items to complete this lab session. (The standard kit for this book, available from *http://www.thehomescientist.com*, includes the items listed in the first group.)

MATERIALS FROM KIT

- Goggles
- Beaker, 250 mL
- Cassia oil
- Centrifuge tubes, 15 mL
- Coverslips
- Forceps
- Glycerol
- Hydrochloric acid
- Hydrogen peroxide, 30%
- Pipettes
- Slides, flat
- Spatula

MATERIALS YOU PROVIDE

- Gloves
- Ammonia, household non-sudsy
- Camera with microscope adapter (optional)
- Collection containers (small plastic bags, etc.)
- Immersion oil (with 1000X microscope only)
- Microscope (ideally 1000X)
- Paper
- Pen
- Refrigerator
- Test tube rack
- Thermos bottle (optional)
- Water, boiling tap
- Water, distilled or deionized
- Specimens, sediment (see text)

WARNING

Concentrated hydrogen peroxide is corrosive and a strong oxidizer. Concentrated hydrochloric acid is corrosive and emits strong fumes. Dilute aqueous ammonia is irritating and emits strong fumes. Work outdoors or under an exhaust hood, and wear splash goggles, gloves, and protective clothing.

BACKGROUND

Diatoms are tiny aquatic algae that have a rigid external skeleton, called a frustule, made up of silica (sand). Diatoms are ubiquitous, found nearly anywhere where saltwater, freshwater, or simply moisture exists. Any specimen of seawater, standing freshwater, or soil contains numerous diatoms. Figures XI-2-1 and XI-2-2 show examples of seawater diatoms and freshwater diatoms, respectively. Figure XI-2-3 shows detail of the silica cell wall of the freshwater diatom *Didymosphenia geminata*, gathered in Rio Espolon, Chile.

Figure XI-2-1: *Assorted seawater diatoms*

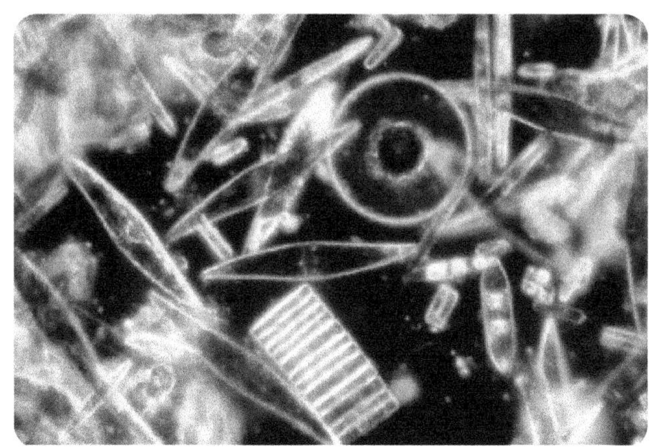

Figure XI-2-2: *Assorted freshwater diatoms*

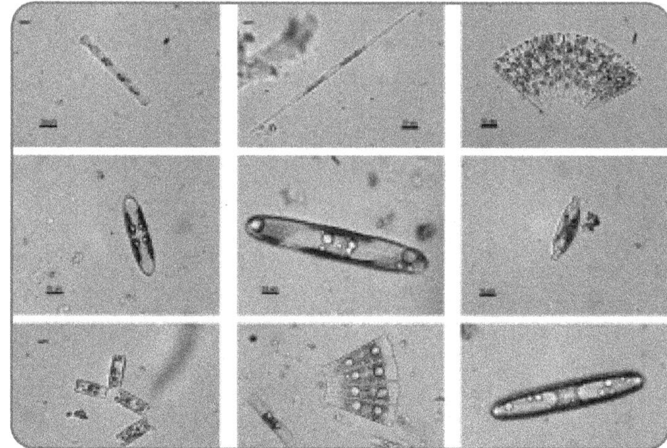

Figure XI-2-3: *Detail of the silica cell wall of the diatom Didymosphenia geminata*

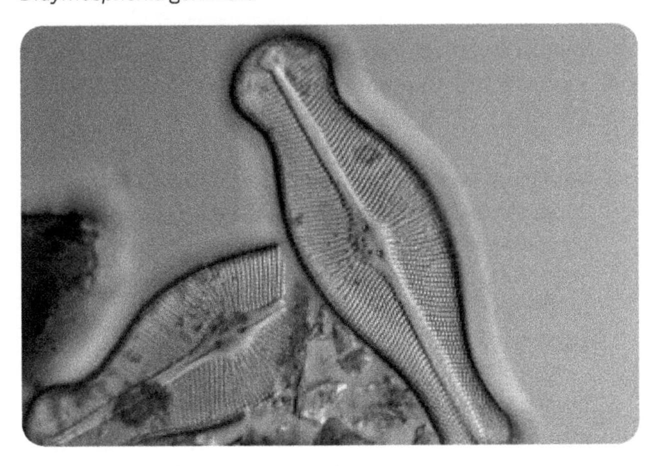

There are thousands upon thousands of diatom species, both living and long extinct. Some are found only in certain isolated locations and others are very widely distributed. Like pollen grains from different plant species, each diatom species has a characteristic size and morphology that allows it to be identified unambiguously. Diatom sizes are similar to those of pollen grains, with typical diatoms ranging from 20 μm to 200 μm in diameter or length, and the largest up to 2,000 μm (2 mm) long. Diatom morphology covers a broad range, from nearly circular to flattened ovals to rectangular to needle-like. One characteristic shared by most diatoms is the bifurcation from which the name "di-atom" originated.

Their siliceous skeletons mean that diatoms are extremely durable. When the algae themselves die, their skeletons remains. In effect, a living diatom is a fossil, in the sense that it is already mineralized. Because diatoms do not degrade with the passage of time, they can be a reliable source of forensic evidence no matter how much time has elapsed since a crime occurred.

The earliest forensic use of diatom evidence was in examining victims who had drowned in questionable circumstances. Microscopic examination of the water found in a drowning victim's lungs allows a medical examiner or forensic biologist to determine the type of water in which the victim drowned. For example, if a man claims that his wife drowned while swimming in a pond, but examination finds no diatoms in the victim's lungs, the examiner may reasonably conclude that the husband probably drowned his wife in the bathtub and later moved her body to the pond.

More recently, diatom analysis has been used as an adjunct to soil analysis. Isolating the diatoms that are often present in soils and sediments provides another way to characterize soil or sediment specimens. Identifying the diatom species present and their relative distribution may allow a specimen to be individualized as originating only within a very small area. In this lab session, we'll isolate the diatoms present in various sediment specimens and attempt to match a questioned specimen against various known specimens, based on the types and numbers of diatoms present in the specimens.

For the known sediment specimens, obtain five small (about 0.5 mL is sufficient, but it does no harm to obtain more) sediment samples from sources as diverse as possible. You can obtain specimens from anywhere there is standing or slowly running water, including lakes, ponds, streams, birdbaths, planters and pots, and so on. Small vials or plastic bags are ideal containers. For the questioned specimen, ask a friend or lab partner to transfer a small amount of one of your five known specimens to a container labeled "Questioned."

This lab session has two procedures. In the first, we'll use a standard protocol to digest sediment specimens in hydrogen peroxide to destroy any organic material present. In the second, we'll isolate and mount the diatoms and compare our questioned specimen against our known specimens to determine if the questioned specimen is consistent with a known specimen.

PROCEDURE XI-2-1: DIGEST DIATOM SPECIMENS

Before you begin this procedure, fill a Thermos bottle with boiling tap water. If you don't have a Thermos bottle, simply keep a pot of water simmering on a stove or hotplate.

The sediment specimens contain soil and various organic matter in addition to the diatoms. In this part of the lab session, we'll digest our specimens in concentrated hydrogen peroxide to oxidize the organic matter, leaving only the diatoms and mineral matter in the specimens. Diatoms are light enough to remain suspended in water for several minutes to several hours, but heavy enough that they eventually settle out, leaving only lighter materials suspended in the supernatant liquid.

After treating the sediment specimens with hydrogen peroxide, we'll use several wash/settle passes to remove everything but the diatoms and mineral matter from the sediment. During the final wash pass, we'll add aqueous ammonia to keep any colloidal clay particles present suspended in the supernatant liquid. By this means, we'll produce a treated sediment that contains only diatoms and relatively massive mineral matter, which we can easily separate from the diatoms in the following procedure.

This procedure requires several days to complete if you follow the instructions as written. If you have access to a centrifuge, you can complete this part of the lab session in an hour or so by centrifuging the specimens rather than allowing them to settle out overnight by gravity.

1. If you have not already done so, put on your splash goggles, gloves, and protective clothing.
2. Label six 15 mL centrifuge tubes, "Q" for the questioned specimen and "K1" through "K5" for the known specimens.
3. Transfer a small amount of each sediment specimen to the corresponding tube. Use the corner of the flat end of the spatula to pick up an amount of sediment about the size of a grain or two. Make sure the specimen is at the bottom of the tube.
4. Working in a well-ventilated area, use a clean pipette to carefully transfer about 0.75 mL (the third line up from the tip of a plastic pipette) of 30% hydrogen peroxide to each of the six tubes. Some specimens may react vigorously, bubbling or even foaming; others may show no obvious signs of reaction.
5. For those specimens that react vigorously, wait for the vigorous reaction to cease and then add further hydrogen peroxide dropwise until no further vigorous reaction is evident.
6. Fill the 250 mL beaker about half full of hot water from the Thermos bottle. Cap the tubes loosely, and immerse all six of them in the hot water bath. The tubes will float, but their tips will be immersed in the water.
7. Every 10 to 15 minutes, remove the tubes from the hot water bath, empty the now-warm water from the beaker, refill the beaker from the Thermos bottle, and then re-immerse the tubes. If the hydrogen peroxide evaporates, add more hydrogen peroxide dropwise to return the liquid to its original level.
8. Repeat step 7 until the specimens have digested for two to four hours.
9. Remove the tubes from the hot water bath and place them in the rack. Add hydrochloric acid dropwise to each of the tubes until foaming ceases. (This may require only a drop or two.)
10. Fill each tube to about the 10 mL line with distilled or deionized water, swirl to mix the contents, cap the tube, and then place it in the rack.
11. Place the rack in the refrigerator and allow the contents to settle overnight.

Ordinarily, we stress keeping lab items and kitchen items completely separated. In this case, however, there is no risk in placing the digested specimen in a kitchen refrigerator. Digestion in hydrogen peroxide destroys all organic matter, including potentially harmful bacteria.

12. Using a clean pipette for each tube, carefully draw up as much of the supernatant liquid as possible from each tube without disturbing the sediment. Discard the liquid.

13. Repeat steps 10 through 12 four times, for a total of five washes. Add two or three drops of dilute aqueous ammonia to the final wash water to help keep clays suspended and prevent the diatoms from clumping. Recap each of the tubes.

At this point, the contents of the tubes are stable. You can do the next procedure immediately, or wait until the next lab session.

PROCEDURE XI-2-2: MOUNT AND OBSERVE DIATOMS

1. If you have not already done so, put on your splash goggles, gloves, and protective clothing.

2. Place six coverslips on a sheet of paper, keeping them well separated. Label each coverslip with the corresponding specimen number, "Q" and "K1" through "K5."

3. With swirling, add about 2 mL of distilled or deionized water to the Q specimen tube. Swirl the tube to suspend the solid matter and then allow the solids to settle for 30 seconds to a minute.

4. Using a clean pipette, withdraw about 0.5 mL of the liquid diatom suspension from the "Q" tube and add it dropwise to the "Q" coverslip until the liquid covers the surface of the coverslip. Use the tip of the pipette to spread the liquid across the surface of the coverslip.

5. Repeat steps 3 and 4 for each of the known liquid diatom suspensions, as shown in Figure XI-2-4.

6. Allow the coverslips to remain undisturbed overnight or until the liquid has evaporated from all of the coverslips. To prevent contamination, you can cover the paper with a cardboard box, plastic kitchen container, or similar object.

7. Label six microscope slides, "Q" for the questioned specimen and "K1" through "K5" for the known specimens.

8. Place one drop of mounting fluid in the center of the "Q" slide.

Figure XI-2-4: *Transferring diatom suspensions to coverslips*

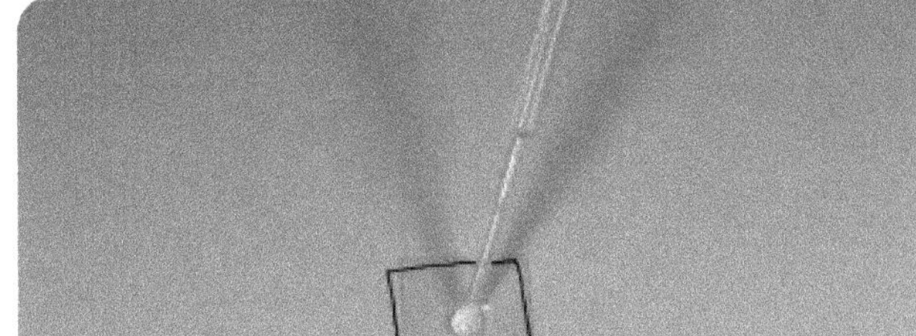

Laboratory supply vendors sell special high–refractive index mounting fluids (sold under trade names such as Naphrax and Zrax) that are designed to optimize the visibility of wet-mounted diatoms. If you have such a mounting fluid, use it, making sure to follow the directions provided with the mountant. Otherwise, use whatever high-RI mounting fluid you have available. The cassia oil included in the FK01 Forensic Science Kit has an RI of about 1.6, less than the ~1.7 RI of the special mountants, but still close enough to the RI of silica to show high relief and excellent contrast.

9. Using the forceps, pick up the "Q" coverslip from the sheet of paper, invert it, and then place it diatom-side down on top of the drop of mounting fluid on the "Q" slide.

10. Repeat steps 8 and 9 for each of the known specimens.

11. Observe the questioned specimen at 40X and 100X to perform a survey of the types and relative numbers of diatoms present, making sure to observe the entire area under the coverslip. If necessary, use 400X and (if available) 1000X magnification to view fine details of diatoms to identify their types unambiguously. Note the appearance and relative numbers of the various diatoms present in your lab notebook. If you have the necessary equipment, shoot an image of the specimen and include it in your lab notebook.

> Although we were able to view diatom details visually, we found it almost impossible to photograph the diatoms at 400X and particularly at 1000X, possibly because we didn't use special diatom mounting fluid. If you cannot photograph the diatoms, make sketches of at least the predominant types in your lab notebook.

12. Repeat step 11 for each of the known specimens, and attempt to determine which of the known specimens is consistent with the questioned specimen.

REVIEW QUESTIONS

Q1: **How much diversity in diatom populations did you find in your various specimens?**

Q2: **Were you able to determine that your questioned specimen was consistent with only one of your known specimens?**

Extract, Isolate, and Visualize DNA

Lab XI-3

EQUIPMENT AND MATERIALS

You'll need the following items to complete this lab session. (The standard kit for this book, available from *http://www.thehomescientist.com*, includes the items listed in the first group.)

MATERIALS FROM KIT

- Goggles
- Beaker, 100 mL
- Beaker, 250 mL
- Centrifuge tubes, 15 mL
- Centrifuge tubes, 50 mL
- Coverslips
- Methylene blue stain
- Microscope slides (flat)
- Papain powder
- Pipettes
- Stirring rod
- Test tubes
- Test tube rack

MATERIALS YOU PROVIDE

- Gloves
- Alcohol (see text)
- Cheesecloth (or muslin, etc.)
- Chopsticks, disposable wooden (optional)
- Dishwashing liquid (Dawn or similar)
- Freezer
- Microscope
- Saucer
- Table salt
- Teaspoon
- Toothpicks
- Specimens: raw and cooked ground beef
- Specimens: vegetables (tomato paste, etc.)

WARNING

None of the materials used in this lab session is particularly hazardous, but as a matter of good practice, you should always wear splash goggles, gloves, and protective clothing when working in the lab.

BACKGROUND

All organisms are made up of cells, from the tiniest species such as single-cell bacteria and protozoa to the largest animals and plants, which are made up of trillions of individual cells. Every cell contains DNA, which is the hereditary genetic material that allows cells and organisms to function and to reproduce themselves.

> **DENNIS HILLIARD COMMENTS**
>
> One exception in humans is that red blood cells do not contain DNA, since they have no nucleus. Fortunately, there are plenty of white blood cells present from which DNA can be obtained.

DNA taken from any cell of a particular individual of any species is identical to that taken from any other cell from that individual, and is unique to that individual. DNA found in different members of the same species—for example, you and your best friend or two rosebushes—is very similar but not identical. DNA found in different but closely related species—for example, humans and chimpanzees or wolves and dogs—has greater differences, but is still extremely similar. DNA found in different, unrelated species—for example, a human and a cucumber—has greater differences still, but remains closely similar.

In this lab session, we'll obtain DNA from various eukaryotic (plant and animal) sources, and examine that DNA. We'll do this in three steps:

Extract the DNA

To extract DNA from cells, we must first *lyse* (break open) the outer cell membranes to release the DNA and other internal cell contents. We'll do this by using ordinary dishwashing liquid—which contains a detergent that breaks down the lipids that make up the outer cell membrane—and ordinary table salt, which aids in the lysing process.

Although this first lysing step releases the DNA from the interior of the cell, eukaryotic DNA is tightly bound to protein molecules called *histones*, so an additional protein digestion step is necessary to release the DNA from the bound histones. To do that, we'll use an enzyme-based meat tenderizer. That enzyme, called papain, breaks down the histones and other proteins present and releases the DNA molecules.

Isolate the DNA

After we lyse the cells with detergent and digest them with papain, the liquid contains not just the DNA we want to isolate, but a complex mixture of proteins, protein fragments, and many other compounds. To isolate and purify the DNA, we'll take advantage of the fact that DNA is relatively soluble in water but nearly insoluble in alcohol, while most of the other compounds present in the liquid are readily soluble in alcohol.

Visualize the DNA

After we extract and isolate the DNA, we'll stain it using methylene blue and observe it under the microscope.

For specimens, you'll need small amounts of uncooked and cooked ground beef. It's best to use fresh ground beef from a butcher or fresh market. Supermarket ground beef can be used but may yield inferior results, depending on its age. For other specimens, you can use any raw fruit or vegetable, such as apple, pear, carrot, corn, onion, or peas. We used canned corn. You can use several different vegetable specimens if you want to see whether the results differ. It's also interesting to compare a raw vegetable (such as a raw tomato) against that same vegetable in processed or cooked form (such as tomato paste or ketchup).

This lab session has three procedures. In the first, we'll extract DNA from our specimens. In the second, we'll isolate that DNA from the other components. In the third, we'll stain our DNA specimens with methylene blue to visualize them and then examine them under the microscope.

PROCEDURE XI-3-1: EXTRACT DNA

Before you begin this procedure, add a teaspoon of table salt to about 100 mL of tap water in the 250 mL beaker, and stir until the salt dissolves. Add about 10 mL (2 teaspoons) of dishwashing liquid and stir gently to mix the solution without creating excess foam.

1. Place a teaspoon-size specimen of fresh, uncooked ground beef in the saucer.

2. Add 15 mL (three teaspoons or one tablespoon) of the salt/detergent solution to the saucer, and use the teaspoon to grind and mash the ground beef thoroughly to produce a suspension of beef cells in the salt/detergent solution.

3. Carefully pour as much as possible of the liquid in the saucer through cheesecloth, muslin or other loosely woven cloth into the 100 mL beaker. Avoid transferring solids from the saucer to the beaker.

4. Pour the liquid from the beaker into a labeled 15 mL centrifuge tube. Transfer 12 to 13 mL of the liquid from the beaker to the tube. If there is excess liquid in the beaker, discard it. If there is insufficient liquid to fill the tube to the 12 to 13 mL level, don't worry about it. Just transfer as much liquid as you can. Cap the tube and set it aside.

5. Repeat steps 1 through 4 with your cooked (browned) ground beef specimen and your vegetable specimens.

6. Prepare a fresh solution of papain enzyme by stirring about half a teaspoon of papain powder or papain-based meat tenderizer into about 10 mL of water and allowing any excess powder to settle.

7. Using a clean pipette, transfer about 1 mL (the line on the pipette tube nearest the bulb) of the papain solution to each of your specimen tubes. Recap each tube and invert it several times to mix the solutions.

> If you want to break the lab session at this point, you can store the liquid DNA specimens in the refrigerator overnight or longer.

PROCEDURE XI-3-2: ISOLATE DNA

Before you begin this procedure, transfer about 40 mL of alcohol to a 50 mL centrifuge tube. Cap the tube and place it in the freezer to cool. You can use either ethanol or isopropanol. Concentrated alcohols (95% ethanol or 99% isopropanol) work best, but even the 70% concentrations work acceptably.

1. Transfer about 5 mL of the raw ground-beef liquid specimen to a test tube.

2. Hold the test tube that contains the first specimen at about a 45° angle and use a pipette to trickle about 5 mL of ice-cold alcohol to the tube, slowly and carefully. Allow the alcohol to run gently down the inside of the tube so that it forms a separate layer rather than mixing with the DNA solution that is already in the tube.

3. Carefully insert a glass or wooden stirring rod into the tube, submerging it in the liquid so that it penetrates the interface between the two liquid layers but does not touch the bottom of the test tube. Dip and withdraw the rod into the lower layer several times, until you see white solid DNA begin to accumulate on the rod, as shown in Figure XI-3-1.

> For the wooden stirring rods, we used disposable chopsticks. You can instead use a glass stirring rod, although the solid DNA adheres better to a wooden rod.

Figure XI-3-1: *Spooling DNA from alcohol layer*

The white flecks visible in Figure XI-3-1 in the top layer and the white cloud at the interface between the upper and lower layers are bits of DNA. DNA is relatively soluble in the lower (aqueous) layer, but nearly insoluble in the upper (alcohol) layer. Moving the rod in and out of the lower layer transfers a small amount of DNA solution into the upper alcohol layer, where it precipitates onto the rod.

1. Twirl the rod gently between your thumb and forefinger to spool solid DNA onto the rod. Continue doing so until you have accumulated a significant amount of DNA on the rod and the deposition rate slows or stops.

2. Withdraw the rod from the liquid and rinse the DNA twice with a few milliliters of cold alcohol.

3. Place the rod aside and allow the DNA to dry completely.

4. If you do not intend to proceed immediately to the following lab session, after each DNA specimen has dried completely, scrape it off the rod and store it in a labeled small paper envelope or other porous container.

5. Repeat steps 1 through 7 for each of your other liquid DNA specimens.

Retain your unused liquid specimens in case you want to repeat the extractions to isolate more DNA. You can store them for a week or so in the refrigerator.

PROCEDURE XI-3-1: VISUALIZE DNA

1. Use a toothpick to transfer a small amount of the dry raw ground beef DNA specimen to a microscope slide.

2. Place a drop or two of 1% methylene blue solution on the specimen and place a coverslip over it.

3. Observe the DNA at 40X, 100X, and 400X magnification. Record your observations in your lab notebook.

4. Repeat steps 1 through 3 for each of your other DNA specimens.

REVIEW QUESTIONS

Q1: **Can you differentiate visually among the different types of DNA you isolated?**

Q2: **Propose an explanation for why we extracted DNA from both raw and cooked ground beef.**

DNA Analysis by Gel Electrophoresis

Lab XI-4

EQUIPMENT AND MATERIALS

You'll need the following items to complete this lab session. (The standard kit for this book, available from *http://www.thehomescientist.com*, includes the items listed in the first group.)

MATERIALS FROM KIT

- Goggles
- Agar powder
- Beaker, 250 mL
- Buffer, loading, 6X concentrate
- Buffer, running, 20X concentrate
- Cylinder, graduated, 10 mL
- Cylinder, graduated, 100 mL
- Leads, alligator clip (2)
- Methylene blue stain
- Pipettes
- Reaction plate, 24-well
- Ruler

MATERIALS YOU PROVIDE

- Gloves
- Aluminum foil
- Balance (optional)
- Batteries, 9V or DC power supply (see text)
- Gel casting comb materials (see text)
- Marking pen (Sharpie or similar)
- Microwave oven
- Plastic containers (see text)
- Scissors
- Soft drink bottle (clean and empty)
- Tape (electrical or masking)
- Toothpicks, plastic
- Water, distilled or deionized
- Specimens: DNA (from Lab XI-3)

WARNING

None of the chemicals used in this lab session present any special hazard, but it is still good practice to wear goggles and gloves while working with any chemical. The battery stack used in the gel electrophoresis exposes potential dangerous voltage, and presents a fire hazard if you allow the positive and negative leads to contact each other. Making up agar gels involves handling very hot liquids.

BACKGROUND

In the preceding lab session, we extracted and isolated DNA from various sources. In this lab session, we'll analyze those DNA specimens using a technique called *gel electrophoresis*.

Gel electrophoresis is conceptually similar to chromatography, but with a slightly different goal. Ordinarily, we use chromatography to separate different compounds from a mixture. With DNA gel electrophoresis, the goal is to separate DNA fragments of different sizes and masses, which are produced by treating a DNA specimen with restriction enzymes to cleave it at known locations into fragments.

To imagine the shape and structure of DNA, think of a standard ladder, with two sides and rungs joining them. Twist the tops of the sides of the ladder to form two counter-rotating, interlocked spirals, with the rungs still joining them. In DNA, the sides of the ladder are made up of molecules of a sugar called 2-deoxyribose, with those sugar molecules bonded together by phosphate groups. The rungs, called *base pairs*, are formed by bonded pairs of four amino acid bases: adenine (abbreviated A), cytosine (C), guanine (G), and thymine (T). Adenine bonds only with thymine, forming an AT base pair, and cytosine only with guanine, forming a CG base pair.

To visualize the complexity of DNA, imagine that your ladder, rather than having only a few rungs, has millions or billions of rungs. Genomic human DNA, for example, has about 3.2 billion base pairs. Untwisted and stretched out, a single strand of genomic human DNA would form a "ladder" about 2.4 μm wide and 2 meters long, with 3.2 billion rungs.

The number of base pairs in genomic DNA is unrelated to the complexity of the organism. For example, while honeybee DNA has only about half as many base pairs as human DNA (1.77 billion versus 3.2 billion), the fruit fly has about 130 million, and E. coli bacteria about 4.6 million, the marbled lungfish has about 130 billion base pairs—more than 40 times the human count—and DNA from one species of amoeba has 670 billion base pairs.

Restriction enzymes function like tiny scissors, cutting the DNA into fragments for subsequent separation by gel electrophoresis. In effect, a restriction enzyme "looks for" a specific base sequence, such as 3 AGG'CCT. When the restriction enzyme "finds" that specific base sequence, it cuts the DNA at that point.

Because that specific short base sequence inevitably occurs many times in any individual DNA specimen, treatment with that one restriction enzyme cuts the DNA into many fragments. Because repeated occurrences of that base sequence may be relatively close together or far apart on the DNA molecule, the fragments are of differing lengths, typically from hundreds to thousands of base pairs.

The specific size distribution of the DNA fragments obtained by treating DNA with a particular restriction enzyme varies according to the positions of the target base sequence in that particular DNA specimen. So, for example, treating your DNA with a particular restriction enzyme results in a different distribution of fragment sizes than treating your lab partner's DNA with the same restriction enzyme. The distribution of fragment sizes in your DNA specimen is unique to you, just as everyone else's distribution is unique to them.

Once the DNA has been cut into fragments by the restriction enzyme, gel electrophoresis is used to analyze the DNA by producing a map of the fragment sizes present in the specimen. A sample of the fragmented DNA is placed in a small well in a gel, which is then immersed in a buffer solution and subjected to DC electric current. Because DNA fragments are negatively charged, they are attracted to the positive electrode, which is positioned at the far end of the gel from the wells that contain the DNA solution.

The gel selectively retards the migration of the DNA fragments toward the positive electrode. Small DNA fragments pass through the gel relatively unhindered, and so reach the positive electrode quickly. Larger fragments move proportionally more slowly because the gel provides more resistance to their progress. If the current is applied until the smallest fragments just reach the end of the gel nearest the positive electrode, larger fragments are strung out along the length of the gel, with the largest fragments barely clear of the well where they originated. The positions of the various fragments provide a graphical map of the fragment size distribution in the specimen.

But DNA fragments are colorless, so it's impossible to track the progress of the electrophoresis visually. For that reason, a *marker dye* is added to the DNA specimen before the electrophoresis run. The marker dye is chosen on the basis of how fast it migrates through the gel. By using a marker dye that moves about the same speed as the smallest DNA fragments of interest, electrophoresis can simply be discontinued when the visible marker dye approaches the positive electrode, at which point the smallest DNA fragments have made about the same amount of progress through the gel.

At this point, the only visible change to the gel is a band of marker dye near the positive electrode end. To visualize the bands of colorless DNA fragments, they're stained using dyes that are selectively attracted to the fragments. The most common stain used in professional laboratories is ethidium bromide, which bonds to the DNA fragments and fluoresces under ultraviolet light to reveal the DNA fragments as bright bands against the dark background of the gel.

DNA gel electrophoresis as practiced by professional forensics labs uses rigidly standardized procedures, equipment, and supplies, and yields predictable and repeatable results. Getting good results with DNA gel electrophoresis with makeshift equipment in a home lab setting is considerably more challenging. In adapting the gel electrophoresis procedure for this lab session, here are some of the issues we encountered and how we decided to handle them:

Restriction enzymes

Although restriction enzymes for DNA analysis are commercially available from many lab vendors such as Carolina Biological Supply, they are relatively expensive and must be shipped and stored refrigerated. Although we obtained and used restriction enzymes for initial testing of this lab session, for the final lab session write-up, we decided to do without them in the interests of economy and convenience for our readers.

We were able to do without restriction enzymes because, although they are the only way to cut DNA into predictable fragments, they are by no means the only way to cut DNA. DNA can be cut into fragments by such ordinary activities as physical handling and cooking.

In general, the more processed a food product is, the shorter the DNA fragments will be. For example, bovine DNA from raw ground beef typically has DNA fragments averaging about 30,000 base pairs (BP). Ground beef heated to 100°C for 10 minutes has an average fragment size of about 1,000 BP. Heating ground beef to 120°C for 30 minutes yields average fragment sizes of about 300 BP. By using both raw and cooked ground beef, and both raw and cooked vegetables, our gel electrophoresis results should show significant variation.

Gel type

Professional forensics labs use agarose gel for electrophoresis. Unfortunately, electrophoresis-grade agarose is relatively expensive, about $1.50 to $3 per gram when purchased in small quantities. One gel may require 1 g or more of agarose, and we intend to run multiple gels, so we ruled out using agarose on cost grounds.

Ordinary unflavored gelatin might seem to be a reasonable alternative. It's cheap and readily available. Unfortunately, gelatin-based gels don't work. They're too weak, too sticky, and yield very poor separations. We decided to use agar, from which agarose is derived, and which is also commonly used in biology to make gels for culturing bacteria and other microorganisms.

Ordinary food-grade agar sells for a few dollars an ounce, and is used as a thickening agent in many Chinese and Japanese food recipes. (You can buy agar in most specialty food stores that carry foods from around the world; look for the pure powder version that does not contain any flavorings or other ingredients.) Agar is a combination of agarose and other compounds. Those other compounds do not take part in the gelling process, but their presence affects the permeability and other characteristics of gels made with them.

> Because of its variability, if you use food-grade agar, we strongly recommend that you run one or more test gels before making up a large batch. Agarose gels are normally used at 0.5% to 3% or higher concentration (0.5 g to 3 g of agarose per 100 mL of water). Agar gels must be more concentrated because not all of the agar is agarose, and the nonagarose components do not contribute to gelling. If you are using food-grade agar, concentrations from about 1 g to 7 g per 100 mL are useful. The concentration affects both the physical characteristics of the gel and its permeability during electrophoresis.

Electrophoresis apparatus

An electrophoresis apparatus is conceptually simple. It consists of a gel chamber surrounded by a buffer chamber, with provisions for connecting electrodes at opposite ends. In use, a comb is suspended in the gel chamber and the chamber is filled with liquid gel. When the gel solidifies, the comb is removed, leaving wells into which samples can be placed. The wells are filled with micropipettes, which deliver small and very precise amounts of the sample into each well. The prepared gel (in its chamber) is then immersed in a buffer solution contained in the buffer chamber and voltage is applied to the electrodes.

Commercial electrophoresis apparatuses are usually made from Plexiglas or a similar plastic, and are quite expensive for what you get, from $100 to $500 or more, depending on size and features. Fortunately, you can make your own gel electrophoresis apparatus inexpensively from readily available materials.

> For instructions on making a simple gel electrophoresis apparatus from materials found around the home, see MAKE: 07: Backyard Biology, Page 65. For instructions on making a professional-grade gel electrophoresis chamber as a weekend project from hardware store parts, see *http://learn.genetics.utah.edu/content/labs/gel/gelchamber/*.

Power supply

Professional gel electrophoresis apparatuses use expensive DC power supplies that can be adjusted over a wide range of voltages, typically 2VDC to 350VDC or more. For a home gel electrophoresis setup, the most convenient alternate power source is three to nine 9V transistor batteries wired in series to provide 27VDC to 81VDC.

To use this method, simply make a battery stack by connecting the crimped terminal of one battery to the smooth terminal of the next, as shown in Figure XI-4-1, and use two alligator clip leads to connect the two open terminals on the stack to your gel electrophoresis apparatus.

Figure XI-4-1: *A 9V transistor battery stack*

Loading buffer and marker dyes

The *loading buffer* is a relatively viscous solution that is mixed with the DNA solution before it is loaded into the well. Its purpose is to keep the DNA solution in the well and force it to migrate through the gel rather than simply diffusing into the pool of running buffer that covers the gel.

The migration rate of DNA fragments through the gel is determined by many factors, most important the size of the fragments, the voltage applied to the gel, the concentration of the gel, and the concentration and makeup of the running buffer. Because DNA fragments are colorless, there's no way visually to determine their position in the gel at any given time. To provide a visual indication of progress, most loading buffers incorporate one or more *marker dyes*, which migrate through the gel at known rates that correspond to specific DNA fragment sizes. For example, the common indicator dye bromophenol blue migrates through a 0.5% to 1.5% agarose gel at about the same rate as DNA fragments of 400 to 500 base pairs, and cresol red at about the same rate as 1,000 to 2,000 bp fragments. The loading buffer supplied with the FK01 Forensic Science Kit uses bromophenol blue.

Whoever named bromophenol blue must have been color blind. It's actually bright yellow in acid solution and deep purple in base solution.

Running buffer

The *running buffer* is a dual-purpose solution, used both to dissolve the powder used to make up the gel and as an immersion bath for the gel while it is being run. Essentially, the running buffer is just a dilute solution of ionic salts that maintains the necessary pH and provides electrical conductivity through the gel between the two electrodes of the power source. The running buffer supplied with the FK01 Forensic Science Kit is a 20X concentrate, which must be diluted with 19 parts of distilled or deionized water to make a working solution.

Visualization stains

Because DNA fragments are colorless, some means must be used to visualize them after the gel is run. Professional forensics labs often use ethidium bromide, which fluoresces under UV light, but it's very expensive and extremely dangerous to handle. Fortunately, a dilute solution of non-toxic methylene blue stain is both inexpensive and reasonably effective as a DNA stain. It selectively stains DNA fragments while only minimally staining the gel bed, making the DNA bands visible as blue bands in the whitish gel under white light.

If you would like to delve further into gel electrophoresis and DNA analysis, one of the best sources of equipment and supplies is Edvotek (*http://www.edvotek.com*). Edvotek packages many different kits designed to illustrate various aspects of DNA analysis in a classroom setting.

In this lab session, we'll use gel electrophoresis to analyze the DNA specimens we produced in the preceding lab session. We'll use a loading buffer that contains bromophenol blue as a marking dye so that we can visually observe progress of the electrophoresis run. When the run is complete, we'll stain the resulting gel with methylene blue to visualize the DNA fragments distributed across the gel.

FORMULARY

If you don't have the FK01 Forensic Science Kit, you can make up the necessary chemicals yourself, as follows:

Loading buffer (6X concentrate) Dilute 6 mL of glycerol with distilled or deionized water to a volume of 20 mL. Dissolve 0.05 g of bromophenol blue sodium salt in the glycerol/water solution. Dilute one volume of this 6X concentrate with five volumes of distilled or deionized water to make up working loading buffer.

Alternatively, mix 0.5 mL of glycerol with 9.5 mL of commercial 0.04% bromophenol blue indicator solution to make up 10 mL of working loading buffer.

Running buffer (20X concentrate) Dissolve 38.17 g of sodium borate tetrahydrate and 33.g of boric acid in about 900 mL of distilled or deionized water and make up the volume to 1,000 mL. Dilute one volume of this 20X concentrate with 19 volumes of distilled or deionized water to make up working running buffer.

Methylene blue stain (50X concentrate) Dissolve 1 g of water-soluble methylene blue stain in 100 mL of distilled or deionized water to make a 1% w/v aqueous solution. Dilute one volume of this concentrate with 49 volumes of distilled or deionized water to make up the 0.02% working-strength visualization stain.

PROCEDURE XI-3-1: BUILD A GEL ELECTROPHORESIS APPARATUS

There's a time-honored custom in science: we scientists build our own apparatus. Not always, of course, nor even most of the time. Nowadays, anyway. Sometimes it's cheaper, easier, and faster just to buy what we need. But when we need something to complete an experiment, and that something isn't available or we don't have the budget to buy it, we make do. If that involves designing and building a piece of equipment, so be it. For this lab session, we needed a gel electrophoresis apparatus and power supply, and we didn't feel like spending $300 or $400 to buy one. So we designed and built our own, at a total cost of about $10.

Like paper chromatography, gel electrophoresis is used to separate mixtures of chemical compounds. (In biology, gel electrophoresis is often used to separate proteins or DNA fragments.) The similarities do not end there. In paper chromatography, the fixed phase (matrix) is a strip of paper; in gel electrophoresis, the matrix is a bed or column of gel. In both cases, the matrices provide resistance to the movement of molecules. In paper chromatography, capillary action (wicking) provides the force required to move the mobile phase (solvent) and analyte(s) through the matrix; in gel electrophoresis, the force is supplied by an external electrical current, with positively charged molecules attracted to the negative electrode, and vice versa.

Although gel electrophoresis is simple conceptually, commercial gel electrophoresis setups sell for $300 and up, and the gels, buffers, and other supplies they require are not cheap. If you can afford a commercial apparatus and supplies, there's no question you'll find it easier and faster to run gels. However, most homeschoolers and hobbyists can't justify spending several hundred dollars on a gel electrophoresis apparatus and supplies. (Check with your homeschool group or co-op, which may have such an apparatus available to borrow.) Fortunately, with a few minutes' work, you can assemble your own apparatus and supplies for only a few dollars, using items from the kit and some common household items. Here's what you'll need:

Shallow plastic container

You'll need a shallow, easily cut plastic container to serve as the gel casting bed. We used the base of a one-quart Gladware container, trimmed to about 1 cm tall. You can also use a soap dish or similar container. If you want to cast multiple gels in one pass, make several of these gel bed containers.

Deeper container

You'll need another plastic container that's larger and deeper than the gel bed. We used a sandwich container, but you can substitute any suitable container. This container needs to be at least as long and wide as the gel bed container—although, ideally, not all that much longer and wider—and at least a centimeter or so taller.

Tape

We'll cut the ends of the gel bed container off, but we need to be able to seal them temporarily back in place while casting gels. We used plastic electrical tape and ordinary masking tape on different gel casting trays; both worked fine. Once the gel solidifies, you simply peel off the tape to expose both ends of the gel. Tape may also be useful in making the comb that's used to produce wells in the gel.

Comb

No, not the kind you use on your hair. A gel electrophoresis comb is a comb-like assembly with teeth that are placed in the gel casting container to produce small wells in the gel as it solidifies. You can buy a commercial comb for an outrageous price, or you can improvise your own with common household materials.

We've made combs from (a) cardboard and aluminum foil or disposable coffee stirrers, (b) wooden paint stirrers and nails or disposable chopsticks, and (c) Lego blocks. The only important things are that the prongs be relatively smooth (so as not to tear up the solidified gel when they're removed), roughly 3 to 4 mm in diameter (or the equivalent area if the teeth are square or rectangular), long enough to reach the bottom of the gel casting container (or nearly so), and spaced at intervals of roughly 1 cm across the width of the gel casting container. Use your imagination and materials at hand.

Aluminum foil

The apparatus requires an electrode at each end of the larger container. We'll use two pieces of aluminum foil to make these electrodes. You may also use the aluminum foil to make the teeth in the comb used to produce wells in the gel.

9V transistor batteries

The apparatus requires a DC power supply, which can be anything from 30 or 40VDC to 150VDC or more. The lower the voltage, or the larger the gels, the longer they take to run.

Five 9V transistor batteries connected in series provide 45VDC, which is enough to run small gels in a reasonable time. Using seven 9V batteries for 63VDC or nine 9V batteries for 81VDC lets you run gels faster, or larger gels in the same amount of time.

For reasonably fast run times, use at least 3V per centimeter of gel length. For example, if your gel bed is 15 cm long, your power supply should be at least 45VDC. (With these parameters, running a gel may require two or three hours or more, depending on the type and concentration of the gel.)

WARNING

Be careful with 9V batteries connected in series. Depending on your skin resistance and other factors, even 45VDC may produce a painful and potentially dangerous electrical shock. **Never** touch the power leads of the battery assembly or touch the liquid in the gel electrophoresis apparatus while power is connected to it.

Never close the circuit (that is, once they are connected to the battery stack, never allow the free ends of the positive (red) and negative (black) leads to contact each other. That short circuit can rapidly weld the connection and cause the batteries to overheat and explode. (It's safe to connect the positive terminal of one battery to the negative terminal of the next because the circuit has not been closed unless you allow the final open positive and negative terminals to contact each other, such as by allowing the free end of the connected red lead to contact the free end of the connected black lead.)

If you run many gels, the cost of 9V batteries starts to add up fast, so it's cheaper in the long run to buy a DC power supply. For smaller gels, any DC power supply that provides fixed-, multiple-, or variable-output DC voltages in the range of 50 to 70VDC or higher at a few hundred mA will suffice. We've seen suitable single-voltage units on eBay for $30 to $90, depending on voltage.

1. To begin, place the marking pen flat on a table and measure the distance between the table surface and the point of the pen. If it is less than 1 cm, place magazines or other spacers between the table surface and the pen until the tip of the pen is about 1 cm above the table surface.

2. Holding the pen firmly in place, place the smaller plastic container flat of the table, and bring its surface into contact with the tip of the pen. Move the plastic container to draw a line around its exterior that is 1 cm above its bottom, as shown in Figure XI-4-2.

Figure XI-4-2: *Marking the gel casting tray for depth*

3. Use scissors to trim the excess from the plastic container, converting it to a shallow plastic tray.

4. Use the scissors to remove most of both ends of the tray, as shown in Figure XI-4-3. Retain these cut ends for future use.

Figure XI-4-3: *Removing the ends from the gel casting tray*

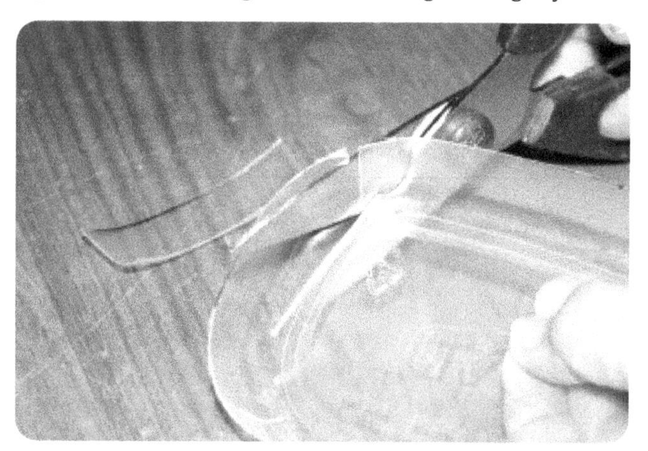

5. Hold the ends of the tray in position, and tape them securely back in place, as shown in Figure XI-4-4. These taped ends will hold the liquid gel in the tray until it's solidified. Before running the gel, you'll remove the ends to allow current to flow through the gel.

Figure XI-4-4: *Taping the ends of the gel casting tray into place*

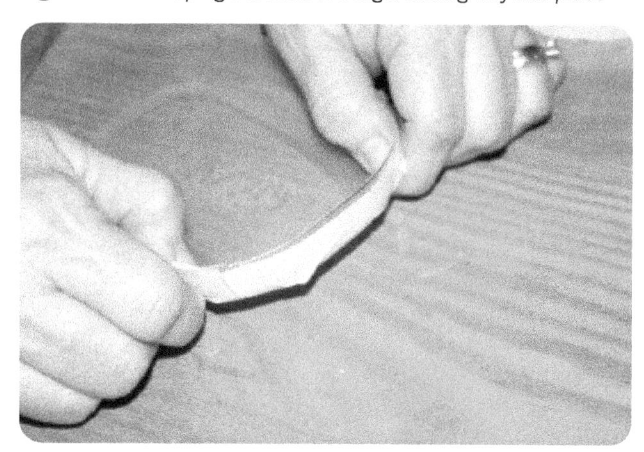

> Calculate the volume of the tray by measuring and multiplying its three dimensions. For example, if your tray is 7 cm wide by 10 cm long by 1 cm deep, its total volume is (7 * 10 * 1) = 70 cm^3, which is the same as 70 mL. When you cast gels, you'll fill the tray to about 0.6 cm (6 mm) to 0.8 cm deep, so each gel tray requires something between 42 mL (0.6 * 70 mL) and 56 mL of gel.

Place the gel tray aside for now. If you want to cast multiple gels in one pass, make several of these trays before continuing. The next step is to make the comb. Here's one way to do it (but feel free to experiment with other methods...).

6. Cut a strip of stiff cardboard about 3 cm wide and 2 cm longer than the width of your gel tray. Draw a centered line the length of the strip. Place tick marks on that line, beginning 1.5 cm from each end of the strip and then evenly spaced at about 1 to 1.5 cm intervals across the width of the gel bed container.

7. Twist pieces of aluminum foil to form stiff cylinders about 2.5 cm long by 3 mm in diameter. (The number you need varies by the width of your gel tray; make one of these cylinders for each of the tick marks across the tray width.) Try to make the surface of the teeth as smooth as possible.

8. Make a 90° bend in each cylinder 1 cm from one end, forming an "L" shape, with one leg 1 cm long and the other 1.5 cm long.

9. Place each cylinder with the bend on a tick mark, and press the longer leg flat against the cardboard. Securely tape the longer leg of each cylinder to the cardboard, with the 1 cm–long leg projecting vertically from the surface of the cardboard.

With the gel-casting tray(s) and comb(s) built, the next task is to assemble the apparatus.

10. Fold aluminum foil over each end of the larger container and connect one end of an alligator clip lead to each of the foil electrodes. (Don't connect the other ends of the alligator clip leads to the battery stack until you're actually ready to run a gel.)

11. Place the gel bed container inside the larger container, centered within the larger container. Verify that there is at least a few millimeters of clearance between each end of the gel bed container and the corresponding electrode on the larger container. Also verify that all sides of the larger container are a centimeter or so higher than the sides of the gel bed container.

At this point, you're ready to prepare your DNA specimens and to cast and run gels, which you'll do in the following procedures. Incidentally, believe it or not, the apparatus you've just built is fully capable of running gels just as good as those produced by an expensive professional apparatus. It's clumsier to use, certainly, but if you use the same agarose and stains that are used on professional apparatus, the resulting gels will be indistinguishable.

PROCEDURE XI-3-2: PREPARE DNA SPECIMENS

1. Make up 500 mL of working-strength running buffer by adding 25 mL of the 20X running buffer concentrate to 475 mL of distilled or deionized water. (You'll need only a small amount of the running buffer for this procedure; you'll use the excess in the following procedures.)

2. Use a clean plastic toothpick to transfer a small amount of your first DNA specimen to the first well in the reaction plate.

3. Add 10 drops of running buffer and 2 drops of the 6X loading buffer to the well, as shown in Figure XI-4-5, and use the toothpick to stir the contents of the well.

4. Repeat steps 2 and 3 for your remaining DNA specimens.

5. Cover the reaction plate and allow it to sit undisturbed overnight.

Figure XI-4-5: *Adding loading buffer to a DNA specimen*

PROCEDURE XI-3-3: PREPARE AND CAST GEL(S)

Before you can run a gel, you need to make up liquid gel solution and cast the gel. If you have more than one gel casting chamber and comb, you can make up multiple gels in advance. Make up no more gels than you plan to use in one or two lab sessions. The gels can be stored overnight in the refrigerator, but you'll get better results with fresh gels.

1. If you have not done so already, calculate the volume of agar gel you need to cast your gel. Multiply the length and width of the gel chamber in centimeters by the desired thickness of the gel to determine how many cubic centimeters (mL) of gel you'll need. For example, if your gel chamber is 7 cm by 10 cm, and you want a 7 mm thick gel, you'll need (7 * 10 * 0.7) = 49 cubic centimeters = 49 mL of gel. If you want to make up two of these gels at one time, you'll need about 100 mL of gel.

> The FK01 Forensic Science Kit includes 10 g of agar, which is sufficient to prepare 500 mL of 2% gel, or enough to cast about 700 cm^2 of gel 0.7 cm thick. If you need more agar, you can use agar that is sold by specialty grocery stores and some Chinese and Japanese restaurants. If you use agar from the grocery store, we recommend using a 3% concentration as a starting point.

2. Transfer the calculated volume of room-temperature running buffer to the beaker.

3. Stir in about 2 g (1/2 to 5/8 teaspoon) of agar powder per 100 mL of running buffer. (The exact amount of agar is not critical.)

4. Carefully heat the liquid in the microwave just until it begins to foam slightly. Keep a very close eye on the beaker as you heat it.

WARNING

The first time we made up 100 mL of agar in the microwave, we figured that one minute would be about right to get the stuff hot but not yet boiling. We intended to heat it for that one minute and then blip it for 5 or 10 seconds more at a time until it boiled.

At about the 50 second mark, the contents of the beaker erupted volcanically, with agar foam flowing up out of the beaker and down its side. We opened the door immediately, but we still ended up with 25 mL of our agar on the microwave tray.

The moral here is that if you use the microwave to heat your agar, do so in very short bursts. It's not actually necessary for the agar solution to boil. Getting it up near the boiling point but not actually boiling the solution is a good way to prevent messy accidents.

5. Using a towel or oven mitt to protect your hand from the heat, remove the agar liquid from the microwave, stir it as shown in Figure XI-4-6 to make sure it is thoroughly mixed, and set it aside to cool. While the agar cools, set up your gel electrophoresis apparatus on a level surface. Verify that the tips of the comb do not contact the bottom of the gel casting tray.

Figure XI-4-6: *Mixing the agar gel*

6. Once the agar has cooled to the point where it feels very warm or hot to the touch (about 50 to 60°C), pour the agar liquid into the gel casting tray to a depth of roughly 0.7 cm, as shown in Figure XI-4-7. If there are any bubbles, use the stirring rod to eliminate them. While the agar is still hot and flowing freely, place the comb near one end of the casting tray with its teeth immersed in the liquid agar.

Figure XI-4-7: *Pouring a gel*

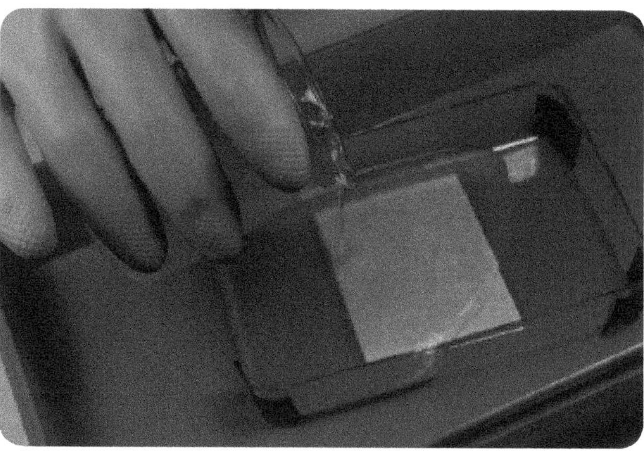

7. Allow the agar to cool and gel completely. Once it has set up, carefully remove the comb. Try to avoid tearing the agar gel when you do so. The goal is to have a solid, flat gel with a series of small neat holes near one end. Remove the ends from the gel casting tray.

PROCEDURE XI-3-4: LOAD AND RUN THE DNA SPECIMENS

1. Using a clean pipette, transfer as much as possible of your first DNA specimen to the first well in the gel, as shown in Figure XI-4-8, making sure not to overfill the well.

Figure XI-4-8: *Loading a DNA specimen into a well*

2. Repeat step 1 for each of your other DNA specimens, transferring each to a separate well. If you have more wells than specimens, alternate wells, leaving an empty well between specimens.

3. Insert the gel chamber into the electrophoresis apparatus, making sure that the end of the gel with the wells is located at the end with the negative electrode.

4. Carefully fill the outer chamber with running buffer until the surface of the gel is entirely immersed in running buffer, with the level of the running buffer a few millimeters above the surface of the gel. Do not pour running buffer directly onto the gel surface, or you may rinse the DNA specimens out of the wells.

5. Connect power to the electrophoresis apparatus, making sure to connect the positive power supply lead to the positive terminal on the electrophoresis apparatus and the negative power supply lead to the negative terminal.

WARNING

Depending on the power supply you use, there may be enough voltage exposed on the running buffer surface to give you a severe shock. Never touch the apparatus (and particularly the running buffer) while power is connected to the apparatus. You can tell it's working because bubbles appear at the positive electrode.

6. Within a few minutes, you should be able to see the marker dye migrating from the wells toward the far end of the gel (the negative terminal side). Continue observing progress periodically. When the marker dye band has nearly reached the far end of the gel, disconnect the power.

Depending on the size and concentration of your gel and the voltage you use, it may take anything from 30 minutes to several hours for the run to complete. Gels run at high voltage may become extremely warm. Allow them to cool to room temperature before handling them.

7. If you have another gel or gels to run, disconnect the power, carefully remove the gel chamber from the electrophoresis apparatus and repeat steps 1 through 6 with a new gel. Unless you are running very high voltage, you can use the same running buffer (in the outer tray) for a second gel.

You can retain developed gels overnight or longer in the refrigerator, but to keep the gels from drying out, it's best to visualize them as soon as possible after you run them and allow them to cool.

PROCEDURE XI-3-5: STAIN AND VISUALIZE THE GEL(S)

With the gel run completed, the gel still looks pretty much the same as it did before the run, except that there is now a band of bromophenol blue dye near the end of the gel farthest from the wells. Never fear. DNA fragments are separated and spread across the gel. We just can't see them until we visualize them by staining them. To do that, we'll use a very dilute solution of methylene blue dye.

1. Make up sufficient 0.02% methylene blue solution to cover the surface of the gel in the electrophoresis apparatus or another container. If you are using the 1% aqueous methylene blue stain from the FK01 Forensic Science Kit, dilute one volume of stain with 49 volumes of distilled or deionized water to make working-strength stain.

> We bought 4 ounces (~118 mL) of 2.303% aqueous methylene blue solution at a pet store. That concentration needs to be diluted about 115:1 water:stain to yield a 0.02% solution of the stain. We made up a supply of 0.02% methylene blue solution by transferring 17.4 mL of the 2.303% methylene blue to an empty 2 L soda bottle and filling it with water. If you use a methylene blue solution of different concentration, adjust the proportions accordingly.

2. With the gel chamber in place in the electrophoresis apparatus or a similar container, fill the outer tray with 0.02% methylene blue until the surface of the gel is completely covered.

3. Allow the gel to sit undisturbed at room temperature overnight.

4. Carefully remove the gel from the apparatus and rinse it with water. You can dispose of the used methylene blue solution by flushing it down the sink or retain it for staining a second gel.

REVIEW QUESTIONS

Q1: **Can you differentiate visually among the different types of DNA you separated by gel electrophoresis?**

Q2: **How, if at all, does the appearance of your gels differ from what you expected?**

Index

B

C

E

F

G

H

I

J

K

M

N

O

P

R

S

T

U

V

W

X

Z

www.ingramcontent.com/pod-product-compliance
Ingram Content Group UK Ltd.
Pitfield, Milton Keynes, MK11 3LW, UK
UKHW050747250525
458861UK00006B/508